Harmonic Analysis and Partial Differential Equations

Chicago Lectures in Mathematics Series
Robert J. Zimmer, series editor
J. P. May, Spencer J. Bloch, Norman R. Lebovitz, and Carlos E. Kenig, editors

Harmonic Analysis and Partial Differential Equations

Essays in Honor of Alberto P. Calderón

EDITED BY
Michael Christ, Carlos E. Kenig, and Cora Sadosky

The University of Chicago Press

Chicago and London

The University of Chicago Press, Chicago 60637
The University of Chicago Press, Ltd., London
© 1999 by The University of Chicago
All rights reserved. Published 1999
Paperback edition 2001
Printed in the United States of America

08 07 06 05 04 03 02 01   2 3 4 5

ISBN: 0-226-10456-7 (cloth)
ISBN: 0-226-10455-9 (paperback)

Library of Congress Cataloging-in-Publication Data

Harmonic analysis and partial differential equations : essays in honor of
    Alberto P. Calderón / edited by Michael Christ, Carlos E. Kenig, and
    Cora Sadosky.
        p.   cm. — (Chicago lectures in mathematics series).
    Includes bibliographic references.
    ISBN: 0-226-10456-7 (cloth : alk. paper)
    1. Harmonic analysis. 2.Differential equations, partial.
I. Calderón, Alberto P. II. Christ, Francis Michael, 1955– .
III. Kenig, Carlos E., 1953– . IV. Sadosky, Cora, 1940– .
V. Series: Chicago lectures in mathematics.
QA403.H225 1999
515'.2433—dc21   99-20385
                                                                         CIP

Contents

Preface

Alberto P. Calderón turned seventy-five in September 1995, and an international conference in his honor was convened at the University of Chicago in February 1996, with the participation of more than three hundred mathematicians.

Professor Calderón was among this century's leading analysts, and his work is characterized by its great originality and depth. Calderón's contributions have been of wide scope, and changed the way researchers approach and think of a wide variety of areas in mathematics and its applications. His fundamental influence is felt strongly in harmonic analysis, partial differential equations, and complex analysis, as well as in such applied areas as signal processing, geophysics, and tomography.

The Chicago conference was a joyous occasion to celebrate Calderón's outstanding achievements. Many of his friends, students, and collaborators from Europe and North and South America came to join him in the celebration. This volume contains the contributions of nineteen of the speakers, including a personal account of the history of the Chicago School of analysis by Elias M. Stein.

The mathematical community was deeply saddened by the loss of Alberto P. Calderón, who died in Chicago on April 16, 1998, after a short illness. We have included, in an introduction, a short account of his life and mathematical career. This volume, originally conceived to mark a joyous milestone, has now also become a memorial tribute.

Introduction

Michael Christ, Carlos E. Kenig, and Cora Sadosky

Alberto P. Calderón was born in Mendoza, Argentina, on September 14, 1920. His father, a physician, instilled in him the love for mathematics and music. Impressed by Alberto's early interest in everything mechanical, he sent him at age twelve to school in Switzerland, in preparation for the ETH, the leading engineering school. This was not to be, and after two years in Zurich, Calderón was called back to Mendoza, where he finished high school and then attended the University of Buenos Aires, graduating as a civil engineer in 1947. Throughout his studies he had been increasingly interested in mathematics, but at the time engineering seemed a more viable career.

Upon graduation as an engineer, Calderón took a job in a geophysics research lab at YPF, the national oil corporation of Argentina. He enjoyed his work, where he dealt with mathematical problems arising from the design of tools for oil prospecting. He was always proud of the problems he solved there and frequently remarked with satisfaction that some problems he could not solve then are still open. But he joked how fortunate it was that his YPF supervisor made his life there difficult, since otherwise he would have remained in Argentina as a state employee until retirement!

This did not come to pass, and Calderón's close relation with the few active mathematicians then at the University of Buenos Aires was instrumental to the breakthrough that led to his mathematical celebrity. In 1948 two concurrent events changed his future: he resigned his position at YPF, and Antoni Zygmund, one of the world's leading analysts, and a professor at the University of Chicago, visited the Institute of Mathematics of the University of Buenos Aires.

At that time the Institute of Mathematics—a single room plus a small good library—was an appendix of the School of Engineering, with Julio Rey Pastor, the Goettingen-trained Spanish mathematician who introduced modern mathematics to Argentina, as its sole professor, and young Dr. Alberto Gonzalez Dominguez as Rey's only assistant. Still, its seminar was active, with the participation of serious young mathematicians, including Spanish

Republic refugees like Luis Santalo, and Calderón attended it. In fact, while still an engineering student, Calderón had attended the advanced calculus courses of Rey Pastor. It was as a student that Calderón met Gonzalez Dominguez, a man with a passion for mathematics, who became first his mentor and then his lifelong friend. And it was Gonzalez Dominguez who, through his studies with Tamarkin at Cornell University, had acquired a strong interest in Fourier analysis by reading Zygmund's 1935 fundamental treatise, "Trigonometrical Series."

In the years immediately after World War II, the U.S. Department of State had a very active visitors program that sent prominent scientists to Latin America. Thus, Adrian Albert, Marshall Stone, and George Birkhoff visited Buenos Aires, and Gonzalez Dominguez arranged through them the visit of Zymund, whose work on Fourier series he so much admired. At the Institute of Mathematics, Zygmund gave a two-month seminar on topics in analysis, based on his book. This seminar was attended by Gonzalez Dominguez, Calderón, Mischa Cotlar, and three other young Argentine mathematicians. Each of the participants had to discuss a portion of the text. Calderón's assignment was to present the Marcel Riesz theorem on the continuity of the Hilbert transform in L^p. According to Cotlar's vivid recollection of the event, Calderón's exposition was entirely acceptable to the junior audience, but not to Zygmund, who appeared agitated and grimaced all the time. Finally, he interrupted Calderón abruptly to ask where had he read the material he was presenting, and a bewildered Calderón answered that he had read it in Zygmund's book. Zygmund vehemently informed the audience that this was not the proof in his book, and after the lecture took Calderón aside and quizzed him about the new short and elegant proof. Calderón confessed that he had first tried to prove the theorem by himself, and then thinking he could not do it, had read the beginning of the proof in the book; but after the first couple of lines, instead of turning the page, had figured out how the proof would finish. In fact, he had found himself an elegant new proof of the Riesz theorem! Zygmund immediately recognized Calderón's power and then and there decided to invite him to Chicago to study with him.

This anecdote illustrates one of Calderón's main characteristics: he always sought his own proofs, developed his own methods. From the start, Calderón worked in mathematics that way: he rarely read the work of others farther than the statements of theorems and after grasping the general nature of the problem went ahead by himself. In this process, Calderón not only rediscovered results but added new insights to the subject. According to Cotlar, while still in Buenos Aires, Calderón had arrived by himself at something very close to the notion of distributions. This came as a consequence of Calderón's interest in quantum mechanics, which had been sparked by Guido Beck, the Italian physicist exiled in Argentina.

The two-month visit of Zygmund to Buenos Aires resulted in two joint papers and an invitation to Calderón to work with him in Chicago. Calderón arrived there in 1949 as a Rockefeller Fellow. The transition was hard because Calderón felt awed when suddenly confronted with the world of top-class mathematics. At the time, the Department of Mathematics at the University of Chicago was reputedly the world's best, and its faculty included, besides Zygmund, Adrian Albert, Shiing S. Chern, Lawrence Graves, Saunders McLane, Marshall Stone (chairman), and André Weil as professors, and Paul Halmos, Irving Kaplansky, Irving Segal, and Edwin Spanier as assistant professors. Calderón recalled he was so shocked that he wanted to return to Argentina, and it required all of Zygmund's power to persuade him to stay in Chicago. Calderón's interests were very broad from the start. He was fascinated by André Weil and established a dialogue with the great French master. Calderón loved all mathematics, and his penchant for concrete analysis did not inhibit his fascination with abstraction. He was interested in the Bourbaki approach and asked Weil for open problems. Weil complied; Calderón later said that these problems eventually sparked some of his ideas on pseudodifferential operators.

As a Rockefeller Fellow, Calderón had gone to Chicago to work with Zygmund, not in pursuit of a degree. But the intervention of Marshall Stone (a most visionary chairman) pushed him to obtain a doctorate, without which Calderón's academic career would have been hindered. To make this possible Stone persuaded him to "staple together" three separate papers into a dissertation. Thus Calderón was able to obtain his Ph.D. in mathematics under Zygmund's supervision in 1950, only a year after arriving in Chicago. The dissertation proved momentous: each of the three papers solved a long-standing open problem in ergodic theory or harmonic analysis. The two on the behavior of harmonic functions at the boundary opened up the possibility of bypassing complex methods in dealing with fundamental questions of harmonic analysis in the circle, thus leading to the extensions to the Euclidean n-dimensional space that were at the core of Zygmund's program for the future.

The collaboration begun by Zygmund and Calderón in 1948 reached fruition in the Calderón-Zygmund theory of singular integrals and ended only with Zygmund's death in 1992. Their memoir, "On singular integrals," published in *Acta Mathematica* in 1952, continues to be one of the most influential papers in the modern history of analysis. Teacher and student not only forged a major mathematical theory but cofounded what became internationally known as the Calderón-Zygmund school of analysis. Their methods have far-reaching consequences in many different branches of mathematics. A prime example of such a general method is one of their first joint results, the famous "Calderón-Zygmund decomposition lemma," invented to prove the "weak-type" continuity of singular integrals of integrable functions, which is now widely used throughout analysis and probability theory.

Although the theory of singular integrals seemed at first quite arcane to most analysts, by the mid sixties its popularity was established thanks to the epoch-making contributions to the theory of differential equations made by Calderón using singular integral operators. His proof of the uniqueness in the Cauchy problem, using algebras of singular integral operators, his existence and uniqueness theory for hyperbolic problems, his reduction of elliptic boundary value problems to singular integral equations on the boundary (the method of the Calderón projector), and the crucial role played by algebras of singular integrals (through the work of Calderón's student R. Seeley) in the proof of the Atiyah-Singer index theorem, gained Calderón international fame. The development of pseudodifferential operators, by Kohn-Nirenberg and Hörmander, is a direct consequence of Calderón's work on the applications of algebras of singular integrals to partial differential equations. Indeed, classical pseudodifferential operators form an algebra which includes the "variable coefficient" singular integral operators with kernels infinitely differentiable outside the diagonal, and partial differential operators with smooth coefficients. Calderón himself, in part through his collaborations with R. Vaillancourt and J. Alvarez-Alonso, contributed significantly to the theory of pseudodifferential operators. Nevertheless, Calderón insisted that the earlier point of view of algebras of singular integral operators with nonsmooth kernels should provide a tool to solve actual problems arising in physics and engineering, in which lack of smoothness is a natural feature. Moreover, he saw this greater generality as a means to "prepare the ground for applications to the theory of quasilinear and nonlinear differential operators." This point of view led to what is now referred to as the "Calderón program," the first success of which was Calderón's proof of the boundedness of the "first commutator." Further impetus in this program came from Calderón's fundamental study of the Cauchy integral on Lipschitz curves (1978). Many important works, like those of Coifman-McIntosh-Meyer, Bony, Verchota, David-Journé and many others followed from this. Calderón's short comment on the need still to address the open problems in this theory appears at the end of this volume.

A commentary on the mathematical significance of some of Calderón's contributions is given in the article by Elias M. Stein that begins the mathematical body of this volume.

It should not be forgotten that Calderón retained his interest in the applications of mathematics all his life, and some items in his bibliography stem from that interest: e.g., the papers on the phase problem for three-dimensional Fourier expansions and on the Radon transform. He was thrilled by the influence his work has had in applied areas, such as signal processing, geophysics and tomography, and he was intrigued by its relationship to wavelet theory.

After receiving his Ph.D. in mathematics from the University of Chicago, Calderón went to teach at Ohio State University for two years. He was a member of the Institute for Advanced Study at Princeton from 1953 to 1955, and an associate professor at the Massachusetts Institute of Technology from 1955 to 1959, returning to the University of Chicago as a professor from 1959 to 1972, and, after three more years at MIT, as a University Professor of Mathematics from 1975 until his retirement in 1985. He served as chairman of the department for two years in the seventies and returned to it permanently in 1989. Calderón was a Honorary Professor at the University of Buenos Aires from 1975 and the Director of the Instituto Argentino de Matematica (IAM) for several years during the seventies and early eighties.

Calderón always kept close links with Argentina and had a strong influence on the development of mathematics in his native country. In particular, seven of the sixteen Ph.D. students he had while at the University of Chicago came from the University of Buenos Aires. This happened in great measure through Zygmund's initiative. As a witness of the extraordinary growth of mathematics in Poland, his native country, during the twenty years of independence from the great powers between the two World Wars, Zygmund was a strong advocate of developing local mathematical talent in noncentral countries. He rightly claimed that mathematics, needing no major investment in either buildings or laboratories, could initiate the uplifting of a whole scientific establishment. On his 1948 trip to Buenos Aires when he "discovered" Calderón, Zygmund also found a group of eager young researchers, among them Mischa Cotlar. After Cotlar had also been "discovered" by George Birkhoff, Marshall Stone intervened to have him admitted to the University of Chicago's doctoral program, even though he had no prior schooling. He became Zygmund's next Ph.D. student after Calderón. It was Zygmund's second visit to Buenos Aires in 1959 that started a series of visits there by Calderón, and it was Zygmund who personally encouraged a group of undergraduate students and young researchers to attend Chicago to work with Calderón. The first to come were Agnes Benedek and Rafael Panzone (who already had obtained their Ph.D. degrees with Gonzalez Dominguez and Cotlar, respectively), who developed with Calderón the theory of Banach space valued singular integrals, and Steven Vagi and Evelio Oklander, who became Calderón's Ph.D. students at the University of Chicago. More students followed, and as was mentioned earlier, all in all seven students from the University of Buenos Aires obtained their Ph.D. degrees under Calderón's supervision at the University of Chicago.

Calderón and Zygmund also took an active interest in the development of mathematics in Spain. It is remarkable that the happy circumstances of the democratic transition after Franco's disappearance allowed the blooming of a Spanish school of real analysis which sprang from the activities of the lone Calderón-Zygmund student from Madrid, Miguel de Guzman, a lifelong friend of both. Miguel had the luck—which eluded his Argentine classmates—to return to his country at a time when it flourished, and his energy and generous vision opened the door for Spain to assume a major role in the international scene of mathematical analysis.

Calderón was recognized all over the world for his outstanding contributions to mathematics. He was a member of the U.S. National Academy of Sciences; the National Academies of Argentina, Spain, and France (corresponding); the Latin American Academy of Sciences; Academy of Sciences of the Third World, and a Fellow of the American Academy of Arts and Sciences. He received honorary doctorates from the University of Buenos Aires, the Technion (Israel), the Ohio State University, and the Universidad Autonoma de Madrid. He gave many invited addresses to universities and to learned societies, and he addressed the International Congress of Mathematicians as invited lecturer in Moscow in 1966 and as plenary lecturer in Helsinki in 1978. He was awarded the 1979 Bocher Prize and the 1989 Steele Prize by the American Mathematical Society. He was also the recipient of the Wolf Prize in Mathematics (Israel, 1989).

In 1991, Calderón was honored with the National Medal of Science, the highest award bestowed by the United States for scientific achievement. In that exceptional recognition he followed his teacher Antoni Zygmund, who had received the National Medal of Science in 1986. This national recognition of the Calderón-Zygmund school's value is testimony to how well the United States can integrate immigrants and benefit from their contributions.

Besides his remarkable research accomplishments, Calderón was a gifted lecturer and an inspiring teacher. But he was uneven; he could teach magnificent courses, but he could also lecture obscurely, depending on the degree of his current involvement with the subject he was teaching. Calderón supervised twenty-seven Ph.D. students —five while at MIT, sixteen at the University of Chicago, and six at the University of Buenos Aires. His students and collaborators had the opportunity to work with a formidable mathematician, with an array of hard and interesting problems to suggest, and they all could profit from his unusual openness in sharing ideas.

Outside of scientific endeavors, Calderón's interests ranged widely. He was fluent in several languages, loved music, played the piano, and danced the tango beautifully. He had a natural talent at fixing all sorts of appliances and always retained the mechanical interests of his early years. To each of his activities he brought the same enthusiasm and the same fresh outlook.

In 1950, Calderón married Mabel Molinelli Wells, a mathematics graduate whom he had met while both were students at the University of Buenos Aires and who was to have a very strong influence on his life. They had a daughter, Maria Josefina, and a son, Pablo, also a mathematician. In the late sixties Calderón's family settled in Buenos Aires, where Mabel died in 1985 after a long illness. Calderón's children and his three grandchildren reside in the

United States, as does Calixto P. Calderón, one of his brothers and also a mathematician; his remaining family is in Argentina.

During a visit to MIT, Calderón had shared an office with the Rumanian-born mathematician Alexandra (Bagdasar) Bellow, an ergodic theorist of world renown, currently Emeritus Professor at Northwestern University. In 1989, Alexandra and Alberto Calderón were married. At the end of her 1991 Noether Lecture, Alexandra described their shared personal and professional fulfillment, saying "life offered (me) peace and happiness after fifty, and mathematics became an asset in the human aspects of my personal life." Calderón had been interested in ergodic theory throughout his career, and his last two papers are joint with his wife.

Alberto P. Calderón, Emeritus Professor of Mathematics of the University of Chicago and Honorary Professor of the University of Buenos Aires, died at the age of seventy-seven on April 16 of 1998, after a short illness. He was one of the greatest mathematicians of the second half of the twentieth century. Those whose lives he entered will long cherish his memory, and his mathematics will live for generations to come.

Ph.D. Students of Alberto Calderón

Robert T. Seeley (1958)
Irwin S. Bernstein (1959)
I. Norman Katz (1959)
Jerome H. Neuwirth (1959)
Earl Robert Berkson (1960)
Evelio Oklander (1964)
Cora S. Sadosky (1965)
Stephen Vagi (1965)
Umberto Neri (1966)
John C. Polking (1966)
Nestor Marcelo Riviere (1966)
Carlos Segovia Fernandez (1967)
Miguel S. J. de Guzman (1968)

Daniel Fife (1968)
Alberto Torchinsky (1971)
Keith W. Powls (1972)
Josefina Dolores Alvarez Alonso (1976)
Telma Caputti (1976)
Robert Richard Reitano (1976)
Carlos E. Kenig (1978)
Angel Bartolome Gatto (1979)
Cristian Enrique Gutierrez (1979)
Kent G. Merryfield (1980)
Michael Christ (1982)
Gerald M. Cohen (1982)
Maria Amelia Muschietti (1984)
Marta Susana Urciolo (1985)

Chapter 1

Calderón and Zygmund's Theory of Singular Integrals

Elias M. Stein

At the occasion of the conference in honor of Alberto Calderón's seventy-fifth birthday, it is most fitting that we celebrate the mathematical achievements for which he is so much admired. Chief among these is his role in the creation of the modern theory of singular integrals. In that great enterprise he had the good fortune of working with the mathematician who had paramount influence on his scientific life: Antoni Zygmund, at first his teacher, and later his mentor and collaborator. So any account of the modern development of that theory must be in large part the story of the efforts of both Zygmund and Calderón. I will try to present this, following roughly the order of events as they unfolded. My aim will be to explore the goals that inspired and motivated them, describe some of their shared accomplishments and later work, and discuss briefly the wide influence of their achievements.

1.1   Zygmund's Vision: 1927–1949

In the first period his scientific work, from 1923 to the middle 1930s, Zygmund devoted himself to what is now called "classical" harmonic analysis, that is, Fourier and trigonometric series of the circle, related power series of the unit disk, conjugate functions, Riemannian theory connected to uniqueness, lacunary series, etc. An account of much of what he did, as well as the work of his contemporaries and predecessors, is contained in his famous treatise, "Trigonometrical Series," published in 1953. The time in which this took place may be viewed as the concluding decade of the brilliant century of classical harmonic analysis: the approximately hundred-year span which began with Dirichlet and Riemann, continued with Cantor and Lebesgue among others, and culminated with the achievements of Kolmogorov, M. Riesz, and Hardy and Littlewood.

It was during that last decade that Zygmund began to turn his attention from the one-dimensional situation to problems in higher dimensions. At first this represented merely an incidental interest, but then later he followed it with increasing dedication, and eventually it was to become the main focus of his scientific work. I want now to describe how this point of view developed with Zygmund.

In outline, the subject of one-dimensional harmonic analysis as it existed in that period can be understood in terms of what were then three closely interrelated areas of study, and which in many ways represented the central achievements of the theory: real-variable theory, complex analysis, and the behavior of Fourier series. Zygmund's first excursion into questions of higher dimensions dealt with the key issue of real-variable theory—the averaging of functions. The question was as follows.

The classical theorem of Lebesgue guaranteed that, for almost every x,

$$\lim_{\substack{x \in I \\ \text{diam}(I) \to 0}} \frac{1}{|I|} \int_I f(y)dy = f(x), \tag{1.1}$$

where I ranges over intervals, and when f is an integrable function on the line $\mathbb{R}^1$. In higher dimensions it is natural to ask whether a similar result held when the intervals I are replaced by appropriate generalizations in $\mathbb{R}^n$. The fact that this is the case when the I's are replaced by balls (or more general sets with bounded "eccentricity") was well known at that time. What must have piqued Zygmund's interest in the subject was his realization (in 1927) that a paradoxical set constructed by Nikodym showed that the answer is irretrievably false when the I's are taken to be rectangles (each containing the point in question), but with arbitrary orientation. To this must be added the counterexample found by Saks several years later, which showed that the desired analogue of (1.1) still failed even if we now restricted the rectangles to have a fixed orientation (e.g., with sides parallel to the axes), as long as one allowed f be a general function in L^1.

It was at this stage that Zygmund effectively transformed the subject at hand by an important advance: he proved that the wished-for conclusion (when the sides are parallel to the axes) held if f was assumed to belong to L^p, with $p > 1$. He accomplished this by proving an inequality for what is now known as the "strong" maximal function. Shortly afterward in Jessen, Marcinkiewicz, and Zygmund [31] this was refined to the requirement that f belong to $L(\log L)^{n-1}$ locally.

This study of the extension of (1.1) to $\mathbb{R}^n$ was the first step taken by Zygmund. It is reasonable to guess that it reinforced his fascination with what was then developing as a long-term goal of his scientific efforts, the extension of the central results of harmonic analysis to higher dimensions. But a great obstacle stood in the way: it was the crucial role played by complex function theory in the whole of one-dimensional Fourier analysis, and for this there was no ready substitute.

In describing this special role of complex methods we shall content ourselves with highlighting some of the main points.

1. *The conjugate function and its basic properties*

   As is well known, the Hilbert transform comes directly from the Cauchy integral formula. We also recall the fact that M. Riesz proved the L^p boundedness properties of the Hilbert transform $f \mapsto H(f) = \frac{p.v.}{\pi} \int\limits_{-\infty}^{\infty} f(x-y)\frac{dy}{y}$ by applying a contour integral to $(F)^p$, where F is the analytic function whose boundary limit has f as its real part.

2. *The theory of the Hardy Spaces H^p*

   These arose in part as substitutes for L^p, when $p \leq 1$, and were by their very nature complex-function-theory constructs. (It should be noted, however, that for $1 < p < \infty$ they were essentially equivalent with L^p by Riesz's theorem.) The main tool used in their study was the Blaschke product of their zeroes in the unit disk. Using it, one could reduce to elements $F \in H^p$ with no zeroes, and from these one could pass to $G = F^{p/2}$; the latter was in H^2 and hence could be treated by more standard (L^2) methods.

3. *The Littlewood-Paley theory*

   This proceeded by studying the dyadic decomposition in frequency space and had many applications; among them was the Marcinkiewicz multiplier theorem. The theory initiated and exploited certain basic "square functions," and these we originally studied by complex-variable techniques closely related to what were used in H^p spaces.

4. *The boundary behavior of harmonic functions*

   The main result obtained here (Privalov [1923], Marcinkiewicz and Zygmund [1938], and Spencer [1943]) stated that for any harmonic function $u(re^{i\theta})$ in the unit disk, the following three properties are equivalent for almost all boundary points $e^{i\theta}$:

   $$u \text{ has a nontangentive limit at } e^{i\theta}, \tag{1.2}$$

   $$u \text{ is nontangentially bounded at } e^{i\theta}, \tag{1.3}$$

the "area integral" $(S(u)(\theta))^2 = \iint\limits_{\Gamma(e^{i\theta})} |\nabla u(z)|^2 dx dy$ is finite ,   (1.4)

where $\Gamma(e^{i\theta})$ is a nontangential approach region with vertex $e^{i\theta}$.

The crucial first step in the proof was the application of the conformal map (to the unit disk) of the famous sawtooth domain (which is pictured in Zygmund [74], vol. 2, p. 200).

This mapping allowed one to reduce the implication (1.3) ⇒ (1.2) to the special case of bounded harmonic functions in the unit disk (Fatou's theorem), and it also played a corresponding role in the other parts of the proof.

It is ironic that complex methods with their great power and success in the one-dimensional theory actually stood in the way of progress to higher dimensions, and appeared to block further progress. The only way past, as Zygmund foresaw, required a further development of "real" methods. Achievement of this objective was to take more than one generation, and in some ways is not yet complete. The mathematician with whom he was to initiate the effort to realize much of this goal was Alberto Calderón.

1.2   Calderón and Zygmund: 1950–1957

1. Zygmund spent a part of the academic year 1948–49 in Argentina, and there he met Calderón. Zygmund brought him back to the University of Chicago, and soon thereafter (in 1950), under his direction, Calderón obtained his doctoral thesis. The dissertation contained three parts, the first about ergodic theory, which will not concern us here. However, it is the second and third parts that interest us, and these represented breakthroughs in the problem of freeing oneself from complex methods and, in particular, in extending to higher dimensions some of the results described in (4) above. In a general way we can say that his efforts here already typified the style of much of his later work: he begins by conceiving some simple but fundamental ideas that go to the heart of the matter and then develops and exploits these insights with great power.

In proving (1.3) ⇒ (1.2) we may assume that u is bounded inside the sawtooth domain Ω that arose in (4) above: this region is the union of approach regions $\Gamma(e^{i\theta})$ ("cones"), with vertex $e^{i\theta}$, for points $e^{i\theta} \in E$, and E a closed set. Calderón introduced the auxiliary harmonic function U, with U the Poisson integral of χ_{cE}, and observed that all the desired facts flowed from the dominating properties of U: namely, u could be split as $u = u_1 + u_2$, where u_1 is the Poisson integral of a bounded function (and hence has nontangential limits a.e.), while by the maximum principle, $|u_2| \leq cU$, and therefore u_2 has (nontangential) limits $= 0$ at a.e. point of E.

The second idea (used to prove the implication (1.2) $\Rightarrow$ (1.4)) has as its starting point the simple identity

$$\Delta u^2 = 2|\nabla u|^2 \tag{1.5}$$

valid for any harmonic function. This will be combined with Green's theorem

$$\iint\limits_{\Omega} (B\Delta A - A\Delta B)dxdy = \int\limits_{\partial\Omega} \left(B\frac{\partial A}{\partial n} - A\frac{\partial B}{\partial n} \right) d\sigma \ ,$$

where $A = u^2$, and B is another ingeniously chosen auxiliary function depending on the domain Ω only. This allowed him to show that

$$\iint\limits_{\Omega} y|\nabla u|^2 dxdy < \infty,$$

which is an integrated version of (1.4).

It may be noted that the above methods and the conclusions they imply make no use of complex analysis, and are very general in nature. It is also a fact that these ideas played a significant role in the later real-variable extension of the H^p theory.

2. Starting in the year 1950, a close collaboration developed between Calderón and Zygmund which lasted almost thirty years. While their joint research dealt with a number of different subjects, their preoccupying interest and most fundamental contributions were in the area of singular integrals. In this connection the first issue they addressed was—to put the matter simply—the extension to higher dimensions of the theory of the Hilbert transform. A real-variable analysis of the Hilbert transform had been carried out by Besicovitch, Titchmarsh, and Marcinkiewicz, and this is what needed to be extended to the $\mathbb{R}^n$ setting.

A reasonable candidate for consideration presented itself. It was the operator $f \mapsto Tf$, with

$$T(f)(x) = p.v. \int\limits_{\mathbb{R}^n} K(y)f(x-y)dy \ , \tag{1.6}$$

when K was homogeneous of degree $-n$, satisfied some regularity, and in addition the cancellation condition $\int\limits_{|x|=1} K(x)d\sigma(x) = 0$.

Besides the Hilbert transform (which is the only real example when $n = 1$), higher-dimensional examples include the operators that arise as second derivatives of the fundamental solution operator for the Laplacian, (which can be written as $\frac{\partial^2}{\partial x_2 \partial x_j}(\Delta)^{-1}$), as well as the related Riesz transforms, $\frac{\partial}{\partial x_j}(-\Delta)^{-1/2}$.

All of this is the subject matter of their historic memoir, "On the existence of singular integrals," which appeared in the *Acta Mathematica* in 1952. There is probably no paper in the last fifty years which had such widespread influence in analysis. The ideas in this work are now so well known that I will only outline its contents. It can be viewed as having three parts.

First, there is the Calderón-Zygmund lemma, and the corresponding Calderón-Zygmund decomposition. The main thrust of the former is as a substitute for F. Riesz's "rising sun" lemma, which had implicitly played a key role in the earlier treatment of the Hilbert transform. Second, using their decomposition, they then proved the weak-type L^1, and L^p, $1 < p < \infty$, estimates for the operator T in (1.6). As a preliminary step they disposed of the L^2 theory of T using Plancherel's theorem. Third, they applied these results to the examples mentioned above, and in addition they proved a.e. convergence for the singular integrals in question.

It should not detract from one's great admiration of this work to note two historical anomalies contained in it. The first is the fact that there is no mention of Marcinkiewicz's interpolation theorem, or to the paper in which it appeared (Marcinkiewicz [1939a]), even though its ideas play a significant role. In the Calderón-Zygmund paper, the special case that is needed is in effect reproved. The explanation for this omission is that Zygmund had simply forgotten about the existence of Marcinkiewicz's note. To make amends he published (in 1956) an account of Marcinkiewicz's theorem and various generalizations and extensions he had since found. In it he conceded that the paper of Marcinkiewicz ". . . seems to have escaped attention and does not find allusion to it in the existing literature."

The second point, like the first, also involves some very important work of Marcinkiewicz. He had been Zygmund's brilliant student and collaborator until his death at the beginning of World War II. It is a mystery why no reference was made to the paper Marcinkiewicz [40] and the multiplier theorem in it. This theorem had been proved by Marcinkiewicz in an n-dimensional form (as a product "consequence" of the one-dimensional form). As an application, the L^p inequalities for the operators $\frac{\partial^2}{\partial x_i \partial x_j}(\Delta^{-1})$ were obtained;[1] these he had proved at the behest of Schauder.

3. As has already been indicated, the n-dimensional singular integrals had its main motivation in the theory of partial differential equations. In their further work, Calderón and Zygmund pursued this connection, following the trail that had been explored earlier by Giraud, Tricomi, and Mihlin. Start-

---

[1] In truth, he had done this for their periodic analogues, but this is a technical distinction.

ing from those ideas (in particular the notion of "symbol") they developed their version of the symbolic calculus of "variable-coefficient" analogues of the singular integral operators. To describe these results one considers an extension of the class of operators arising in (1.6), namely of the form,

$$T(f)(x) = a_0(x)f(x) + p.v. \int_{\mathbb{R}^n} K(x,y)f(x-y)dy , \qquad (1.7)$$

where $K(x,y)$ is for each x a singular integral kernel of the type (1.6) in y, which depends smoothly and boundedly on x; also $a_0(x)$ is a smooth and bounded function.

To each operator T of this kind there corresponds its symbol $a(x,\xi)$, defined by

$$a(x,\xi) = a_0(x) + \widehat{K}(x,\xi) , \qquad (1.8)$$

where $\widehat{K}(x,\xi)$ denotes the Fourier transform of $K(x,y)$ in the y-variable. Thus $a(x,\xi)$ is homogeneous of degree 0 in the ξ variable (reflecting the homogeneity of $K(x,y)$ of degree $-n$ in y); and it depends smoothly and boundedly on x. Conversely to each function $a(x,\xi)$ of this kind there exists a (unique) operator (1.7) for which (1.8) holds. One says that a is the symbol of T and also writes $T = T_a$.

The basic properties that were proved were, first, the regularity properties

$$T_a : L_k^p \to L_k^p , \qquad (1.9)$$

where L_k^p are the usual Sobolev spaces, with $1 < p < \infty$.

Also the basic facts of symbolic manipulations

$$T_{a_1} \cdot T_{a_2} = T_{a_1 \cdot a_2} + Error \qquad (1.10)$$

$$(T_a)^* = T_{\bar{a}} + Error \qquad (1.11)$$

where the *Error* operators are smoothing of order 1, in the sense that *Error* : $L_k^p \to L_{k+1}^p$.

A consequence of the symbolic calculus is the factorizability of any linear partial differential operator L of order m,

$$L = \sum_{|\alpha| \le m} a_\alpha(x) \left(\frac{\partial}{\partial x} \right)^\alpha ,$$

where the coefficients a_α are assumed to be smooth and bounded. One can write

$$L = T_a(-\Delta)^{m/2} + (Error)' \qquad (1.12)$$

for an appropriate symbol a, where the operator $(Error)'$ refers to an operator that maps $L_k^p \to L_{k-m+1}^p$ for $k \geq m - 1$. It seemed clear that this symbolic calculus should have wide applications to the theory of partial differential operators and to other parts of analysis. This was soon to be borne out.

1.3   Acceptance: 1957–1965

At this stage of my narrative I would like to share some personal reminiscences. I had been a student of Zygmund at University of Chicago, and in 1956 at his suggestion I took my first teaching position at MIT, where Calderón was at that time. I had met Calderón several years earlier when he came to Chicago to speak about the "method of rotations" in Zygmund's seminar. I still remember my feelings when I saw him there; these first impressions have not changed much over the years: I was struck by the sense of his understated elegance, his reserve, and quiet charisma.

At MIT we would meet quite often and over time an easy conversational relationship developed between us. I do recall that we, in the small group who were interested in singular integrals then, felt a certain separateness from the larger community of analysts—not that this isolation was self-imposed, but more because our subject matter was seen by our colleagues as somewhat arcane, rarefied, and possibly not very relevant. However, this did change, and a fuller acceptance eventually came. I want to relate now how this occurred.

1. Starting from the calculus of singular integral operators that he had worked out with Zygmund, Calderón obtained a number of important applications to hyperbolic and elliptic equations. His most dramatic achievement was in the uniqueness of the Cauchy problem (Calderón [5]). There he succeeded in a broad and decisive extension of the results of Holmgren (for the case of analytic coefficients), and Carleman (in the case of two dimensions). Calderón's theorem can be formulated as follows.

Suppose u is a function which in the neighborhood of the origin in $\mathbb{R}^n$ satisfies the equation of mth order:

$$\frac{\partial^m u}{\partial x_n^m} = \sum_\alpha a_\alpha(x) \frac{\partial^\alpha}{\partial x^\alpha} \,, \tag{1.13}$$

where the summation is taken over all indices $\alpha = (\alpha_1, \ldots, \alpha_n)$ with $|\alpha| \leq n$, and $\alpha_n < m$. We also assume that u satisfies the null initial Cauchy conditions

$$\left.\frac{\partial^j u(x)}{\partial x_n^j}\right|_{x_n=0} = 0 \,, \quad j = 0, \ldots, m - 1 \,. \tag{1.14}$$

Besides (1.13) and (1.14), it suffices that the coefficients a_α belong to $C^{1+\epsilon}$, that the characteristics are simple, and $n \neq 3$, or $m \leq 3$. Under these hypotheses u vanishes identically in a neighborhood of the origin.

Calderón's approach was to reduce matters to a key "pseudo-differential inequality" (in a terminology that was used later). This inequality is complicated, but somewhat reminiscent of a differential inequality that Carleman had used in two dimensions. The essence of it is that

$$\int_0^a \phi_k \left\| \frac{\partial u}{\partial t} + (P + iQ)(-\Delta)^{1/2} u \right\|^2 dt \leq c \int_0^a \phi_k \|u\|^2 dt, \qquad (1.15)$$

where $u(0) = 0$ implies $u \equiv 0$, if (1.15) holds for $k \to \infty$.

Here P and Q are singular integral operators of the type (1.7), with real symbols and P is invertible; we have written $t = x_n$, and the norms are L^2 norms taken with respect to the variables $x_1, \ldots, x_{n-1}$. The functions ϕ_k are meant to behave like t^{-k}, which when $k \to \infty$ emphasizes the effect taking place near $t = 0$. In fact, in (1.13) we can take $\phi_k(t) = (t + 1/k)^{-k}$.

The proof of assertions like (1.15) is easier in the special case when all the operators commute; their general form is established by using the basic facts (1.10) and (1.11) of the calculus.

The paper of Calderón was, at first, not well received. In fact, I learned from him that it was rejected when submitted to what was then the leading journal in partial differential equations, *Commentaries of Pure and Applied Mathematics*.

2. At about that time, because of the applicability of singular integrals to partial differential equations, Calderón became interested in formulating the facts about singular integrals in the setting of manifolds. This required the analysis of the effect coordinate changes had on such operators. A hint that the problem was tractable came from the observation that the class of kernels, $K(y)$, of the type arising in (1.6), was invariant under linear (invertible) changes of variables $y \mapsto L(y)$. (The fact that $K(L(y))$ satisfied the same regularity and homogeneity that $K(y)$ did, was immediate; that the cancellation property also holds for $K(L(y))$ is a little less obvious.)

R. Seeley was Calderón's student at that time, and he dealt with this problem in his thesis (see Seeley [59]). Suppose $x \mapsto \psi(x)$ is a local diffeomorphism, then the result was that modulo error terms (which are "smoothing" of one degree) the operator (1.7) is transformed into another operator of the same kind,

$$T'(f)(x) \cdot a_0'(x) = f(x) + \text{p.v.} \int K'(x, y) f(x - y) dy,$$

but now $a_0'(x) = a_0(\psi(x))$, and $K'(x, y) = K'(\psi(x), L_x(y))$, where L_x is the linear transformation given by the Jacobian matrix $\frac{\partial \psi(x)}{\partial x}$. On the level of symbols this meant that the new symbol a' was determined by the old symbol according to the formula

$$a'(x, \xi) = a(\psi(x), L'_x(\xi)) \ ,$$

with L'_x the transpose-inverse of L_x. Hence the symbol is actually a function on the cotangent space of the manifold.

The result of Seeley was not only highly satisfactory as to its conclusions, but it was also very timely in terms of events that were about to take place. Following an intervention by Gelfand [1960], interest grew in calculating the "index" of an elliptic operator on a manifold. This index is the difference of the dimension of the null-space and the codimension of the range of the operator, and is an invariant under deformations. The problem of determining it was connected with a number of interesting issues in geometry and topology. The result of the "Seeley calculus" proved quite useful in this context: the proofs proceeded by appropriate deformations and matters were facilitated if these could be carried out in the more flexible context of "general" symbols, instead of restricting attention to the polynomial symbols coming from differential operators. A contemporaneous account of this development (during the period 1961–64), may be found in the notes of the seminar on the Atiyah-Singer index theorem (see Palais [51]); for an historical survey of some of the background, see also Seeley [61].

3. With the activity surrounding the index theorem, it suddenly seemed as if everyone was interested in the algebra of singular integral operators. However, one further step was needed to make this a household tool for analysts: it required a change of point of view. Even though this change of perspective was not major, it was significant psychologically and methodologically, since it allowed one to think more simply about certain aspects of the subject and because it suggested various extensions.

The idea was merely to change the role of the definitions of the operators, from (1.7) for singular integrals to pseudo-differential operators

$$T_a(f)(x) = \int_{\mathbb{R}^n} a(x, \xi) \hat{f}(\xi) e^{2\pi i x \cdot \xi} d\xi \ , \tag{1.16}$$

with symbol a. (Here $\hat{f}$ is the Fourier transform, $\hat{f}(\xi) = \int_{\mathbb{R}^n} e^{-2\pi i x \cdot \xi} f(x) dx$.)

Although the two operators are identical (when $a(x, \xi) = a_0(x) + \widehat{K}(x, \xi)$), the advantage lies in the emphasis in (1.16) on the L^2 theory and Fourier transform, and the wider class of operators that can be considered, in particular, including differential operators. The formulation (1.16) allows one to deal more systematically with the composition of such operators and incorporate the lower-order terms in the calculus.

To do this, one might adopt a wider class of symbols of "homogeneous-type": roughly speaking, $a(x, \xi)$ belongs to this class (and is of *order m*) if $a(x, \xi)$ is for large ξ, asymptotically the sum of terms homogeneous in ξ of degrees $m - j$, with $j = 0, 1, 2, \ldots$.

The change in point of view described above came into its full flowering with the papers of Kohn and Nirenberg [36] and Hörmander [30], (after some work by Unterberger and Bokobza [70] and Seeley [60]). It is in this way that singular integrals were subsumed by pseudo-differential operators. Despite this, singular integrals, with their formulation in terms of kernels, still retained their primacy when treating real-variable issues, issues such as L^p or L^1 estimates (and even for some of the more intricate parts of the L^2 theory). The central role of the kernel representation of these operators became, if anything, more pronounced in the next twenty years.

1.4 Calderón's New Theory of Singular Integrals: 1965–

In the years 1957–58 there appeared the fundamental work of DeGiorgi and Nash, dealing with smoothness of solutions of partial differential equations, with minimal assumptions of regularity of the coefficients. One of the most striking results—for elliptic equations—was that any solution u of the equation

$$L(u) \equiv \sum_{i,j} \frac{\partial}{\partial x_i} \left(a_{ij}(x) \frac{\partial u}{\partial x_j} \right) = 0 \tag{1.17}$$

in an open ball satisfies an a priori interior regularity as long as the coefficients are uniformly elliptic, i.e.,

$$c_1 |\xi|^2 \leq \sum_{i,j} a_{ij}(x) \xi_i \xi_j \leq c_2 |\xi|^2 . \tag{1.18}$$

In fact, no regularity is assumed about the a_{ij} except for the boundedness implicit in (1.18), and the result is that u is Hölder continuous with an exponent depending only on the constants c_1 and c_2.

Calderón was intrigued by this result. He initially expected, as he told me, that one could obtain such conclusions and others by refining the calculus of singular integral operators (1.15), making minimal assumptions of smoothness on $a_0(x)$ and $K(x, y)$. While this was plausible—and indeed in his work with Zygmund they had already derived properties of the operators (1.7) and their calculus when the dependence on x was, for example, of class $C^{1+\epsilon}$—this hope was not to be realized. Further understanding about these things could be achieved only if one were ready to look in a somewhat different direction. I want to relate now how this came about.

1. The first major insight arose in answer to the following:

Question: Suppose M_A is the operator of multiplication (by the function A),

$$M_A : f \mapsto A \cdot F .$$

What are the least regularity assumptions on A needed to guarantee that the commutator $[T, M_A]$ is bounded on L^2, whenever T is of order 1? In $\mathbb{R}^1$, if T happens to be $\frac{d}{dx}$, then $[T, M_A] = M_{A'}$, and so the condition is exactly

$$A' \in L^\infty(\mathbb{R}^1) . \tag{1.19}$$

In a remarkable paper, Calderón [1965], showed that this is also the case more generally. The key case, containing the essence of the result he proved, arose when $T = H\frac{d}{dx}$, with H the Hilbert transform. Then T is actually $\left|\frac{d}{dx}\right|$, its symbol is $2\pi|\xi|$. and $[T, M_A]$ is the "commutator" C_1,

$$C_1(f)(x) = \frac{1}{\pi} p.v. \int\limits_{-\infty}^{\infty} \frac{A(x) - A(y)}{(x - y)^2} f(y) dy . \tag{1.20}$$

Calderón proved that $f \mapsto C_1(f)$ is bounded on $L^2(\mathbb{R})$ if (1.19) held.

There are two crucial points that I want to emphasize about the proof of this theorem. The first is the reduction of the boundedness of the bilinear term $(f, g) \to \langle C_1(f), g \rangle$ to a corresponding property of a particular bilinear mapping, $(F, G) \to B(F, G)$, defined for (appropriate) holomorphic functions in the upper half-plane $\{z = x + iy, y > 0\}$ by

$$B(F, G)(x) = i \int_0^\infty F'(x + iy) G(x + iy) dy . \tag{1.21}$$

This B is a primitive version of a "para-product" (in this context, the justification for this terminology is the fact that $F(x) \cdot G(x) = B(F, G)(x) + B(G, F)(x)$). It is, in fact, not too difficult to see that $f \mapsto C_1(f)$ is bounded on $L^2(\mathbb{R}^1)$ if B satisfies the Hardy-space estimate

$$\|B(F, G)\|_{H^1} \le c\|F\|_{H^2}\|G\|_{H^2} . \tag{1.22}$$

The second major point in the proof is the assertion need to establish (1.22). It is the converse part of the equivalence

$$\|S(F)\|_{L^1} \sim \|F\|_{H^1} , \tag{1.23}$$

for the area integral S (which appeared in (1.4)).

The theorem of Calderón, and in particular the methods he used, inspired a number of significant developments in analysis. The first came because of the enigmatic nature of the proof: a deep L^2 theorem had been established by methods (using complex function theory) that did not seem susceptible of a general framework. In addition, the nontranslation-invariance character of the operator C_1 made Plancherel's theorem of no use here. It seemed likely that a method of "almost-orthogonal" decomposition—pioneered by Cotlar for the classical Hilbert transform—might well succeed in this case also. This lead to a reexamination of Cotlar's lemma (which had originally applied to the case of commuting self-adjoint operators). A general formulation was obtained as follows: Suppose that on a Hilbert space, $T = \sum T_j$, then

$$\|T\|^2 \leq \sum_k \sup_j \{\|T_j T_{j+k}^\star\| + \|T_j^\star T_{j+k}\|\}. \tag{1.24}$$

Despite the success in proving (1.24), this alone was not enough to reprove Calderón's theorem. As understood later, the missing element was a certain cancellation property. Nevertheless, the general form of Cotlar's lemma, (1.24), quickly led to a number of highly useful applications, such as singular integrals on nilpotent group (intertwining operators), pseudo-differential operators, etc.

Calderón's theorem also gave added impetus to the further evolution of the real-variable H^p theory. This came about because the equivalence (1.23) and its generalizations allowed one to show that the usual singular integrals (1.6) were also bounded on the Hardy space H^1 (and in fact on all H^p, $0 < p < \infty$). Taken together with earlier developments and some later ideas, the real-variable H^p theory reached its full-flowering a few years later. One owes this long-term achievement to the work of G. Weiss, C. Fefferman, Burkholder, Gundy, and Coifman, among others.

2. It became clear after a time that understanding the commutator C_1 (and its "higher" analogues) was in fact connected with an old problem that had been an ultimate, but unreached, goal of the classical theory of singular integrals: the boundedness behavior of the Cauchy-integral taken over curves with minimal regularity. The question involved can be formulated as follows: in the complex plane, for a contour γ and a function f defined on it, form the Cauchy integral

$$F(z) = \frac{1}{2\pi i} \int_\gamma \frac{f(\zeta)}{\zeta - z} d\zeta \,,$$

with F holomorphic outside γ. Define the mapping $f \to \mathcal{C}(f)$ by $\mathcal{C}(f) = F_+ + F_-$, where F_+ are the limits of F on γ approached from either side. When γ is the unit circle, or real axis, then $f \to \mathcal{C}(f)$ is essentially the Hilbert transform. Also when γ has some regularity (e.g., γ is in $C^{1+\epsilon}$), the expected properties of $\mathcal{C}$ (i.e., L^2, L^p boundedness, etc.) are easily obtained from the Hilbert transform. The problem was what happened when, say, γ was less regular, and here the main issue that presented itself was the behavior of the Cauchy integral when γ was a Lipschitz curve.

If γ is a Lipschitz graph in the plane, $\gamma = \{x + iA(x), x \in R\}$, with $A' \in L^\infty$, then up to a multiplicative constant,

$$\mathcal{C}(f)(x) = p.v. \int_{-\infty}^{\infty} \frac{1}{x - y + i(A(x) - A(y))} f(y)b(y)dy, \qquad (1.25)$$

where $b = 1 + iA'$. The formal expansion

$$\frac{1}{x - y + iA(A(y))} = \frac{1}{x - y} \cdot \sum_{k=0}^{\infty} (-i)^k \left(\frac{A(x) - A(y)}{x - y} \right)^k \qquad (1.26)$$

then makes clear that the fate of Cauchy integral $\mathcal{C}$ is inextricably bound up with that of the commutator C_1 and its higher analogues C_k given by

$$C_k f(x) = \frac{p.v.}{\pi} \int_{-\infty}^{\infty} \left(\frac{A(x) - A(y)}{x - y} \right)^k \frac{f(y)}{x - y} dy \ .$$

The further study of this problem was begun by Coifman and Meyer in the context of the commutators C_k, but the first breakthrough for the Cauchy integral was obtained by Calderón [9] (using different methods), in the case the norm $\|A'\|_{L^\infty}$ was small. His proof made decisive use of the complex-analytic setting of the problem. It proceeded by an ingenious deformation argument, leading to a nonlinear differential inequality; this nonlinearity accounted for the limitation of small norm for A' in the conclusion. But even with this limitation, the conclusion obtained was stunning.

The crowning result came in 1982, when Coifman and Meyer having enlisted the help of McIntosh, and relying on some of their earlier ideas, together proved the desired result without limitation on the size of $\|A'\|_{L^\infty}$. The method (in Coifman, McIntosh, and Meyer [20]) was operator-theoretic, emphasizing the multilinear aspects of the C_k, and in distinction to Calderón's approach was not based on complex-analytic techniques.

3. The major achievement represented by the theory of the Cauchy integral led to a host of other results, either by a rather direct exploitation of the conclusions involved, or by extensions of the techniques that were used. I will briefly discuss two of these developments.

The first was a complete analysis of the L^2 theory of "Calderón-Zygmund operators." By this terminology is meant operators of the form

$$T(f)(x) = \int_{\mathbb{R}^n} K(x,y)f(y)dy \qquad (1.27)$$

initially defined for test-function $f \in \mathcal{S}$, with the kernel K a distribution. It is assumed that away from the diagonal K agrees with a function that satisfies familiar estimates such as

$$|K(x,y)| \le A|x-y|^{-n}, \qquad |\nabla_{x,y}K(x,y)| \le A|x-y|^{-n-1}. \qquad (1.28)$$

The main question that arises (and is suggested by the commutators C_k) is what are the additional conditions that guarantee that T is a bounded operator on $L^2(\mathbb{R}^n)$ to itself. The answer, found by David and Journé [23] is highly satisfying: A certain "weak boundedness" property, namely $|(Tf,g)| \le Ar^n$ wherever f and g are suitably normalized bump functions, supported in a ball of radius r; also that both $T(1)$ and $T^\star(1)$ belong to BMO. These conditions are easily seen to be also necessary.

The argument giving the sufficiency proceeded in decomposing the operator into a sum, $T = T_1 + T_2$, where for T_1 the additional cancellation condition $T_1(1) = T_1^\star(1) = 0$ held. As a consequence the method of almost-orthogonal decomposition, (1.24), could be successfully applied to T_1. The operator T_2 (for which L^2 boundedness was proved differently) was of para-product type, chosen so as to guarantee the needed cancellation property.

The conditions of the David-Journé theorem, while applying in principle to the Cauchy integral, are not easily verified in that case. However, a refinement (the "$T(b)$ theorem"), with $b = 1 + iA'$, was found by David, Journé, and Semmes, and this does the job needed.

A second area that was substantially influenced by the work of the Cauchy integral was that of second-order elliptic equations in the context of minimal regularity. Side by side with the consideration of the divergence-form operator L in (1.17) (where the emphasis is on the minimal smoothness of the coefficients), one was led to study also the potential theory of the Laplacian (where now the emphasis was on the minimal smoothness of the boundary). In the latter setting, a natural assumption to make was that the boundary is Lipschitzian. In fact, by an appropriate Lipschitz mappings of domains, the situation of the Laplacian in a Lipschitz domain could be realized as a special case of the divergence-form operator (1.17), where the domain was smooth (say, a half-space).

The decisive application of the Cauchy integral to the potential theory of the Laplacian in a Lipschitz domain was in the study of the boundedness of the double layer potential (and the normal derivative of the single layer potential). These are $n-1$ dimensional operators, and they can be realized by applying the "method of rotations" to the one-dimensional operator (1.25). One should mention that another significant aspect of Laplacians on Lipschitz domains was the understanding brought to light by Dahlberg of the nature of harmonic measure and its relation to A_p weights. These two strands, initially independent, have been linked together, and with the aid of further ideas a rich theory has developed, owing to the added contributions of Jerison, Kenig, and others.

Finally, we return to the point where much of this began—the divergence-form equation (1.17). Here the analysis growing out of the Cauchy integral also had its effect. Here I will mention only the usefulness of multilinear analysis in the study of the case of "radially independent" coefficients; also in the work on the Kato problem—the determination of the domain of $\sqrt{L}$ in the case the coefficients can be complex-valued.

1.5   Some Perspectives on Singular Integrals: Past, Present, and Future

The modern theory of singular integrals, developed and nurtured by Calderón and Zygmund, has proved to be a very fruitful part of analysis. Beyond the achievements described above, a number of other directions have been cultivated with great success, with work being vigorously pursued up to this time; in addition, here several interesting open questions present themselves. I want to allude briefly to three of these directions and mention some of the problems that arise.

1. *Method of the Calderón-Zygmund lemma.* As is well known, this method consists of decomposing an integrable function into its "good" and "bad" parts; the latter being supported on a disjoint union of cubes, and having mean-value zero on each cube. Together with an L^2 bound and estimates of the type (1.28), this leads ultimately to the weak-type (1,1) results, etc.

It was recognized quite early that this method allowed substantial extension. The generalizations that were undertaken were not so much pursued for their own sake but rather were motivated in each case by the interest of the applications. Roughly, in order of appearance, here were some of the main instances:

(i) *The heat equation and other parabolic equations.* This began with the work of F. Jones [32] for the heat equation, with the Calderón-Zygmund cubes replaced by rectangles whose dimensions reflected the homogeneity of the heat operator. The theory was extended by Fabes and Rivière to encompass more general singular integrals respecting "nonisotropic" homogeneity in Euclidean spaces.

(ii) *Symmetric spaces and semisimple Lie groups.* To be succinct, the crucial point was the extension to the setting of nilpotent Lie groups with dilations ("homogeneous groups"), motivated by problems connected with Poisson integrals on symmetric spaces, and construction of intertwining operators.

(iii) *Several complex variables and subelliptic equations.* Here we return again to the source of singular integrals, complex analysis, but now in the setting of several variables. An important conclusion attained was that for a broad class of domains in $\mathbb{C}^n$, the Cauchy-Szegö projection is a singular integral, susceptible to the above methods. This was realized first for strongly pseudo-convex domains, next weakly pseudo-convex domains of finite type in $\mathbb{C}^2$; and more recently, convex domains of finite-type in $\mathbb{C}^n$. Connected with this is the application of the above ideas to the $\bar{\partial}$-Neumann problem, and its boundary analogue for certain domains in $\mathbb{C}^n$, as well as the study solving operators for subelliptic problems, such as Kohn's Laplacian, Hörmander's sum of squares, etc.; these matters also involved using ideas originating in the study of nilpotent groups as in (ii).

The three kinds of extensions mentioned above are prime examples of what one may call "one-parameter" analysis. This terminology refers to the fact that the cubes (or their containing balls), which occur in the standard $\mathbb{R}^n$ set-up, have been replaced by suitable one-parameter family of generalized "balls," associated to each point. While the general one-parameter method clearly has wide applicability, it is not sufficient to resolve the following important question:

Problem: Describe the nature of the singular integrals operators which are given by Cauchy-Szegö projection, as well as those that arise in connection with the solving operators for the $\bar{\partial}$ and $\bar{\partial}_b$ complexes for general smooth finite-type pseudo-convex domains in $\mathbb{C}^n$.

Some speculation about what may be involved in resolving this question can be found below.

2. *The method of rotations.* The method of rotations is both simple in its conception, and far-reaching in its consequences. The initial idea was to take the one-dimensional Hilbert transform, induce it on a fixed (subgroup) $\mathbb{R}^1$ of $\mathbb{R}^n$, rotate this $\mathbb{R}^1$ and integrate in all directions, obtaining in this way the singular integral (1.6) with odd kernel, which can be written as

$$T_\Omega(f)(x) = p.v. \int\limits_{\mathbb{R}^n} \frac{\Omega(y)}{|y|^n} f(x-y)dy, \qquad (1.29)$$

where Ω is homogeneous of degree 0, integrable in the unit sphere, and odd. In much the same way the general maximal operator

$$M_\Omega(f)(x) = \sup_{r>0} \frac{1}{r^n} \left| \int\limits_{|y| \leq r} \Omega(y)f(x-y)dy \right| \tag{1.30}$$

arises from the one-dimensional Hardy-Littlewood maximal function.

This method worked very well for L^p, $1 < p$ estimates, but not for L^1 (since the weak-type L^1 "norm" is not subadditive). The question of what happens for L^1 was left unresolved by Calderón and Zygmund. It is now to a large extent answered: we know that both (1.29) and (1.30) are indeed of weak-type (1,1) if Ω is in $L(\log L)$. This is the achievement of a number of mathematicians, in particular Christ and Rubio de Francia.

When the method of rotations is combined with the singular integrals for the heat equation (as in 1(i) above), one arrives at the "Hilbert transform on the parabola." Consideration of the Poisson integral on symmetric spaces leads one also to inquire about some analogous maximal functions associated to homogeneous curves. The initial major breakthroughs in this area of research were obtained by Nagel, Rivière, and Wainger. The subject has since developed into a rich and varied theory: beginning with its translation invariant setting on $\mathbb{R}^n$ (and its reliance on the Fourier transform), and then prompted by several complex variables, to a more general context connected with oscillatory integrals and nilpotent Lie groups, where it was rechristened as the theory of "singular radon transforms."

A common unresolved enigma remains about these two areas which have sprung out of the method of rotations. This is a question which has intrigued workers in the field, and whose solution, if positive, would be of great interest.

Problem:

(a) Is there an L^1 theory for (1.29) and (1.30) if Ω is merely integrable?[2]

(b) Are the singular Radon transforms, and their corresponding maximal functions, of weak-type (1,1)?

3. *Product theory and multiparameter analysis.* To oversimplify matters, one can say that "product theory" is that part of harmonic analysis in $\mathbb{R}^n$ which is invariant with respect to the n-fold dilations: $x = (x_1, x_2, \ldots, x_n) \to (\delta_1 x_1, \delta_2 x_2, \ldots, \delta_n x_n)$, $\delta_j > 0$. Another way of putting it is that its initial concern is with operators that are essentially products of operators acting on each variable separately, and then more generally with operators (and associated function spaces) which retain some of these characteristics. Related

---

[2] For (1.29) we assume also that Ω is odd.

to this is the multiparameter theory, standing part-way between the one-parameter theory discussed above and product theory: here the emphasis is on operators which are "invariant" (or compatible with) specified subgroups of the group of n-parameter dilations.

The product theory of $\mathbb{R}^n$ began with Zygmund's study of the strong maximal function, continued with Marcinkiewicz's proof of his multiplier theorem, and has since branched out in a variety of directions where much interesting work has been done. Among the things achieved are an appropriate H^p and BMO theory, and the many properties of product (and multiparameter) singular integrals which have came to light. This is due to the work of S.-Y. Chang, R. Fefferman, and J.-L. Journé, to mention only a few of the names.

Finally, I want to come to an extension of the product theory (more precisely, the induced "multiparameter analysis") in a direction which has particularly interested me recently. Here the point is that the underlying space is no longer Euclidean $\mathbb{R}^n$, but rather a nilpotent group or another appropriate generalization. On the basis of recent, but limited, experience I would hazard the guess that multiparameter analysis in this setting could well turn out to be of great interest in questions related to several complex variables. A first vague hint that this may be so, came with the realization that certain boundary operators arising from the $\bar{\partial}$-Neumann problem (in the model case corresponding to the Heisenberg group) are excellent examples of multiple-parameter singular integrals (see Müller, Ricci, and Stein [46]). A second indication is the description of Cauchy-Szegö projections and solving operators for $\bar{\partial}_b$ for a wide class of quadratic surfaces of higher codimension in $\mathbb{C}^n$, in terms of appropriate quotients of products of Heisenberg groups (see Nagel, Ricci, and Stein [48]). And even more suggestive are recent calculations (made jointly with A. Nagel) for such operators in a number of pseudo-convex domains of finite type. All this leads one to hope that a suitable version of multiparameter analysis will provide the missing theory of singular integrals needed for a variety of questions in several complex variables. This is indeed an exciting prospect.

1.6 Bibliographical Notes

I wish to provide here some additional citations of the literature closely connected to the material I have covered. However, these notes are not meant to be in any sense a systematic survey of relevant work.

1.1 Zygmund [74] is a greatly revised and expanded second edition of his 1935 book. His initial work on the strong maximal function is in Zygmund [71]. His views about the central role of complex methods in

Fourier analysis are explained in Zygmund [72]. A historical survey of square functions and an account of Zygmund's work in this area can be found in Stein [65].

1.2.2 For the real variable theory of the Hilbert transform see Besicovitch [1], Titchmarsh [69], and Marcinkiewicz [39].

1.2.3 Two papers of Calderón and Zygmund dealing with the symbolic calculus of operators (1.7) ([12] and [13]).

1.3.1 Further work of Calderón, applying singular integrals to partial differential equations, is contained in [6] and [7].

1.4.1 The theory of para-products was developed later in Bony [2]. The general form of Cotlar's lemma, (1.24), as well as the application to intertwining operators, may be found in Knapp and Stein [35]; the application to pseudo-differential operators is in Calderón and Vaillancourt [14]. The relation between the boundedness of the usual singular integrals on Hardy spaces and equivalences like (1.23) is in Stein [63], chapter 7. An account of the real-variable H^p theory can be found in Stein [66], chapters 3 and 4.

1.4.2 For systematic presentations of topics such as commutators, the Cauchy integral, multilinear analysis, and the $T(b)$ theorem, the reader should consult Coifman and Meyer [21], Meyer [44], and Meyer and Coifman [45].

1.4.3 In Kenig [34] the reader will find an exposition of the area dealing with the operator (1.17) as well as the Laplacian on domains with Lipschitz boundary.

1.5.1 In connection with (ii), the reader may consult Stein [64]. For the occurrence of Calderón-Zygmund-type singular integrals on strictly-pseudo-convex domains, see Koranyi-Vági [37], C. Fefferman [27], and Folland and Stein [28]. Some corresponding results for domains in $\mathbb{C}^2$ of finite type may be found in Christ [17], Machedon [38], Nagel, Rosay, Stein, and Wainger [48]. For the Cauchy-Szegö projection on convex domains in $\mathbb{C}^n$, see McNeal and Stein [43].

Regarding the Calderón-Zygmund lemma, two further sources should be cited. In Coifman and Weiss [22] the use of this method on spaces of a general character is systematized. The work of C. Fefferman [26] contains an important departure regarding the Calderón-Zygmund method, involving certain additional L^2 arguments, and allowing him to prove a number of subtle weak-type results. This method has proved to be relevant in various other instances, in particular to the study of operators of the type (1.29) and (1.30).

1.5.2 The method of rotations originates in Calderón and Zygmund [11]. For (1.29) and (1.30) see also Seeger [58] and Tao [68]. For singular Radon transforms, see Stein and Wainger [67], Phong and Stein [52], and Christ, Nagel, Stein, and Wainger [18], where other references can be found.

1.5.3 Among the papers that may be consulted for the product and multi-parameter theory in the Euclidean set-up are: S.-Y. Chang and R. Fefferman [15], Journé [33], Ricci and Stein [66].

References

[1] A. Besicovitch, *Sur la nature des fonctions à carré sommable mésurables,* Fund. Math. 4 (1926), 172–195.

[2] J. M. Bony, *Calcul symbolique et propagation des singularités pour les equations aux dérivées partielles non linéares,* Ann. Sci. École Norm. Sup. 14 (1981), 209–256.

[3] A. P. Calderón, *On the behavior of harmonic functions near the boundary,* Trans. Amer. Math. Soc. 68 (1950), 47–54.

[4] ———, *On a theorem of Marcinkiewicz and Zygmund,* Trans. Amer. Math. Soc. 68 (1950), 55–61.

[5] ———, *Uniqueness in the Cauchy problem for partial differential equations,* Amer. J. of Math. 80 (1958), 16–36.

[6] ———, *Existence and uniqueness theorems for systems of partial differential equations,* Proc. Symp. U. of Maryland 1961 (1962), 147–195, Gordon and Breach, New York.

[7] ———, *Boundary value problems for elliptic equations,* Joint Soviet-American Symposium on partial differential equations (1964), Novosibirsk.

[8] ———, *Commutators of singular integral operators,* Proc. Nat. Acad. Sci. 53 (1965), 1092–1099.

[9] ———, *Cauchy integrals on Lipschitz curves and related operators,* Proc. Nat. Acad. Sci. U.S.A. 74 (1977), 1324–1327.

[10] A. P. Calderón and A. Zygmund, *On the existence of certain singular integrals,* Acta Math. 88 (1952), 85–139.

[11] ———, *On singular integrals,* Amer. J. Math. 78 (1956), 289–309.

[12] ———, *Algebras of certain singular integrals,* Amer. J. Math. 78 (1956), 310–320.

[13] ———, *Singular integral operators and differential equations,* Amer. J. Math. 79 (1957), 801–821.

[14] A. P. Calderón and R. Vaillancourt, *A class of bounded pseudo-differential operators,* Proc. Nat. Acad. Sci. U.S.A. 69 (1972), 1185–1187.

[15] S.-Y. A. Chang and R. Fefferman, *A continuous version of the duality of H^1 and BMO on the bidisk,* Ann. of Math. 112 (1982), 179–201.

[16] ———, *Some recent developments in Fourier and analysis and H^p theory on product domains,* Bull. Amer. Math. Soc. 12 (1985), 1–43.

[17] M. Christ, *Regularity properties of the $\overline{\partial}_b$ equation on weakly pseudoconvex CR manifolds of dimension 3,* J. Amer. Math. Soc. 1 (1998), 587–646.

[18] M. Christ, A. Nagel, E. M. Stein, and S. Wainger, *Singular and Maximal Radon Transforms,* Ann. of Math., to appear 1999.

[19] M. Christ and J. L. Rubio de Francia, *Weak-type $(1,1)$ bounds for rough operators II,* Invent. Math. 93 (1988), 225–237.

[20] R. R. Coifman, A. McIntosh, and Y. Meyer, *L'intégrale de Cauchy sur les courbes lipschitziennes,* Ann. of Math. 116 (1982), 361–387.

[21] R. R. Coifman and Y. Meyer, *Au-delà des opèrateurs pseudo-différentiels,* Astérisque no. 57, Société Math. de France.

[22] R. R. Coifman and G. Weiss, *Analyse harmonique non-commutative sur certains éspaces homogenes,* Lecture Notes in Math. no. 242, Springer-Verlag, 1974.

[23] G. David and J.-L. Journé, *A boundedness criterion for generalized Calderón-Zygmund operators,* Ann. of Math. 120 (1984), 371–397.

[24] E. DeGiorgi, *Sulla differenziabilità ed analiticità delle estremali degli integrali multipli regolari,* Mem. Acad. Sci. Torino 3 (1957), 25–43.

[25] E. Fabes, M. Jodeit, Jr., and N. Rivière, *Potential techniques for boundary value problems on C^1 domains*, Acta Math. 141 (1978), 165–186.

[26] C. Fefferman, *Inequalities for strongly singular convolution operators*, Acta Math. 124 (1970), 9–36.

[27] ———, *The Bergman kernel and biholomorphic mappings of pseudoconvex domains*, Inv. Math. 26 (1974), 1–65.

[28] G. B. Folland and E. M. Stein, *Estimates for the $\bar{\partial}_b$-complex and analysis on the Heisenberg group*, Comm. Pure Appl. Math. 27 (1974), 429–522.

[29] I. M. Gelfand, in Russian, Math. Surveys 15 (1960), 103–113.

[30] L. Hörmander, *Pseudo-differential operators*, Comm. Pure Appl. Math. 18 (1965), 501–507.

[31] B. Jessen, J. Marcinkiewicz, and A. Zygmund, *Note on the differentiability of multiple integrals*, Fund. Math. 25 (1935), 217–234.

[32] B. F. Jones, Jr., *A class of singular integrals*, Amer. J. of Math. 86 (1964), 441–462.

[33] J.-L. Journé, *Calderón-Zygmund operators on product spaces*, Rev. Mat. Iberoamer. 1 (1985), 55–91.

[34] C. E. Kenig, *Harmonic analysis techniques for second order elliptic boundary value problems*, CBMS, Regional Conference Series in Mathematics no. 83 A.M.S., 1994.

[35] A. W. Knapp and E. M. Stein, *Intertwining operators for semi-simple groups*, Ann. of Math. 93 (1971), 489–578.

[36] J. J. Kohn and L. Nirenberg, *An algebra of pseudo-differential operators*, Comm. Pure Appl. Math. 18 (1965), 269–305.

[37] A. Korányi and S. Vági, *Singular integrals in homogeneous spaces and some problems of classical analysis*, Ann. Scuola Norm. Sup. Pisa 25 (1971), 575–648.

[38] M. Machedon, *Szegö kernels on pseudo-convex domains with one degenerate eigenvalue*, Ann. of Math. 128 (1988), 619–640.

[39] J. Marcinkiewicz, *Sur les séries de Fourier*, Fund. Math. 27 (1936), 38–69.

[40] ———, *Sur l'interpolation d'opérations*, C. R. Acad. Sci. Paris 208 (1939a), 1272–1273.

[41] ———, *Sur les multiplicateurs des séries de Fourier*, Studia Math. 8 (1939b), 78–91.

[42] J. Marcinkiewicz and A. Zygmund, *On a theorem of Lusin*, Duke Math. J. 4 (1938), 473–485.

[43] J. McNeal and E. M. Stein, *The Szegö projection in convex domains*, Math. Zeit 224 (1997), 519–553.

[44] Y. Meyer, *Ondelettes et Opérateurs* (vols. I and II), Hermann, 1990.

[45] Y. Meyer and R. R. Coifman, *Ondelettes et Opérateurs* (vol. III), Hermann, 1991.

[46] D. Müller, F. Ricci, and E. M. Stein, *Marcinkiewicz multipliers and multi-parameter structure on Heisenberg groups I*, Inv. Math. 119 (1995), 199-233.

[47] A. Nagel, F. Ricci, and E. M. Stein, 1999, in preparation.

[48] A. Nagel, J. P. Rosay, E. M. Stein, and S. Wainger, *Estimates for the Bergman and Szegö kernels in $\mathbb{C}^2$*, Ann. of Math. 129 (1989), 113–149.

[49] J. Nash, *Continuity of solutions of parabolic and elliptic equations*, Amer. J. of Math. 80 (1958), 931–954.

[50] O. Nikodym, *Sur les ensembles accessibles*, Fund. Math. 10 (1927), 116-168.

[51] R. Palais, *Seminar on the Atiyah-Singer index theorem*, Ann. of Math. Study no. 15, Princeton Univ. Press, 1965.

[52] D. H. Phong and E. M. Stein, *Hilbert integrals, singular integrals and Radon transforms I*, Acta Math. 157 (1986), 99–157.

[53] I. Privalov, *Sur une généralisation du théorème de Fatou*, Mat. Sbornik 3 (1923), 232–235.

[54] M. Riesz, *Sur les fonctions conjuguées*, Mat. Zeit. 27 (1927), 218–244.

[55] F. Ricci and E. M. Stein, *Multiparameter singular integrals and maximal functions*, Ann. Inst. Fourier 42 (1992), 637–670.

[56] L. P. Rothschild and E. M. Stein, *Hypoelliptic differential operators and nilpotent groups,* Acta Math. 137 (1976), 247–320.

[57] S. Saks, *On the strong derivatives of functions of intervals,* Fund. Math. 25 (1935), 235–252.

[58] A. Seeger, *Singular integral operators with rough kernels,* Jour. of the A.M.S. 9 (1996), 95–106.

[59] R. T. Seeley, *Singular integrals on compact manifolds,* Amer. J. of Math. 81 (1959), 658–690.

[60] ———, *Refinement of the functional calculus of Calderón and Zygmund,* Indag. Math 27 (1965), 167–204.

[61] ———, *Elliptic singular integrals,* Proc. Symp. Pure Math. 10 (1967), 308–315.

[62] D. C. Spencer, *A function-theoretic identity,* Amer. J. of Math. 65 (1943), 147–160.

[63] E. M. Stein, *Singular Integrals and Differentiability Properties of Functions,* Princeton University Press, 1970.

[64] ———, *Some problems in harmonic analysis suggested by symmetric spaces and semisimple groups,* in Proc. Int. Congress Math. Nice I, 173–189, Gauthier-Villars, 1970.

[65] ———, *The development of square functions in the work of A. Zygmund,* Bull. Amer. Math. Soc. 7 (1982), 359–376.

[66] ———, *Harmonic Analysis.* Princeton University Press, 1992.

[67] E. M. Stein and S. Wainger, *Problems in harmonic analysis related to curvature,* Bull. Amer. Math. Soc. 84 (1978), 1239–1295.

[68] T. Tao, *Three regularity results in harmonic analysis,* Ph.D. dissertation, Princeton University.

[69] E. C. Titchmarsh, *On conjugate functions,* Proc. London Math. Soc. 29 (1929), 49–80.

[70] J. Unterberger and J. Bokobza, *Les opérateurs de Calderón-Zygmund précisés,* C. R. Acad. Sci. Paris 259 (1964), 1612–1614.

[71] A. Zygmund, *On the differentiability of multiple integrals,* Fund. Math. 23 (1934), 143–149.

[72] ———, *Complex methods in the theory of Fourier Series,* Bull. A.M.S. 49 (1943), 805–822.

[73] ———, *On a theorem of Marcinkiewicz concerning interpolation of operations,* Jour. de Math. 35 (1956), 223–248.

[74] ———, *Trigonometric Series,* 2d ed., Cambridge Univ. Press, 1953.

Chapter 2

Transference Principles in Ergodic Theory

Alexandra Bellow

The purpose of this paper is not to give a survey, but rather to discuss a few selected topics having to do with transference in Ergodic Theory.

2.1 Classical Transference Principle

Let me first give some historical background.

Birkhoff's proof of the Pointwise Ergodic Theorem appeared in 1931 [1] and received a great deal of attention. There have been many proofs of Birkhoff's Pointwise Ergodic Theorem, in particular many proofs via the Maximal Ergodic Inequality. Let me mention two of the early ones, both going back to 1939: the Kakutani-Yosida paper [2] giving a sharp form of the Maximal Ergodic Inequality and Wiener's celebrated paper on the Dominated Ergodic Theorem [3] giving the dominated estimate in L^p for the Ergodic Maximal Function. More on Wiener's paper later.

Now there is another celebrated paper that preceded Birkhoff's Proof of the Ergodic Theorem by one year, namely the Hardy-Littlewood paper in *Acta Mathematica*, 1930 [4] (the years 1930, 1931 were spectacular years for analysis). In their fundamental paper, Hardy and Littlewood introduced what is now known as the Hardy-Littlewood Maximal Function and proved various maximal inequalities, dominated estimates in L^p, $1 < p < \infty$, and much more. Having some knowledge of the game of cricket helps in developing a feeling for the "basic inequality" in the early part of the paper. For

Research partially supported by NSF grant DMS-9303327.

it was the game of cricket—Hardy's "other" great passion— that motivated this "basic inequality." For Hardy, who practiced Mathematics for the internal esthetics of Mathematics, this may be the closest he ever came to practical applications in his mathematical work. At any rate, one quickly realizes that this "basic inequality" implies what is now known as the Hardy-Littlewood Weak-Type Maximal Inequality, even though the Weak-Type Maximal Inequality is not explicitly stated as such in the Hardy-Littlewood paper.

If one thinks of the set of integers $\mathbb{Z}$ as a totally σ-finite measure space with mass 1 at each point, then translation by 1 is a measure-preserving transformation and the Hardy-Littlewood Maximal Function is precisely the Maximal Ergodic Function. Thus the Weak-Type Inequality and the Dominated Estimates in L^p for the Hardy-Littlewood Maximal Function follow from the corresponding estimates for the Ergodic Maximal Function.

The remarkable fact is that the converse is also true: The Maximal Ergodic Inequality and the Dominated Ergodic Estimates in L^p can be derived from the corresponding results for the Hardy-Littlewood Maximal Function. This is now known as the basic "transference principle" in Ergodic Theory. To the best of my knowledge, the story of the transference principle begins in 1939 with N. Wiener's paper on the Dominated Ergodic Theorem. Wiener first proves a version of the Hardy-Littlewood Maximal Inequality using a Vitali-type covering argument and then uses a transference argument to derive the corresponding Maximal Inequality in the abstract dynamical system. The technique is roughly this: consider the function along the orbit, one orbit at a time, then integrate and apply Fubini's Theorem making use of the measure-preserving character of the transformations. My guess is that this was a natural approach for Wiener to try. As a very young man, he had spent some time at Cambridge University, where he came under Hardy's influence. Wiener greatly admired Hardy, whom he considered his "master in mathematical training." So it should not come as a surprise that when he learned about the Maximal Ergodic Inequality, he tried to derive it from the Hardy-Littlewood Inequality.

It appears that Aurel Wintner was also aware of the possibility of transferring a deterministic inequality to the ergodic setting. Wintner suggested the idea to Hartman who implemented it in yet another proof of the Maximal Ergodic Inequality [5]; in fact, he derived the Maximal Ergodic Inequality from F. Riesz's Sunrise Lemma, via a transference argument.

Finally, in 1968, Alberto Calderón formulated the general Transference Principle in Ergodic Theory in his beautiful paper "Ergodic Theory and translation-invariant operators" [6]. His motivation was to derive the Ergodic Hilbert Transform of Cotlar from the ordinary Hilbert Transform.

People in Harmonic Analysis were quick to grasp the Calderón Transference Principle. Motivated also by the Calderón-Zygmund work on singular integrals (in particular the rotation method that allowed the transference of results about the Hilbert Transform from the 1-dimensional to the n-dimensional setting), and beginning in the early 1970s, R. Coifman and G. Weiss embarked on a remarkable program to extend the framework of the transference principle further: they replaced the group $\mathbb{Z}$ or $\mathbb{R}$ by a more general (locally compact, amenable) group G and the measure-preserving action by a continuous representation of G acting on some Banach space. This has been a very active area of research, in which Earl Berkson and his collaborators also figure prominently. But I shall not dwell on this, because this is somewhat outside the scope of the paper. I would like to recall, however, that Coifman and Weiss [7] showed how to transfer the strong-type (p,p) maximal inequality in the case when the operators of the representation are positivity-preserving, using the lovely transference argument of A. de la Torre [8]. For transference of strong-type (p,p) maximal inequalities, respectively weak-type maximal inequalities in the case of separation-preserving representations, see [9] (this paper also contains some interesting counterexamples), respectively [10]; see also [11].

It is a curious fact that the Calderón Transference Principle took longer to reach the ergodic circles per se. There were still books on Ergodic Theory appearing in the 1980s—excellent books otherwise—that barely mentioned Calderón's paper, unaware of the transference principle. In the 1980s I recall hearing J. Bourgain lecture on the proof of the remarkable ergodic theorems along the sequence of squares and along the sequence of primes ([12], [13], [14]) and making use of transference arguments—that was a novelty in those days. I learned about the Calderón Transference Principle in 1987. Calderón gave me a reprint of his paper and said: "You may find this of some interest." Did I ever! By now the Calderón Transference Principle has become a standard tool, a classic in Ergodic Theory, and I am happy to report that singular integrals are a "household word" in Ergodic Theory as well.

Transference techniques, in various guises, have become common practice in Ergodic Theory. Let me give a small sample. In [15], Jones and Olsen show how to transfer a maximal inequality from one positive invertible isometry on L^p to any other, while later, in [16], Jones, Olsen, and Wierdl show how to transfer oscillation (variational) inequalities from the integers to positive invertible isometries. This allows them, using Akcoglu's Dilation Theorem, to extend various pointwise ergodic theorems, known in the measure-preserving case, to Dunford-Schwartz operators [15], respectively, to positive contractions on L^p [16]. Another interesting ergodic theorem obtained via transference arguments is given in [17]. See also the important, beautiful recent work on ergodic theorems in the case of non-abelian groups of Nevo [18] and Nevo and Stein [19].

I shall now state the Calderón Transference Principle not in its full generality, but in the discrete case, in a form which suffices for our purposes.

Let μ be a probability on $\mathbb{Z}$. If $\varphi \in \ell^1(\mathbb{Z})$ define

$$(\mu\varphi)(p) = \sum_{j \in \mathbb{Z}} \mu(j)\varphi(p+j), \text{ for } p \in \mathbb{Z}.$$

Let $(X, \mathcal{A}, m, \tau)$ be an abstract dynamical system (that is, $(X, \mathcal{A}, m)$ is a probability space and $\tau : X \to X$ an invertible measure-preserving transformation). If $f \in L^1 = L^1(X)$ define

$$(\mu f)(x) = \sum_{j \in \mathbb{Z}} \mu(j)f(\tau^j x), \text{ for } x \in X.$$

Note that we use Greek letters to denote functions on $\mathbb{Z}$ and ordinary letters to denote functions on X.

Theorem 2.1 (Transference Principle). *Let (μ_n) be a sequence of probability measures on $\mathbb{Z}$. Consider the assertions:*
(i) There is a constant $C > 0$ such that

$$\# \left\{ j \in \mathbb{Z} : \sup_n (\mu_n \varphi)(j) > \lambda \right\} \leq \frac{C}{\lambda} \|\varphi\|_1 \text{ for all } \varphi \in \ell^1_+(\mathbb{Z}), \lambda > 0$$

(that is, we have a weak-type $(1,1)$ inequality on $\ell^1(\mathbb{Z})$).
(ii) There is a constant $C > 0$ such that for every abstract dynamical system $(X, \mathcal{A}, m, \tau)$ we have

$$m \left\{ x \in X : \sup_n (\mu_n f)(x) > \lambda \right\} \leq \frac{C}{\lambda} \|f\|_1 \text{ for all } f \in L^1_+(X), \lambda > 0$$

(that is, we have a weak-type $(1,1)$ inequality in the ergodic context, i.e., on $L^1(X)$).
Then (i) $\Longrightarrow$ (ii).

Comments:

1. The constant C is the same in (i) and (ii).

2. One can remove the restriction on the abstract dynamical system that the total mass $m(X) = 1$; one can allow σ-finite measure spaces $(X, \mathcal{A}, m)$ and one obtains an equivalent statement.

3. The Transference Principle also applies if we replace the weak-type $(1,1)$ estimate by a "weak-type (p,p) estimate" (respectively a "strong-type (p,p) estimate") for $1 < p < \infty$. The moral of the story is that if we have the estimate for the dynamical system of the integers with translations, we can derive it for any other dynamical system.

4. Examples.

(a) If one sets $\mu_n = \frac{1}{n}\sum_{j=0}^{n-1}\delta_j$ (one-sided Cesaro averages), then (i) yields the Hardy-Littlewood Maximal Inequality

$$\#\left\{p \in \mathbb{Z} : \sup_n \left|\frac{1}{n}\sum_{j=0}^{n-1}\varphi(p+j)\right| > \lambda\right\} \le \frac{C}{\lambda}\|\varphi\|_1$$

for all $\varphi \in \ell^1(\mathbb{Z})$, $\lambda > 0$. (ii) yields the Maximal Ergodic Inequality

$$m\left\{x \in X : \sup_n \left|\frac{1}{n}\sum_{j=0}^{n-1}f(\tau^j x)\right| > \lambda\right\} \le \frac{C}{\lambda}\|f\|_1$$

for all $f \in L^1(X)$, $\lambda > 0$.

Alternatively, one can use the two-sided Cesaro averages

$$\mu_n = \frac{1}{2n+1}\sum_{j=-n}^{n}\delta_j.$$

(b) Actually the Calderón Transference Principle is general enough to include also the "Ergodic Hilbert Transform"; i.e., Cotlar's Ergodic Hilbert Transform can be derived from the Classical Hilbert Transform. The reason for this is that we need not restrict the μ_n's in the Transference Principle to be probabilities, we can allow real-valued μ_n's. As a matter of fact, the Calderón Transference Principle applies to sublinear operators, as long as they commute with translations and are "semilocal" [6]. In particular, one can obtain square-function estimates and variational inequalities of the type Bourgain considered in his work. But we shall not go into this. For our purposes the above form of the Calderón Transference Principle suffices.

2.2 Applications to Convolution Powers

Calderón has also had a hand in this in recent years. When Alberto Calderón and I got married in 1989, I called my adviser, Professor Kakutani, to let him know. His response was: "That's wonderful news, for Ergodic Theory. This means we may get Alberto Calderón interested in Ergodic Theory again."

Consider now the abstract dynamical system $(X, \mathcal{A}, m, \tau)$, μ a probability on $\mathbb{Z}$, and for $f : X \to \mathbb{R}$ define as before

$$(\mu f)(x) = \sum_{j \in \mathbb{Z}}\mu(j)f(\tau^j x), \; x \in X.$$

This defines a weighted average, a linear operator which is in fact a contraction in every L^p space. Now iterate this operator n times

$$(\mu^n f)(x) = \sum_{j \in \mathbb{Z}} \mu^n(j) f(\tau^j x).$$

This corresponds in the right-hand side to the n-fold convolution μ^n of the probability measure μ. In probabilistic language this means we are considering a summation associated with a random walk on $\mathbb{Z}$, where the position of the particle after n steps is given by μ^n.

There is another reason why the operator $f \to \mu f$ is of interest. In 1951 Kakutani noted that this operator corresponded to a stationary Markov chain with phase space X [20]. In fact, if we set

$$P(n, x, A) = \sum_{j \in \mathbb{Z}} \mu^n(j) \chi_A(\tau^j x) = \int_{\mathbb{Z}} \chi_A(\tau^j x) d\mu^n(j),$$

where $\chi_A =$ the indicator function of $A \in \mathcal{A}$, then

$$(\mu^n f)(x) = \int_X f(y) P(n, x, dy).$$

From the fact that τ preserves the measure m, it follows that

$$\int_X P(1, x, A) dm(x) = m(A).$$

Thus the initial distribution m is invariant under $P(1, x, A)$, and we are dealing with a stationary Markov chain with transition probability $P(1, x, A)$.

However, in order to obtain good convergence properties of the sequence $((\mu^n f)(x))$, we need to impose some restrictions on μ. To illustrate this, take

$$\mu = \frac{1}{2}(\delta_0 + \delta_1).$$

When we iterate this (under convolution) we get the binomial average

$$\mu^n = \frac{1}{2^n} \left(\sum_{k=0}^n \binom{n}{k} \delta_k \right)$$

and hence

$$(\mu^n f)(x) = \frac{1}{2^n} \sum_{k=0}^n \binom{n}{k} f(\tau^k x), \quad x \in X.$$

It is well known and not difficult to see that the Mean Ergodic Theorem holds, i.e., that these binomial averages converge in the mean in $L^p(X)$ for every $1 \le p < \infty$

$$\mu^n f \xrightarrow{L^p} \tilde{f}, \quad \tilde{f}(\tau x) = \tilde{f}(x), \quad x \in X.$$

On the other hand, a.e. convergence breaks down. In fact, it was first observed by J. Rosenblatt [21] that if τ is ergodic, then the "strong sweeping out property" holds, that is: Given any $\varepsilon > 0$, it is possible to find $E \in \mathcal{A}$, $m(E) < \varepsilon$ such that

$$\limsup_n (\mu^n \chi_E)(x) = 1 \text{ a.e. } (m),$$

$$\liminf_n (\mu^n \chi_E)(x) = 0 \text{ a.e. } (m).$$

In fact, as was shown in [22], this phenomenon, the "strong sweeping out property," happens whenever we have a μ with finite second moment

$$\sum_{k \in \mathbb{Z}} k^2 \mu(k) < \infty$$

and

$$E(\mu) = \sum_{k \in \mathbb{Z}} k\mu(k) \neq 0.$$

The probability $\frac{1}{2}\delta_0 + \frac{1}{2}\delta_1$ is just the prototype of this behavior. To rule out this pathological behavior, we must impose some restrictions on μ.

Let μ be a probability on $\mathbb{Z}$. Let $\hat{\mu}(t) = \sum_{k \in \mathbb{Z}} \mu(k) \exp(-2\pi i k t)$ be its Fourier transform. We say that the probability μ has "bounded angular ratio" if the range of the Fourier transform $\{\hat{\mu}(t) : t \in [0, 1)\}$ is contained in some proper Stolz domain. Equivalently, this means that $|\hat{\mu}(t)| < 1$ for all $t \in [0, 1)$, $t \neq 0$ and

$$\sup_{t \in [0,1)} \frac{|1 - \hat{\mu}(t)|}{1 - |\hat{\mu}(t)|} < \infty.$$

Now there was a remarkable ergodic theorem proved in 1961 for the iterates of a self-adjoint operator. The case $p = 2$ was due to Burkholder and Chow [23]; the general case, $1 < p < \infty$, is due to Stein [24]. Let me recall Stein's theorem:

Theorem 2.2 (Stein). *Let S be a linear operator which is simultaneously defined and bounded as an operator $S : L^1(X) \to L^1(X)$ and $S : L^\infty(X) \to L^\infty(X)$. Suppose that*
(1) $\|S\|_1 \leq 1$
(2) $\|S\|_\infty \leq 1$
(3) S is self-adjoint: $S^* = S$ on $L^2(X)$.
Then for $1 < p < \infty$ we have the dominated estimate in L^p; i.e., there is a constant A_p such that

$$\left\| \sup_n |S^n f| \right\|_p \leq A_p \|f\|_p \text{ for } f \in L^p(X).$$

Moreover if S is also positive-definite in the Hilbert space sense, that is $(Sf, f) \geq 0$ for all $f \in L^2(X)$, then

$$\lim_n S^n f(x) \text{ exists a.e. and in the } L^p \text{ norm}$$

The result that follows was inspired by Stein's Theorem:

Theorem 2.3 ([22]). *Assume that the probability μ has bounded angular ratio. Then for $1 < p < \infty$ we have the dominated estimate in L^p; i.e., there exists a constant C_p such that*

$$\left\| \sup_n |\mu^n f| \right\|_p \leq C_p \|f\|_p \text{ for } f \in L^p(X).$$

Moreover, given any $f \in L^p(X)$, there exists a unique τ-invariant function $f^* \in L^p(X)$ such that

$$\lim_n (\mu^n f)(x) = f^*(x) \text{ a.e. and in the } L^p \text{ norm.}$$

Comments:

1. If the probability μ is symmetric, i.e. $\mu(-k) = \mu(k)$ and strictly aperiodic (this means that $|\hat{\mu}(t)| < 1$ for all $t \in [0, 1)$, $t \neq 0$), then the range of the Fourier transform is contained in some interval $[-1 + \delta, 1]$, so we trivially have "bounded angular ratio" and the previous Theorem applies. Actually in this case the operator $Sf = \mu f$ which is an L^1- and L^∞-contraction, is also self-adjoint, $S^* = S$, so we may use Stein's Theorem. Thus Theorem 2.3 above can be regarded as an extension of Stein's Theorem to the non-self-adjoint case.

2. There are probability measures μ that have bounded angular ratios, but are far from being symmetric, in fact have support entirely contained in $\mathbb{Z}^+$, as was observed in [22].

3. The "bounded angular ratio" condition allows us to go over to the Fourier transform side and use the spectral method to get the Dominated Estimate in L^2; then we proceed as in Stein's proof, use his complex interpolation theorem to get the strong maximal inequality in L^p, $p > 1$.

The fact that μ is strictly aperiodic means, by the Choquet-Deny theorem, that μ^n are asymptotically invariant under translation, that is, $\|\mu^n * \delta_1 - \mu^n\|_1 \to 0$. This allows us to obtain easily a dense set of functions in L^p for which we have a.e. convergence, namely (see [22])

$$\{f_1 + (f_2 \circ \tau - f_2); f_1 \in L^p \text{ is } \tau - \text{invariant}, \ f_2 \in L^\infty\}$$

(respectively,

$$\{f_1 + (f_2 \circ \tau - f_2); f_1 \in L^p \text{ is } \tau - \text{invariant}, \ f_2 \in L^\infty \cap L^1\}$$

if the measure is infinite).

Question: Does a.e. convergence also hold for $f \in L^1(X)$ in Theorem 2.3? In particular, is this always the case when the probability μ is symmetric and strictly aperiodic?

An old paper of Oseledets [25] contains a "positive" answer to the latter question, but it appears that the proof is incomplete.

The question is also of interest from the point of view of convolutions on $\mathbb{Z}$. By Sawyer's theorem and the Transference Principle, this is equivalent to asking whether for the convolutions on $\mathbb{Z}$, the corresponding maximal function $M\varphi = \sup_n |\mu^n * \varphi|$ is weak-type $(1,1)$; i.e., does there exist $C = C(\mu) > 0$ such that

$$\# \left\{ p \in \mathbb{Z} : \sup_n |(\mu^n * \varphi)(p)| > \lambda \right\} \leq \frac{C}{\lambda} \|\varphi\|_1 \text{ for all } \varphi \in \ell^1(\mathbb{Z}), \lambda > 0?$$

This is an elementary but very natural question to which we should have an answer.

We know the answer in some special cases.

Theorem 2.4. *If μ is a symmetric probability on $\mathbb{Z}$ and $\mu(k) \geq \mu(k+1)$ for all $k \geq 0$, then the answer to the above Question is yes.*

The reason for this is that in this case μ, and each μ^n can be written as a convex combination of the symmetric version of the usual ergodic averages, the Cesaro averages

$$C_p = \frac{1}{2p+1} \sum_{j=-p}^{p} \delta_j,$$

and hence $\sup_n |\mu^n f|$ is dominated by the Maximal Ergodic function (see [22]).

Theorem 2.5. *Let μ be a probability on $\mathbb{Z}$ which is strictly aperiodic, has expectation zero ($E(\mu) = \sum_k k\mu(k) = 0$), and a finite second moment ($\sum_k k^2\mu(k) < \infty$). Then again the answer to the above Question is yes.*

The proof is not trivial and is given in the next chapter.

Our motivation for looking at probabilities μ with a second moment was an interesting result of K. Reinhold-Larsson [26] (she considered moments $> 2 + \delta$). We also had the mathematical assistance of Pablo Calderón, who prevented us from going awry in this problem. We were trying to prove a similar result for probabilities μ with a moment of order $p > 1$ and $E(\mu) = 0$.

This simply is not true. In the meantime, explicit counterexamples showing this were constructed independently by G. Chistyakov [27] and V. Losert [28]. Thus, if we restrict our attention to probabilities on $\mathbb{Z}$ that have a moment of some order and expectation zero, then $p = 2$ is the best one can do; $p = 2$ is "sharp."

Comments:

1. In the aftermath of Stein's remarkable 1961 Ergodic Theorem for the iterates of a self-adjoint operator, people tried to figure out what happens in the case of L^1 with a.e. convergence. It took a while before a counterexample was found: Ornstein came up with a counterexample in 1968. The self-adjoint operator in Ornstein's counterexample is of the form $S = PQP$, where P and Q are conditional expectation operators [29].

2. The proofs of a.e. convergence for $f \in L^2$ are concise, elegant, and make use of spectral methods. The case of L^1 is quite different. L^1 stands apart from L^2 and L^p, $p > 1$. The techniques required for proving a.e. convergence in L^1 are quite different (see the proof in the next chapter): covering lemmas, Calderón-Zygmund decomposition, careful estimates of Fourier coefficients, etc.

3. If the answer to the above Question is positive, that would be very interesting indeed, because that would provide a natural class of examples for which Stein's Ergodic Theorem extends all the way to L^1.

4. If the answer to the Oseledets Question is negative, that too would be very interesting. In particular, this would provide an alternative proof of the existence of Ornstein's counterexample with conditional expectations. This would follow from a beautiful old theorem of Burkholder based on his notion of "stochastic convexity" [30].

2.3 Transference via Square Functions

There is a class of results that have been obtained recently and that can also be regarded as transference results, but of a different kind: these results relate one process to another via a square function.

I shall illustrate this with one example which in a sense is the archetype: it relates martingales and the standard differentiation operators.

Let
$$X = [0, 1)(\operatorname{mod} 1),$$

$\mathcal{F}_k = \text{ the dyadic } \sigma-\text{field with } 2^k \text{ atoms } \left[0, \frac{1}{2^k}\right), \left[\frac{1}{2^k}, \frac{2}{2^k}\right), \ldots, \left[\frac{2^{k-1}}{2^k}, 1\right),$

for $k \geq 1$. For $f \in L^1(X)$, define the dyadic martingale

$$f_k = E(f|\mathcal{F}_k), \text{ for } k \geq 1.$$

For $f \in L^1(X)$ and extended periodically to $\mathbb{R}$, define the dyadic differentiation operators

$$(D_k f)(x) = 2^k \int_x^{x+\frac{1}{2^k}} f(t)\,dt, \text{ for } k \geq 1.$$

Consider the square function defined by

$$(Sf)(x) = \left(\sum_{n=1}^{\infty} |(D_n f)(x) - f_n(x)|^2 \right)^{\frac{1}{2}}.$$

Theorem 2.6 ([31]). *The (sublinear) map $f \to Sf$ is weak-type $(1,1)$*

$$m\{x \in X : (Sf)(x) > \lambda\} \leq \frac{C}{\lambda}||f||_1 \text{ for all } f \in L^1(X),$$

and, for each $1 < p < \infty$, it is strong type (p,p), that is

$$||Sf||_p \leq C_p||f||_p \text{ for all } f \in L^p(X).$$

This allows one to transfer information from the dyadic martingales to the differentiation averages and vice versa.

References

[1] G. D. Birkhoff, *Proof of the ergodic theorem*, Proc. Nat. Acad. Sci. USA 17 (1931), 656–660.

[2] K. Yosida and S. Kakutani, *Birkhoff's ergodic theorem and the maximal ergodic theorem*, Proc. Imp. Acad. Tokyo 15 (1939), 165–168.

[3] N. Wiener, *The ergodic theorem*, Duke Math. J. 5 (1939), 1–18.

[4] G. H. Hardy and J. E. Littlewood, *A maximal theorem with function-theoretic applications*, Acta Math. 54 (1930), 81–116.

[5] P. Hartman, *On the ergodic theorems*, Amer. J. Math. 69 (1947), 193–199.

[6] A. P. Calderón, *Ergodic theory and translation-invariant operators*, Proc. Nat. Acad. Sci. USA 59 (1968), 349–353.

[7] R. R. Coifman and G. Weiss, *Transference methods in analysis*, CBMS Regional Conf. Series Math. 31, Amer. Math. Soc., Providence, RI, 1977 (reprinted 1986).

[8] A. de la Torre, *A simple proof of the maximal ergodic theorem*, Can. J. Math. 28 (1976), 1073–1075.

[9] N. Asmar, E. Berkson, and T. A. Gillespie, *Transference of the strong-type maximal inequalities by separation-preserving representations*, Amer. J. Math. 113 (1991), 47–74.

[10] ————, *Transference of the weak-type maximal inequalities by distributionally bounded representations*, Quart. J. Math. Oxford (2), 43 (1992), 259–282.

[11] E. Berkson, J. Bourgain, and T. A. Gillespie, *On the almost everywhere convergence of ergodic averages for power-bounded operators in L^p-subspaces*, Integral Equations and Operator Theory 14 (1991), 678–715.

[12] J. Bourgain, *On the maximal ergodic theorem for certain subsets of the integers*, Isr. J. Math. 61 (1988), 39–72.

[13] ————, *On the pointwise ergodic theorem on L^p for arithmetic sets*, Isr. J. Math. 61 (1988), 73–84.

[14] ————, *Pointwise ergodic theorems for arithmetic sets*, IHES Publication, October 1989.

[15] R. L. Jones and J. Olsen, *Subsequence pointwise ergodic theorems for operators in L^p*, Isr. J. Math. 77 (1992), 33–54.

[16] R. L. Jones, J. Olsen, and M. Wierdl, *Subsequence ergodic theorems for L^p contractions*, Trans. Amer. Math. Soc. 331 (1992), 837–850.

[17] R. L. Jones, *Ergodic averages on spheres*, J. Analyse Math. 61 (1993), 29–45.

[18] A. Nevo, *Harmonic analysis and pointwise ergodic theorems for noncommuting transformations*, J. Amer. Math. Soc. 7 (1994), 875–902.

[19] A. Nevo and E. M. Stein, *A generalization of Birkhoff's pointwise ergodic theorem*, Acta Math. 173 (1994), 135–154.

[20] S. Kakutani, *Random ergodic theorems and Markoff processes with a stable distribution*, Proc. Second Berkeley Symp. Math. Stat. and Prob. II, Univ. of California Press (1951), 247–261.

[21] J. Rosenblatt, *Ergodic group actions*, Arch. Math. 47 (1968), 263–269.

[22] A. Bellow, R. Jones, and J. Rosenblatt, *Almost everywhere convergence of convolution powers*, Ergod. Th. & Dynam. Sys. 14 (1994), 415–432.

[23] D. L. Burkholder and Y. S. Chow, *Iterates of conditional expectation operators*, Proc. Amer. Math. Soc. 12 (1961), 490–495.

[24] E. M. Stein, *On the maximal ergodic theorem*, Proc. Nat. Acad. Sci. 47 (1961), 1894–1897.

[25] V. I. Oseledets, *Markov chains, skew products and ergodic theorems for "general" dynamical systems*, Theor. Prob. Appl. 10 (1965), 499–504.

[26] K. Reinhold-Larsson, *Almost everywhere convergence of convolution powers in $L^1(X)$*, Illinois J. Math. 37 (1993), 666–679.

[27] G. Chistyakov, private communication.

[28] V. Losert, *A remark on almost everywhere convergence of convolution powers*, to appear in Illinois J. Math.

[29] D. S. Ornstein, *On the pointwise behavior of iterates of a self-adjoint operator*, J. Math. Mech. 18 (1968), 473–478.

[30] D. L. Burkholder, *Maximal inequalities as necessary conditions for almost everywhere convergence*, Zeit. Wahrsch. 3 (1964), 75–88.

[31] R. L. Jones, R. Kaufman, J. M. Rosenblatt, and M. Wierdl, *Oscillation in Ergodic Theory*, Ergod. Th. and Dynam. Sys. 18 (1998), 889–935.

Chapter 3

A Weak-Type Inequality for Convolution Products

Alexandra Bellow and Alberto P. Calderón

The purpose of this paper is to give a proof of Theorem 2.5 of the previous chapter. The setting is the same as in the previous chapter. We begin with the following:

Theorem 3.1. *Let (μ_n) be a sequence of probability measures on $\mathbb{Z}$ and for $f : X \to \mathbb{R}$ define the maximal operator*

$$(Mf)(x) = \sup_n |(\mu_n f)(x)|, \ x \in X.$$

We assume
(∗) (Regularity of coefficients). There is $0 < \alpha \leq 1$ and $C > 0$ such that for each $n \geq 1$

$$|\mu_n(x+y) - \mu_n(x)| \leq C \frac{|y|^\alpha}{|x|^{1+\alpha}} \ for \ x, y \in \mathbb{Z}, \ and \ \ 0 < 2|y| \leq |x|.$$

Then the maximal operator M is weak-type $(1,1)$; i.e., there is $C' > 0$ such that for any $\lambda > 0$

$$m\{x \in X : (Mf)(x) > \lambda\} \leq \frac{C'}{\lambda} ||f||_1 \ for \ all \ f \in L^1(X).$$

Because of the Calderón Transference Principle, it suffices to know the validity of the theorem for the dynamical system $\mathbb{Z}$ (with the group $\mathbb{Z}$ acting on itself by translations and the counting measure dy; below, for a set $A \subset \mathbb{Z}$ its measure will be denoted by $|A|$), namely,

Theorem 3.2. *Let (μ_n) be a sequence of probability measures on $\mathbb{Z}$ and for $f : \mathbb{Z} \to \mathbb{R}$ define the maximal operator*

$$(Mf)(x) = \sup_n |(\mu_n f)(x)|, \ x \in \mathbb{Z}.$$

We assume

(*) (Regularity of coefficients). There is $0 < \alpha \leq 1$ and $C > 0$ such that, for each $n \geq 1$,

$$|\mu_n(x + y) - \mu_n(x)| \leq C \frac{|y|^\alpha}{|x|^{1+\alpha}} \ for \ x, y \in \mathbb{Z}, \ 0 < 2|y| \leq |x|.$$

Then the maximal operator M is weak-type $(1,1)$; i.e., there is a $C' > 0$ such that for any $\lambda > 0$

$$|\{x \in \mathbb{Z} : (Mf)(x) > \lambda\}| \leq \frac{C'}{\lambda} \|f\|_1 \ for \ all \ f \in \ell^1 = \ell^1(\mathbb{Z}).$$

Comments:

1. We proved Theorem 3.2 several years ago, unaware of the existence of Zo's paper (see [2], also [1], 292–295). It turns out that Theorem 3.2 above is a special case of Zo's Theorem. In fact, it is enough to notice that, for $0 < 2|y| \leq |x|$,

$$\varphi(x, y) = \sup_n |\mu_n(x + y) - \mu_n(x)| \leq C \frac{|y|^\alpha}{|x|^{1+\alpha}}$$

   and that

$$\int_{\{|x| \geq 2|y|\}} C \frac{|y|^\alpha}{|x|^{1+\alpha}} \, dx \leq C(\alpha) < \infty$$

   independently of $y \in \mathbb{R} - \{0\}$. Thus the basic assumption in Zo's Theorem is satisfied. It should be noted also that our proof of Theorem 3.2 and the proof of Zo's Theorem are very similar: they are direct applications of the Calderón-Zygmund decomposition.

2. The maximal operator M in Theorems 3.1 and 3.2 is obviously strong type (∞, ∞) and hence, by the Marcinkiewicz Interpolation Theorem, also strong type (p, p), for $1 < p < \infty$.

For the sake of completeness we reproduce below our proof.

Proof. We start with $f \in \ell^1_+$ and $\lambda > 0$. Apply the Calderón-Zygmund decomposition to f and $\rho = \frac{\lambda}{4}$. We get dyadic intervals (Q_i), which are pairwise disjoint, such that

$$\rho < \frac{1}{|Q_i|} \int_{Q_i} f(y) \, dy \leq 2\rho \ for \ all \ i$$

and
$$0 \le f(x) \le \rho \text{ for } x \notin \cup_i Q_i.$$

Set
$$f_1(x) = \begin{cases} \frac{1}{|Q_i|} \int_{Q_i} f(y) \, dy & \text{for } x \in Q_i \\ f(x) & \text{for } x \notin \cup_i Q_i \end{cases}$$

and
$$f_2 = f - f_1.$$

We have

(1) $\begin{cases} \rho \le f_1(x) \le 2\rho & \text{for } x \in \cup_i Q_i \\ 0 \le f_1(x) \le \rho & \text{for } x \notin \cup_i Q_i, \end{cases}$

(2) $\begin{cases} \frac{1}{|Q_i|} \int_{Q_i} f_2(y) \, dy = 0 & \text{for each } i \\ f_2(x) = 0 & \text{for } x \notin \cup_i Q_i, \end{cases}$

(3) $| \cup_i Q_i | \rho \le \int_{\cup_i Q_i} f(y) \, dy \le \|f\|_1,$

(4) $\|f_1\|_1 \le \|f\|_1$ and $\|f_2\|_1 \le 2\|f\|_1.$

Since $0 \le f_1 \le 2\rho = \frac{\lambda}{2}$, we have $M f_1 \le \frac{\lambda}{2}$. Also since $f = f_1 + f_2$ and M is sublinear we have

$$\{Mf > \lambda\} \le \left\{ M f_1 > \frac{\lambda}{2} \right\} \cup \left\{ M f_2 > \frac{\lambda}{2} \right\} = \left\{ M f_2 > \frac{\lambda}{2} \right\}.$$

Thus we only need to worry about f_2.

Consider the intervals Q_i^* obtained by dilating Q_i by a factor of 5 (but with the same center) and note that by (3)

(5) $| \cup_i Q_i^* | \le \sum_i |Q_i^*| \le \sum_i 5|Q_i| = 5| \cup_i Q_i | \le \frac{5}{\rho} \|f\|_1 = \frac{20}{\lambda} \|f\|_1.$

Thus when checking the weak-type inequality for f_2 it is enough to consider the complement of $\cup_i Q_i^*$. Fix $x \in S = (\cup_i Q_i^*)^c$, and choose $y_i \in Q_i$ for each i. We have, using (2), for each n,

$$\begin{aligned} (\mu_n * f_2)(x) &= \sum_i \int_{Q_i} \mu_n(x-y) f_2(y) \, dy \\ &= \sum_i \int_{Q_i} [\mu_n(x-y) f_2(y) - \mu_n(x-y_i) f_2(y)] \, dy \\ &= \sum_i \int_{Q_i} [\mu_n(x-y) - \mu_n(x-y_i)] f_2(y) \, dy. \end{aligned}$$

But by assumption (∗) (Regularity of coefficients), since $|y_i - y| \leq |Q_i|$ and $|x - y_i| \geq 2|Q_i|$, we have

$$|\mu_n(x - y) - \mu_n(x - y_i)| \leq C\frac{|y_i - y|^\alpha}{|x - y_i|^{1+\alpha}} \leq C\frac{|Q_i|^\alpha}{|x - y_i|^{1+\alpha}}$$

and hence

$$|(\mu_n * f_2)(x)| \leq \sum_i \int_{Q_i} |f_2(y)|\frac{C|Q_i|^\alpha}{|x - y_i|^{1+\alpha}}\, dy = g(x).$$

This function $g(x)$ defined for $x \in S$ does not depend on n and is integrable:

$$
\begin{aligned}
\int_S g(x)\, dx &= C\int_S\left\{\sum_i \int_{Q_i} |f_2(y)|\frac{|Q_i|^\alpha}{|x-y_i|^{1+\alpha}}\, dy\right\}\, dx \\
&= C\sum_i\left\{\int_S dx\left(\int_{Q_i} |f_2(y)|\frac{|Q_i|^\alpha}{|x-y_i|^{1+\alpha}}\, dy\right)\right\} \\
&= C\sum_i\left\{\int_{Q_i} |f_2(y)|\left(\int_S \frac{|Q_i|^\alpha}{|x-y_i|^{1+\alpha}}\, dx\right)\, dy\right\} \\
&\leq C\sum_i\left\{\int_{Q_i} |f_2(y)||Q_i|^\alpha\left(\int_{(Q_i^*)^c} \frac{1}{|x-y_i|^{1+\alpha}}\, dx\right)\, dy\right\}.
\end{aligned}
$$

But, as observed earlier, $|x - y_i| \geq 2|Q_i|$ for $x \in (Q_i^*)^c$, so that

$$\int_{(Q_i^*)^c} \frac{1}{|x - y_i|^{1+\alpha}}\, dx \leq \frac{c}{|Q_i|^\alpha},$$

where $c = c(\alpha)$ depends only on α, and hence

$$\int_S g(x)\, dx \leq cC\sum_i\left\{\int_{Q_i} |f_2(y)|\, dy\right\} = \mathbf{C}\|f_2\|_1 \leq 2\mathbf{C}\|f\|_1.$$

Since

$$Mf_2(x) \leq g(x) \text{ for } x \in S,$$

it follows that

$$\left|\left\{x \in S : (Mf_2)(x) > \frac{\lambda}{2}\right\}\right| \leq \left|\left\{x \in S : g(x) > \frac{\lambda}{2}\right\}\right|$$

$$\leq \frac{2}{\lambda}\int_S g(x)\, dx \leq \frac{4\mathbf{C}}{\lambda}\|f\|_1.$$

This completes the proof of Theorem 3.2.   □

Below we shall use the notation

$$e(s) = \exp(2\pi i s).$$

Lemma 3.3. *There is a constant $C > 0$ such that, for any $x, y \in \mathbb{R}$, we have $0 < 2|y| \le |x|$, and $t \in \mathbb{R}$*

$$\left| \frac{e((x+y)t) - 1}{(x+y)^2} - \frac{e(xt) - 1}{x^2} \right| \le C|t| \frac{|y|}{x^2}.$$

Proof. Note that

$$\frac{d}{du} \left(\frac{e(ut) - 1}{u^2} \right) = \frac{(2\pi i t)e(ut) \cdot u^2 - 2u(e(ut) - 1)}{u^4}$$

$$= (2\pi i t) \cdot \frac{e(ut)}{u^2} - \frac{2(e(ut) - 1)}{u^3}$$

and that

$$|e(ut)| \le 1 \text{ and } |e(ut) - 1| \le 2\pi|t||u|.$$

Hence

$$\frac{e((x+y)t) - 1}{(x+y)^2} - \frac{e(xt) - 1}{x^2} = \int_x^{x+y} \frac{d}{du} \left(\frac{e(ut) - 1}{u^2} \right) du$$

$$= (2\pi i t) \int_x^{x+y} \frac{e(ut)}{u^2} \, du - 2 \int_x^{x+y} \frac{e(ut) - 1}{u^3} \, du.$$

Since $|x + y| \ge |x| - |y| \ge |x| - \frac{1}{2}|x| = \frac{1}{2}|x|$, we have

$$\left| \int_x^{x+y} \frac{e(ut)}{u^2} \, du \right| \le \left| \frac{1}{x} - \frac{1}{x+y} \right| = \frac{|y|}{|x||x+y|} \le 2 \frac{|y|}{x^2},$$

$$\left| \int_x^{x+y} \frac{e(ut) - 1}{u^3} \, du \right| \le 2\pi|t| \left| \int_x^{x+y} \frac{1}{u^2} \, du \right| \le 4\pi|t| \frac{|y|}{x^2},$$

and the Lemma is proved. $\qquad\square$

Corollary 3.4. *Let (μ_n) be a sequence of probabilities on $\mathbb{Z}$ and let $\theta_n(t) = \hat{\mu}_n(t) =$ the Fourier transform of μ_n, $t \in [-\frac{1}{2}, \frac{1}{2})$. We assume that, for each $n \ge 1$, θ_n is twice continuously differentiable and that*

$$\sup_n \int_{-\frac{1}{2}}^{\frac{1}{2}} |\theta_n''(t)||t| \, dt < \infty.$$

Then the μ_n satisfy the regularity assumption (∗) of Theorem 3.2.

Proof. For $u \in \mathbb{Z}$ we have, using integration by parts (with the appropriate antiderivative) and the fact that the functions involved are periodic of period 1:

$$\mu_n(u) = \int_{-\frac{1}{2}}^{\frac{1}{2}} \theta_n(t) e(ut) \, dt$$

$$= \left[\theta_n(t) \frac{e(ut)}{2\pi i u} \right]_{t=-\frac{1}{2}}^{t=\frac{1}{2}} - \int_{-\frac{1}{2}}^{\frac{1}{2}} \theta_n'(t) \cdot \frac{e(ut)}{2\pi i u} \, dt = - \int_{-\frac{1}{2}}^{\frac{1}{2}} \theta_n'(t) \cdot \frac{e(ut)}{2\pi i u} \, dt$$

$$= - \left[\theta_n'(t) \cdot \frac{e(ut) - 1}{(2\pi i)^2 u^2} \right]_{t=-\frac{1}{2}}^{t=\frac{1}{2}} + \int_{-\frac{1}{2}}^{\frac{1}{2}} \theta_n''(t) \cdot \frac{e(ut) - 1}{(2\pi i)^2 u^2} \, dt = \int_{-\frac{1}{2}}^{\frac{1}{2}} \theta_n''(t) \cdot \frac{e(ut) - 1}{(2\pi i)^2 u^2} \, dt.$$

Now for $x, y \in \mathbb{Z}$, $0 < 2|y| \leq |x|$, using Lemma 3.3, we get

$$|\mu_n(x + y) - \mu_n(x)| = \left| \int_{-\frac{1}{2}}^{\frac{1}{2}} \theta_n''(t) \cdot \left(\frac{e((x + y)t) - 1}{(2\pi i)^2 (x + y)^2} - \frac{e(xt) - 1}{(2\pi i)^2 x^2} \right) \, dt \right|$$

$$\leq \int_{-\frac{1}{2}}^{\frac{1}{2}} |\theta_n''(t)| \cdot C|t| \frac{|y|}{x^2} \, dt = C \frac{|y|}{x^2} \int_{-\frac{1}{2}}^{\frac{1}{2}} |\theta_n''(t)| |t| \, dt,$$

finishing the proof. $\qquad\qquad\qquad\qquad\qquad\qquad\qquad\qquad\qquad\qquad\qquad\square$

Corollary 3.5. *Let μ be a probability on $\mathbb{Z}$ which is strictly aperiodic, has expectation zero (that is, $E(\mu) = \sum_k k\mu(k) = 0$), and finite second moment (that is, $\sum_k k^2 \mu(k) < \infty$). Let $\theta(t) = \hat{\mu}(t)$. Then*

$$\sup_n \int_{-\frac{1}{2}}^{\frac{1}{2}} |(\theta^n)''(t)| |t| \, dt < \infty,$$

and hence the sequence (μ^n) satisfies the regularity assumption $(*)$ of Theorem 3.2.

Proof. First note that θ is twice continuously differentiable, that the Fourier transform of μ^n is θ^n, and that

$$(1) \quad (\theta^n)''(t) = n(n - 1)\theta^{n-2}(t) \cdot (\theta'(t))^2 + n\theta^{n-1}(t) \cdot \theta''(t).$$

Since

$$\theta(t) = \sum_k \mu(k) e(-tk),$$

we have

$$\theta'(t) = \sum_k (-2\pi i k) \mu(k) e(-tk) \text{ and } \theta'(0) = -2\pi i \sum_k k\mu(k) = 0,$$

$$\theta''(t) = \sum_k -(4\pi^2)k^2\mu(k)e(-tk) = -(4\pi^2)\left[\sum_k k^2\mu(k)e(-tk)\right],$$

$$\theta''(0) = -(4\pi)^2\left[\sum_k k^2\mu(k)\right] = -a, \text{ where } a = \sum_k 4\pi^2k^2\mu(k) > 0.$$

In particular,

$$|\theta'(t)| = |\theta'(t) - \theta'(0)| = \left|-\sum_k (2\pi ik)\mu(k)[e(-tk) - 1]\right|$$

$$\leq \sum_k 2\pi|k|\mu(k)2\pi|k||t| = \left(\sum_k 4\pi^2k^2\mu(k)\right)|t|$$

so that

$$(2) \quad |\theta'(t)| \leq a|t| \text{ for all } |t| \leq \frac{1}{2}$$

and

$$(3) \quad |\theta''(t)| \leq a \text{ for all } |t| \leq \frac{1}{2}.$$

By Taylor's formula,

$$\theta(t) = 1 - \frac{a}{2}t^2 + o(t^2) \text{ as } t \to 0.$$

Compare $\theta(t)$ with

$$\exp(-\varepsilon t^2) = 1 - \varepsilon t^2 + o(t^2) \text{ as } t \to 0.$$

There is $\delta > 0$ sufficiently small such that for $|t| \leq \delta$ we have

$$|\theta(t)| \leq 1 - \frac{a}{2}t^2 + \frac{a}{4}t^2 = 1 - \frac{a}{4}t^2$$

and

$$\exp(-\varepsilon t^2) \geq 1 - \varepsilon t^2 - \varepsilon t^2 = 1 - 2\varepsilon t^2 \geq |\theta(t)|$$

provided $2\varepsilon \leq \frac{a}{4}$. Since

$$\sup_{\substack{|t| \geq \delta \\ t \in [-\frac{1}{2}, \frac{1}{2})}} |\theta(t)| = b < 1$$

by choosing a conveniently small ε we get

$$(4) \quad |\theta(t)| \leq \exp(-\varepsilon t^2) \text{ for all } |t| \leq \frac{1}{2}.$$

By (1), (2), (3), and (4) we can now estimate

$$|(\theta^n)''(t)| \cdot |t| \leq n(n-1)\exp(-(n-2)\varepsilon t^2) \cdot a^2|t|^3 + n\exp(-(n-1)\varepsilon t^2) \cdot a|t|$$

so we only need to check the boundedness (in n) of the integrals

$$\int_{\mathbb{R}} n(n-1)\exp(-(n-2)\varepsilon t^2) \cdot |t|^3 \, dt$$

and

$$\int_{\mathbb{R}} n\exp(-(n-1)\varepsilon t^2) \cdot |t| \, dt,$$

and this is easily done. $\square$

References

[1] M. de Guzmán, *Real Variable Methods in Fourier Analysis*, North Holland Publ. Co., 1981, Mathematics Studies 46, Amsterdam, New York, Oxford.

[2] F. Zo, *A note on the approximation of the identity*, Studia Math. 55 (1976), 111–122.

Chapter 4

Transference Couples and Weighted Maximal Estimates

Earl Berkson, Maciej Paluszyński, and Guido Weiss

Suppose that $1 < p < \infty$. We apply the notion of transference couple in order to transfer maximal multiplier transforms, along with their bounds, from $L^p(\mathbb{T})$ to $L^p(\omega)$, where ω belongs to a certain class W_p of weight functions on $\mathbb{R}$ which satisfy the A_p condition of Muckenhoupt. The structural simplicity afforded by the use of transference couples permits us to rely on elementary tools throughout, and, in the special case of the weights $\omega \in W_p$, our methods generalize the weighted Carleson-Hunt Theorem in the direction of a wide class of maximal multiplier transforms acting on $L^p(\omega)$. The requisite features of transference couples are summarized in §4.3, where we also indicate how transference methods for square function estimates can be extended to the setting of transference couples.

4.1   Introduction and Notation

It will be convenient in all that follows to adopt the convention that the symbol "K" with a (possibly empty) set of subscripts denotes a nonnegative real constant which depends only on its subscripts, and which can change in value from one occurrence to another. The Fourier transform $\widehat{f}$ of a function $f \in L^1(\mathbb{R})$ will be given by $\widehat{f}(y) \equiv \int_{\mathbb{R}} f(t)\, e^{-2\pi i t y}\, dt$. For $u \in \mathbb{R}$ and

Research by Earl Berkson partially supported by NSF grant DMS 9401009; research by Maciej Paluszyński partially supported by a grant from the Southwestern Bell Company; research by Guido Weiss partially supported by NSF grant DMS 9302828.

$f : \mathbb{R} \to \mathbb{C}$, we define $\tau_u f : \mathbb{R} \to \mathbb{C}$ by writing $(\tau_u f)(x) \equiv f(x+u)$. We begin the discussion by recalling a few background items concerning weights, fixing some notation in the process. In all that follows (except as noted otherwise in §4.3), p will be a fixed index satisfying $1 < p < \infty$. A measurable function $\omega : \mathbb{R} \to [0, \infty]$ such that $0 < \omega < \infty$ a.e. (respectively, a sequence of positive real numbers $\mathfrak{w} \equiv \{\mathfrak{w}_k\}_{k=-\infty}^{\infty}$) will be called a *weight function* on $\mathbb{R}$ (respectively, a *weight sequence*). We shall denote by $A_p(\mathbb{R})$ the class of weight functions on $\mathbb{R}$ satisfying the A_p condition of Muckenhoupt, which is stated as follows.

Definition 4.1. A weight function ω on $\mathbb{R}$ belongs to the class $A_p(\mathbb{R})$ provided that there is a real number C (called *an $A_p(\mathbb{R})$ weight constant for ω*) such that for all compact intervals Q whose length $|Q|$ is positive, we have

$$\left(\frac{1}{|Q|} \int_Q \omega(t)\, dt\right)\left(\frac{1}{|Q|} \int_Q [\omega(t)]^{-1/(p-1)}\, dt\right)^{p-1} \le C. \qquad (4.1)$$

Similarly, we denote by $A_p(\mathbb{Z})$ the class of weight sequences satisfying the discrete counterpart of (4.1).

Definition 4.2. A weight sequence $\mathfrak{w} \equiv \{\mathfrak{w}_k\}_{k=-\infty}^{\infty}$ belongs to the class $A_p(\mathbb{Z})$ provided that there is a real number C (called *an $A_p(\mathbb{Z})$ weight constant for $\mathfrak{w}$*) such that

$$\left(\frac{1}{M-L+1} \sum_{k=L}^{M} \mathfrak{w}_k\right)\left(\frac{1}{M-L+1} \sum_{k=L}^{M} (\mathfrak{w}_k)^{-1/(p-1)}\right)^{p-1} \le C,$$

whenever $L \in \mathbb{Z}$, $M \in \mathbb{Z}$, and $L \le M$.

For an extensive treatment of $A_p(\mathbb{R})$, we refer the reader to [10]. Fundamental facts regarding $A_p(\mathbb{Z})$ can be found in [11].

Let $\omega \in A_p(\mathbb{R})$. We shall symbolize by $M_{p,\omega}(\mathbb{R})$ the class consisting of the multipliers for $L^p(\omega)$. By definition, a multiplier for $L^p(\omega)$ is a function $\phi \in L^\infty(\mathbb{R})$ such that the mapping $f \mapsto \left(\phi \widehat{f}\right)^{\vee}$, defined initially on the Schwartz class $\mathcal{S}(\mathbb{R})$, extends to a bounded linear transformation $\mathrm{T}_\phi^{(p,\omega)}$ of $L^p(\omega)$ into $L^p(\omega)$. In this case we write $\|\phi\|_{M_{p,\omega}(\mathbb{R})} = \left\|\mathrm{T}_\phi^{(p,\omega)}\right\|$. Notice the following immediate consequence of the preceding definitions: if $\psi \in M_{p,\omega}(\mathbb{R})$, and $f \in L^p(\omega) \cap L^2(\mathbb{R})$, then $\mathrm{T}_\psi^{(p,\omega)} f = \left(\psi \widehat{f}\right)^{\vee}$. This handy fact will be used without explicit mention. The functions belonging to $M_{p,\omega}(\mathbb{R})$ (identified modulo equality a.e.) are readily seen to form a normed algebra under pointwise operations and the norm $\|\cdot\|_{M_{p,\omega}(\mathbb{R})}$. (In fact, with this norm, $M_{p,\omega}(\mathbb{R})$ is a Banach algebra [7, §2].)

Suppose next that $\mathfrak{w} \in A_p(\mathbb{Z})$. Let $\ell^p(\mathfrak{w})$ be the corresponding Banach space consisting of all complex-valued sequences $x = \{x_k\}_{k=-\infty}^{\infty}$ such that

$$\|x\|_{\ell^p(\mathfrak{w})} \equiv \left\{ \sum_{k=-\infty}^{\infty} |x_k|^p \mathfrak{w}_k \right\}^{1/p} < \infty.$$

The class $M_{p,\mathfrak{w}}(\mathbb{T})$ consisting of the multipliers for $\ell^p(\mathfrak{w})$ is described as follows. A function $\psi \in L^\infty(\mathbb{T})$ is a *multiplier for* $\ell^p(\mathfrak{w})$ provided that: (i) for each $x \in \ell^p(\mathfrak{w})$, and $j \in \mathbb{Z}$, the series $(\psi^\vee * x)(j) \equiv \{\sum_{k=-\infty}^{\infty} \psi^\vee(j-k)x_k\}$ converges absolutely, and (ii) the mapping $\mathrm{T}_\psi^{(p,\mathfrak{w})} : x \in \ell^p(w) \mapsto \psi^\vee * x$ is a bounded linear mapping of $\ell^p(\mathfrak{w})$ into $\ell^p(\mathfrak{w})$. We then write $\|\psi\|_{M_{p,\mathfrak{w}}(\mathbb{T})} = \left\|\mathrm{T}_\psi^{(p,\mathfrak{w})}\right\|$. After identifying functions modulo equality a.e. on $\mathbb{T}$, we see that $M_{p,\mathfrak{w}}(\mathbb{T})$ is a Banach algebra under pointwise operations and the norm $\|\cdot\|_{M_{p,\mathfrak{w}}(\mathbb{T})}$ [7, §2]. Notice that in the special case where $\omega \equiv 1$ (respectively, $\mathfrak{w} \equiv 1$), $M_{p,\omega}(\mathbb{R})$ (respectively, $M_{p,\mathfrak{w}}(\mathbb{T})$) becomes the usual Banach algebra of Fourier multipliers $M_p(\mathbb{R})$ (respectively, $M_p(\mathbb{T})$).

For $\omega \in A_p(\mathbb{R})$ (respectively, $\mathfrak{w} \in A_p(\mathbb{Z})$), it is well known that the classical Hilbert kernel defines a bounded convolution operator on $L^p(\omega)$ (respectively, on $\ell^p(\mathfrak{w})$) (see [11]), and hence the characteristic functions χ_I, where I runs through all the intervals of $\mathbb{R}$ (respectively, through all the arcs of $\mathbb{T}$) form a bounded subset of $M_{p,\omega}(\mathbb{R})$ (respectively, $M_{p,\mathfrak{w}}(\mathbb{T})$). In fact, the boundedness of the Hilbert transform implies the weighted analogue of Stečkin's Theorem concerning functions of bounded variation—specifically, $\mathrm{BV}(\mathbb{R}) \subseteq M_{p,\omega}(\mathbb{R})$, and $\mathrm{BV}(\mathbb{T}) \subseteq M_{p,\mathfrak{w}}(\mathbb{T})$.

In [8] the notion of transference couple was introduced, and it was shown that under an appropriate subpositivity assumption such couples transfer maximal convolution operators, along with their bounds, from groups to measure spaces (see §4.3 below for precise statements of the definitions and results we shall require concerning transference couples). In the present note, we develop machinery for applying the transference techniques of [8] in order to obtain bounds for maximal multiplier transforms in weighted settings. The advantage of our approach is its structural simplicity; however, our present methods require us to confine attention to those weight functions belonging to $A_p(\mathbb{R})$ which are controlled, in the following sense, by their restrictions to $\mathbb{Z}$.

Definition 4.3. The class W_p consists of the weight functions $\omega > 0$ belonging to $A_p(\mathbb{R})$ which satisfy the following condition:

there is a positive constant ρ such that, for each $k \in \mathbb{Z}$,
$$\rho^{-1}\omega(k) \leq \omega(x) \leq \rho\omega(k), \quad \text{for all } x \in [k, k+1]. \tag{4.2}$$

Some indication of the size of the class W_p can be inferred from the following examples. Suppose that $\mathfrak{w} \in A_p(\mathbb{Z})$. A standard method for using $\mathfrak{w}$ to construct a weight $\omega \in A_p(\mathbb{R})$ was introduced in [11]. Specifically, the corresponding weight ω is defined by taking $\omega = \mathfrak{w}_k$ on the intervals $[k - (1/4), k + (1/4)]$, $k \in \mathbb{Z}$, and requiring ω to be linear on the intermediate intervals $[k + (1/4), k + (3/4)]$, $k \in \mathbb{Z}$. It is elementary to verify that the weight function ω so constructed belongs to W_p. It is also easy to see that W_p contains each real-valued weight function $\omega \in A_p(\mathbb{R})$ such that ω is a strictly positive, even function which is monotone on $[0, \infty)$. In particular, if $0 < \alpha < 1$ and $1 + \alpha < p$, then $\omega_\alpha(x) \equiv |x|^\alpha + 1$ belongs to W_p.

In §4.2 we develop the properties of the class W_p which implement transference methods. Section 4.3 is devoted to a discussion of transference couples needed for our main results on the transference of maximal multiplier transforms in §4.4.

4.2   Properties of the Class W_p

Let $\omega \in W_p$. Here and henceforth we denote by $w \equiv \{w_k\}_{k=-\infty}^{\infty}$ the restriction of ω to $\mathbb{Z}$ (that is, $w_k = \omega(k)$, for all $k \in \mathbb{Z}$). We begin this section with the following theorem, which uses Definition (4.3) to "discretize" calculations in $L^p(\omega)$. This result will enable us to deduce a weighted de Leeuw type theorem which furnishes periodic elements of $M_{p,\omega}(\mathbb{R})$.

Theorem 4.4. *If $\omega \in W_p$, then the following assertions are valid.*

1. *$w \in A_p(\mathbb{Z})$.*

2. *The positive constant ρ in (4.2) has the property that for each measurable function f defined on $\mathbb{R}$ we have*

$$\rho^{-1} \int_0^1 \sum_{k=-\infty}^{\infty} |f(t+k)|^p w_k \, dt \leq \int_{\mathbb{R}} |f(x)|^p \omega(x) \, dx$$
$$\leq \rho \int_0^1 \sum_{k=-\infty}^{\infty} |f(t+k)|^p w_k \, dt.$$

3. *For each $s \in \mathbb{R}$ the corresponding translation operator τ_s is a bounded linear mapping of $L^p(\omega)$ into $L^p(\omega)$, and $\{\tau_u\}_{u \in \mathbb{R}}$ is a strongly continuous one-parameter group of operators on $L^p(\omega)$.*

Proof. (i) is an immediate consequence of (4.2) and the definitions for $A_p(\mathbb{R})$, $A_p(\mathbb{Z})$. To see that (ii) holds, we proceed from the identity

$$\int_{\mathbb{R}} |f(x)|^p \omega(x) \, dx = \sum_{k=-\infty}^{\infty} \int_k^{k+1} |f(x)|^p \omega(x) \, dx.$$

Application of (4.2) together with the the change of variable $t = x - k$ in each summand establishes (ii). To obtain (iii) observe first that since $w \in A_p(\mathbb{Z})$, the sequences $\{w_{k+1}/w_k\}_{k=-\infty}^{\infty}$ and $\{w_k/w_{k+1}\}_{k=-\infty}^{\infty}$ are bounded. It follows by (ii) that the operators τ_1 and τ_{-1} are bounded linear mappings of $L^p(\omega)$ into itself, and hence τ_n is a bounded operator on $L^p(\omega)$ for each $n \in \mathbb{Z}$. It is easy to see with the aid of (i) and (4.2) that

$$\zeta \equiv \sup_{\substack{x \in \mathbb{R} \\ 0 \leq u < 1}} \frac{\omega(x - u)}{\omega(x)} < \infty.$$

Consequently, if $s = n + u$, where $n \in \mathbb{Z}$ and $u \in [0, 1)$, we see that τ_s is a bounded linear map of $L^p(\omega)$ into itself satisfying

$$\|\tau_s\| \leq \zeta^{1/p} \left[\max \{ \|\tau_1\|, \|\tau_{-1}\| \} \right]^{|n|}.$$

Moreover, this shows that, for each compact interval J of $\mathbb{R}$, we now have $\sup_{s \in J} \|\tau_s\| < \infty$. Since the algebra $C_0^\infty(\mathbb{R})$ consisting of the infinitely differentiable, compactly supported functions defined on $\mathbb{R}$ is dense in $L^p(\omega)$, the remaining assertion in (iii) is now evident. $\square$

Our next item serves as a weighted analogue for (half of) de Leeuw's Periodization Theorem [9, Theorem 4.5].

Theorem 4.5. *Suppose that* $\omega \in W_p$, *and* $\Lambda \in M_{p,w}(\mathbb{T})$. *Define* Φ *on* $\mathbb{R}$ *by putting* $\Phi(y) = \Lambda\left(e^{2\pi i y}\right)$, *for all* $y \in \mathbb{R}$. *Then* $\Phi \in M_{p,\omega}(\mathbb{R})$, $\|\Phi\|_{M_{p,\omega}(\mathbb{R})} \leq \rho^{2/p} \|\Lambda\|_{M_{p,w}(\mathbb{T})}$ *(where* ρ *is the constant in (4.2)), and, for each* $f \in L^p(\omega)$,

$$\left(T_\Phi^{(p,\omega)} f \right)(x) = \sum_{k=-\infty}^{\infty} \Lambda^\vee(k) f(x - k), \quad \textit{for almost all } x \in \mathbb{R}, \qquad (4.3)$$

the series on the right converging absolutely for almost all $x \in \mathbb{R}$.

Proof. Let $f \in L^p(\omega)$. By 4.4(ii), for almost all $x \in [0, 1]$, $\{f(x + k)\}_{k=-\infty}^{\infty} \in \ell^p(w)$. It follows from this and the definition of $M_{p,w}(\mathbb{T})$ that, for almost all $x \in [0, 1]$, the series $\sum_{k=-\infty}^{\infty} \Lambda^\vee(k) f(x + m - k)$ converges absolutely for all $m \in \mathbb{Z}$. This shows that the series on the right of (4.3) converges absolutely for almost all $x \in \mathbb{R}$. Define $\mathcal{H}f$ a.e. on $\mathbb{R}$ by writing

$$\left(\mathcal{H}f \right)(x) = \sum_{k=-\infty}^{\infty} \Lambda^\vee(k) f(x - k).$$

Let $\{\kappa_j\}_{j=0}^{\infty}$ be the Fejér kernel for $\mathbb{T}$, and put $\Lambda_j = \kappa_j * \Lambda$, $j \geq 0$. By [6, Theorem (5.2)], for $j \geq 0$, $\Lambda_j \in M_{p,w}(\mathbb{T})$, and

$$\|\Lambda_j\|_{M_{p,w}(\mathbb{T})} \leq \|\Lambda\|_{M_{p,w}(\mathbb{T})}. \qquad (4.4)$$

Put

$$\left(\mathcal{H}_j f\right)(x) = \sum_{k=-\infty}^{\infty} \Lambda_j^{\vee}(k) f(x - k), \quad \text{for all } x \in \mathbb{R}.$$

Since $\left|\Lambda_j^{\vee}(k) f(x - k)\right| \le \left|\Lambda^{\vee}(k) f(x - k)\right|$, we can use dominated convergence in $\ell^1(\mathbb{Z})$ to infer that as $j \to \infty$,

$$\left(\mathcal{H}_j f\right)(x) \to \left(\mathcal{H} f\right)(x), \quad \text{for almost all } x \in \mathbb{R}. \qquad (4.5)$$

Using Theorems 4.4-(ii) and (4.4), we have for $j \ge 0$,

$$\|\mathcal{H}_j f\|_{L^p(\omega)}^p \le \rho \int_0^1 \sum_{k=-\infty}^{\infty} \left| \sum_{m=-\infty}^{\infty} \Lambda_j^{\vee}(m) f(x + k - m) \right|^p w_k \, dx$$

$$\le \rho \|\Lambda_j\|_{M_{p,w}(\mathbb{T})}^p \int_0^1 \sum_{k=-\infty}^{\infty} |f(x + k)|^p w_k \, dx$$

$$\le \rho^2 \|\Lambda\|_{M_{p,w}(\mathbb{T})}^p \|f\|_{L^p(\omega)}^p.$$

From this estimate and (4.5) we infer with the aid of Fatou's Lemma that

$$\|\mathcal{H} f\|_{L^p(\omega)} \le \rho^{2/p} \|\Lambda\|_{M_{p,w}(\mathbb{T})} \|f\|_{L^p(\omega)}, \quad \text{for all } f \in L^p(\omega). \qquad (4.6)$$

Next, observe that for $g \in \mathcal{S}(\mathbb{R})$, and $j \ge 0$, we have $\mathcal{H}_j g \in \mathcal{S}(\mathbb{R})$, and

$$\left(\mathcal{H}_j g\right)^{\wedge}(y) = \sum_{k=-j}^{j} \Lambda_j^{\vee}(k) e^{-2\pi i k y} \widehat{g}(y) = \Lambda_j\left(e^{2\pi i y}\right) \widehat{g}(y), \quad \text{for all } y \in \mathbb{R}.$$

By Lebesgue's Theorem, $\Lambda_j\left(e^{2\pi i y}\right) \to \Phi(y)$, for almost all $y \in \mathbb{R}$. Consequently, $\mathcal{H}_j g \to (\Phi \widehat{g})^{\vee}$ in $L^2(\mathbb{R})$. Combining this with (4.5), we see that $\mathcal{H} g = (\Phi \widehat{g})^{\vee}$, for all $g \in \mathcal{S}(\mathbb{R})$. In view of (4.6), the proof is now complete. $\square$

The following result sets up strong ties between $M_{p,\omega}(\mathbb{R})$ and $M_{p,w}(\mathbb{T})$.

Theorem 4.6. *Suppose that* $\omega \in W_p$, $\Psi \in L^{\infty}(\mathbb{R})$, *and the support of* Ψ *is a subset of* $[-1/2, 1/2]$. *Define* $\psi \in L^{\infty}(\mathbb{T})$ *by writing*

$$\psi\left(e^{2\pi i t}\right) = \Psi(t), \quad \text{for} \quad -\frac{1}{2} \le t < \frac{1}{2}. \qquad (4.7)$$

Then in order that $\Psi \in M_{p,\omega}(\mathbb{R})$ *it is necessary and sufficient that* $\psi \in M_{p,w}(\mathbb{T})$. *If this is the case, then*

$$\eta^{-1} \|\psi\|_{M_{p,w}(\mathbb{T})} \le \|\Psi\|_{M_{p,\omega}(\mathbb{R})} \le \eta \|\psi\|_{M_{p,w}(\mathbb{T})}, \qquad (4.8)$$

where η *is a positive constant depending only on* p *and* ω.

Proof. For the sufficiency proof, let $\Phi(t) = \psi(e^{2\pi i t})$, for all $t \in \mathbb{R}$. By Theorem 4.5, $\Phi \in M_{p,\omega}(\mathbb{R})$, and $\|\Phi\|_{M_{p,\omega}(\mathbb{R})} \leq \rho^{2/p} \|\psi\|_{M_{p,w}(\mathbb{T})}$. Since the characteristic function $\chi_{[-1/2,1/2)} \in M_{p,\omega}(\mathbb{R})$, and $\Psi = \Phi\chi_{[-1/2,1/2)}$ a.e. on $\mathbb{R}$, we see that $\Psi \in M_{p,\omega}(\mathbb{R})$ with $\|\Psi\|_{M_{p,\omega}(\mathbb{R})} \leq K_{p,\omega} \|\psi\|_{M_{p,w}(\mathbb{T})}$.

Conversely, suppose that $\Psi \in M_{p,\omega}(\mathbb{R})$. The reasoning leading up to [7, Theorem (4.17)] readily adapts to the present circumstances so as to show that $\psi \in M_{p,w}(\mathbb{T})$, with $\|\psi\|_{M_{p,w}(\mathbb{T})} \leq K_{p,\omega} \|\Psi\|_{M_{p,\omega}(\mathbb{R})}$. We omit the details for expository reasons. $\square$

Remark. In the unweighted setting, the necessity assertion of Theorem 4.6, together with the left-hand inequality in (4.8), are contained in [3, Theorem 1] and [13, Theorem (2.3)].

It will now be convenient to introduce a further item of notation.

Definition 4.7. Suppose that $\Theta : \mathbb{R} \to \mathbb{C}$ is compactly supported. Define $\Theta^{\#} : \mathbb{R} \to \mathbb{C}$ by writing

$$\Theta^{\#}(x) = \sum_{k=-\infty}^{\infty} \Theta(x-k), \quad \text{for all } x \in \mathbb{R}. \tag{4.9}$$

Using Theorems 4.5 and 4.6, we arrive at the following result.

Theorem 4.8. *Suppose that* $\omega \in W_p$, $\Psi \in M_{p,\omega}(\mathbb{R})$, *and the support of* Ψ *is a subset of* $[-1/2, 1/2]$. *Then* $\Psi^{\#} \in M_{p,\omega}(\mathbb{R})$, *and* $\|\Psi^{\#}\|_{M_{p,\omega}(\mathbb{R})} \leq K_{p,\omega} \|\Psi\|_{M_{p,\omega}(\mathbb{R})}$.

The next section will review the key features of transference couples. Theorem 4.8 will play an instrumental role in setting up transference couples needed for the applications in §4.4.

4.3 Transference Couples

In all that follows G will be a locally compact group with given left Haar measure λ. If $\mathfrak{X}$ is a Banach space, and $1 \leq p < \infty$, we denote by $L^p(\lambda, \mathfrak{X})$ the space of all $\mathfrak{X}$-valued, λ-measurable functions g such that $\int_G \|g(u)\|^p \, d\lambda(u) < \infty$. If $k \in L^1(\lambda)$, $g \in L^p(\lambda, \mathfrak{X})$, then the convolution $f * g$ is defined for λ-a.a. $x \in G$ by

$$(k * g)(x) = \int_G k(xy)g\left(y^{-1}\right) \, d\lambda(y) = \int_G k(y)g\left(y^{-1}x\right) \, d\lambda(y).$$

As is well known, convolution by k is a bounded linear mapping of $L^p(\lambda, \mathfrak{X})$ into itself, since

$$\|k * g\|_p \leq \|k\|_1 \|g\|_p.$$

We shall denote by $N_{p,\mathfrak{X}}(k)$ the norm of convolution by $k \in L^1(\lambda)$ on $L^p(\lambda, \mathfrak{X})$. In the special case when $\mathfrak{X} = \mathbb{C}$, we shall write $N_p(k)$ in place of $N_{p,\mathbb{C}}(k)$. The proof of [5, Lemma (4.2)] (reproduced in [8, p. 77]) shows that if μ is an arbitrary measure, and $\mathfrak{X}$ is a nonzero closed subspace of $L^p(\mu)$, then $N_{p,\mathfrak{X}}(k) = N_p(k)$.

Obviously, in the case of the general Banach space $\mathfrak{X}$, we have $N_{p,\mathfrak{X}}(k) \le \|k\|_{L^1(\lambda)}$, and familiar classical examples show that in general $N_{p,\mathfrak{X}}(k)$ can have a much smaller order of magnitude than $\|k\|_{L^1(\lambda)}$. For this reason, transference methods for transplanting individual convolution operators must aim at preserving convolution norms rather than L^1-norms of convolution kernels.

The remainder of this section will be devoted to a review of some essential background items concerning the notion of transference couple. This notion was introduced in [8], where further details can be found.

Definition 4.9. Let $\mathfrak{B}(\mathfrak{X})$ be the algebra of all bounded linear operators mapping a Banach space $\mathfrak{X}$ into itself. A *transference couple defined on G and acting in $\mathfrak{X}$* is a pair (S, T) of strongly continuous mappings of G into $\mathfrak{B}(\mathfrak{X})$ such that, after writing $S \equiv \{S_u\}_{u \in G}$, $T \equiv \{T_u\}_{u \in G}$, we have:

1. $c_S \equiv \sup \{\|S_u\| : u \in G\} < \infty$;

2. $c_T \equiv \sup \{\|T_u\| : u \in G\} < \infty$;

3. $S_u T_v = T_{uv}$, for all $u \in G$, $v \in G$.

In particular, if R is a strongly continuous, uniformly bounded representation of G in $\mathfrak{X}$, then (R, R) is a transference couple. Consequently, results for transference couples generalize traditional transference methods, which are based on representations of G.

Definition 4.10. Given $k \in L^1(\lambda)$, and a transference couple (S, T) defined on G and acting in a Banach space $\mathfrak{X}$, we use $\mathfrak{X}$-valued Bochner integration to define the *transferred convolution operator* $H_k \in \mathfrak{B}(\mathfrak{X})$ by writing:

$$H_k x = \int_G k(u) T_u x \, d\lambda(u), \quad \text{for all } x \in \mathfrak{X}.$$

Notice that by elementary reasoning we have $\|H_k\| \le c_T \|k\|_{L^1(\lambda)}$. When the group G is amenable, this crude estimate can be considerably improved as follows.

Theorem 4.11 ([8, Theorem (2.7)]). *If G is an amenable group, $k \in L^1(\lambda)$, and (S, T) is a transference couple defined on G and acting in a Banach space $\mathfrak{X}$, then*

$$\|H_k\| \le c_S c_T N_{p,\mathfrak{X}}(k), \quad \text{for } 1 \le p < \infty.$$

In the special case when $\mathfrak{X}$ is a closed subspace of $L^p(\mu)$, where μ is an arbitrary measure and $1 \leq p < \infty$, the conclusion in Theorem 4.11 can be replaced by the estimate $\|H_k\| \leq c_S c_T N_p(k)$. In order to discuss the counterpart of this result for maximal estimates, we first introduce the following notation. Given a sequence $\{k_j\}_{j=1}^{\infty} \subseteq L^1(\lambda)$, we denote by $N_p\left(\{k_j\}_{j=1}^{\infty}\right)$ ($\in [0, \infty]$) the strong type (p, p) norm of the maximal convolution operator on $L^p(\lambda)$ defined by the sequence of convolution kernels $\{k_j\}_{j=1}^{\infty}$. We shall also need the following auxiliary notion.

Definition 4.12. Suppose that (Y, μ) is an arbitrary measure space, $1 \leq p < \infty$, and G is an amenable group. Let $S = \{S_u\}_{u \in G}$ be a strongly continuous mapping of G into $\mathfrak{B}\left(L^p(\mu)\right)$. We say that S is a *subpositive family* provided that there is a family of positive operators $\mathcal{P} = \{\mathcal{P}_u\}_{u \in G} \subseteq \mathfrak{B}\left(L^p(\mu)\right)$ such that

1. for each $u \in G$, and each $f \in L^p(\mu)$, we have $|S_u f| \leq \mathcal{P}_u(|f|)$ μ-a.e. on Y;

2. $c_{\mathcal{P}} \equiv \sup_{u \in G} \|\mathcal{P}_u\| < \infty$.

The transference by couples of the bounds for maximal convolution operators has the following form.

Theorem 4.13. [8, Theorem (2.11)]. *Suppose that μ is an arbitrary measure, $1 \leq p < \infty$, and G is an amenable group. Suppose further that $\{k_j\}_{j=1}^{\infty} \subseteq L^1(\lambda)$, and let (S, T) be a transference couple defined on G and acting in $\mathfrak{X} = L^p(\mu)$. Then if S is a subpositive family, we have (in the notation of (4.9)(ii) and (4.12)(ii)):*

$$\|\mathfrak{M}f\|_{L^p(\mu)} \leq c_{\mathcal{P}} c_T N_p\left(\{k_j\}_{j=1}^{\infty}\right) \|f\|_{L^p(\mu)}, \quad \text{for all } f \in L^p(\mu),$$

where $\mathfrak{M}f = \sup_{j \in \mathbb{N}} \left|H_{k_j} f\right|$.

Although in what follows we shall not need to transfer square function estimates, it seems appropriate to mention here that for subspaces of $L^p(\mu)$, Theorem 4.11 can be generalized to square functions under milder hypotheses on transference couples than those employed in Theorem 4.13. Methods for transferring the bounds associated with square functions defined by sequences of multiplier transforms were initiated in [4] and [1]. The methods used to establish [1, Theorems 2.2 and 2.8] are readily extended to the setting of transference couples, where they furnish the following two results.

Theorem 4.14 (Scholium). *Suppose that μ is an arbitrary measure, $1 \leq p < \infty$, X is a closed subspace of $L^p(\mu)$, and G is an amenable group.*

Suppose further that $\{k_j\}_{j=1}^{\infty} \subseteq L^1(\lambda)$, and let (S,T) be a transference couple defined on G and acting in X. If α is a constant such that

$$\left\|\left\{\sum_{j=1}^{\infty}|k_j * f_j|^2\right\}^{1/2}\right\|_{L^p(\lambda)} \leq \alpha \left\|\left\{\sum_{j=1}^{\infty}|f_j|^2\right\}^{1/2}\right\|_{L^p(\lambda)}, \qquad (4.10)$$

for all sequences $\{f_j\}_{j=1}^{\infty} \subseteq L^p(\lambda)$,

then we have (in the notation of (4.9) and (4.10)):

$$\left\|\left\{\sum_{j=1}^{\infty}|H_{k_j} g_j|^2\right\}^{1/2}\right\|_{L^p(\mu)} \leq \alpha\, c_S\, c_T \left\|\left\{\sum_{j=1}^{\infty}|g_j|^2\right\}^{1/2}\right\|_{L^p(\mu)},$$

for all sequences $\{g_j\}_{j=1}^{\infty} \subseteq X$.

Theorem 4.15 (Scholium). *Assume all the hypotheses of Scholium 4.14 except (4.10). If β is a constant such that*

$$\left\|\left\{\sum_{j=1}^{\infty}|k_j * f|^2\right\}^{1/2}\right\|_{L^p(\lambda)} \leq \beta\, \|f\|_{L^p(\lambda)}, \quad \text{for all } f \in L^p(\lambda),$$

then we have (in the notation of (4.9) and (4.10)):

$$\left\|\left\{\sum_{j=1}^{\infty}|H_{k_j} g|^2\right\}^{1/2}\right\|_{L^p(\mu)} \leq \beta\, c_S\, c_T\, \|g\|_{L^p(\mu)}, \quad \text{for all } g \in X.$$

4.4 Transference of Maximal Estimates from $L^p(\mathbb{T})$ to $L^p(\omega)$, $\omega \in W_p$

Throughout what follows we consider a weight function $\omega \in W_p$, where $1 < p < \infty$. For $u \in \mathbb{R}$, we denote by γ_u the corresponding character of $\mathbb{R}$ specified by $\gamma_u(t) \equiv e^{2\pi i u t}$. Hence by Theorem 4.4(iii), $\gamma_u \in M_{p,\omega}(\mathbb{R})$, with $T\gamma_u^{(p,\omega)} = \tau_u$. Suppose now that $\Psi \in M_{p,\omega}(\mathbb{R})$ and the support of Ψ is a subset of $[-1/2, 1/2]$. For each $u \in \mathbb{R}$ define $\Gamma_u \in M_{p,\omega}(\mathbb{R})$ by writing $\Gamma_u = \gamma_u \Psi$. Invoking Theorem 4.8, we see that $\Gamma_u^{\#} \in M_{p,\omega}(\mathbb{R})$, and

$$\left\|\Gamma_u^{\#}\right\|_{M_{p,\omega}(\mathbb{R})} \leq K_{p,\omega} \|\gamma_u\|_{M_{p,\omega}(\mathbb{R})} \|\Psi\|_{M_{p,\omega}(\mathbb{R})}, \qquad (4.11)$$

where

$$\gamma_{-u}(t)\Gamma_u^{\#}(t) = \sum_{m=-\infty}^{\infty} e^{-2\pi i u m}\Psi(t - m), \quad \text{for all } t \in \mathbb{R}. \qquad (4.12)$$

Notice, in particular, that the mapping $u \in \mathbb{R} \mapsto \gamma_{-u}\Gamma_u^\# \in M_{p,\omega}(\mathbb{R})$ has period 1 on $\mathbb{R}$. Moreover, the strong continuity of $\{\tau_u\}_{u\in\mathbb{R}}$ asserted by Theorem 4.4(iii) shows by Banach-Steinhaus that $\sup_{u\in[-1,1]}\|\gamma_u\|_{M_{p,\omega}(\mathbb{R})} < \infty$. Consequently, we find with the aid of periodicity in u and (4.11) that

$$\sup_{u\in\mathbb{R}}\left\|\gamma_{-u}\Gamma_u^\#\right\|_{M_{p,\omega}(\mathbb{R})} = \sup_{u\in[0,1)}\left\|\gamma_{-u}\Gamma_u^\#\right\|_{M_{p,\omega}(\mathbb{R})} \leq K_{p,\omega}\|\Psi\|_{M_{p,\omega}(\mathbb{R})}. \quad (4.13)$$

Now let $\mathcal{L}$ be the linear manifold consisting of all $g \in L^p(\omega)\cap L^2(\mathbb{R})$ such that $\widehat{g}$ vanishes a.e. in the complement of some corresponding compact set. Since $\omega \in A_p(\mathbb{R})$, $\mathcal{L}$ is dense in $L^p(\omega)$. It is also clear that, for each $u \in \mathbb{R}$, the translate of Ψ by u, $\Psi((\cdot)+u)$, belongs to $M_{p,\omega}(\mathbb{R})$ with corresponding multiplier transform on $L^p(\omega)$ specified by $f \in L^p(\omega) \mapsto \gamma_{-u}T_\Psi^{(p,\omega)}(\gamma_u f)$. It follows from these observations and (4.12) that for each $g \in \mathcal{L}$, the mapping $u \in \mathbb{R} \mapsto T_{\gamma_{-u}\Gamma_u^\#}^{(p,\omega)} g \in L^p(\omega)$ is continuous. An application of uniform boundedness, furnished by (4.13), now shows that the mapping $u \in \mathbb{R} \mapsto T_{\gamma_{-u}\Gamma_u^\#}^{(p,\omega)}$ is continuous with respect to the strong operator topology of $\mathfrak{B}(L^p(\omega))$. For $u \in \mathbb{R}$, and $z = e^{2\pi i u}$, we can rewrite (4.12) in the form

$$\gamma_{-u}(t)\Gamma_u^\#(t) = \sum_{m=-\infty}^{\infty} z^{-m}\Psi(t-m), \quad \text{for all } t \in \mathbb{R}.$$

Assembling the foregoing facts, we arrive at the following theorem.

Theorem 4.16. *Suppose that* $\omega \in W_p$, $\Psi \in M_{p,\omega}(\mathbb{R})$, *and the support of* Ψ *is a subset of* $[-1/2, 1/2]$. *Then for each* $z \in \mathbb{T}$, $M_{p,\omega}(\mathbb{R})$ *contains the function* $\Psi^{(z)}$ *specified by*

$$\Psi^{(z)}(t) = \sum_{m=-\infty}^{\infty} z^{-m}\Psi(t-m), \quad \text{for all } t \in \mathbb{R}.$$

Moreover,

$$\sup_{z\in\mathbb{T}}\left\|\Psi^{(z)}\right\|_{M_{p,\omega}(\mathbb{R})} \leq K_{p,\omega}\|\Psi\|_{M_{p,\omega}(\mathbb{R})}, \quad (4.14)$$

and the mapping $z \in \mathbb{T} \mapsto T_{\Psi^{(z)}}^{(p,\omega)}$ *is continuous with respect to the strong operator topology of* $\mathfrak{B}(L^p(\omega))$.

Theorem 4.16 will be used to generate transference couples defined on $\mathbb{T}$ and acting in $L^p(\omega)$. To begin with, we specialize the assertions in Theorem 4.16 by taking Ψ to be the Fourier transform of the de la Vallée Poussin kernel of order $\pi/2$—in other words, we consider the special case where Ψ is the function b on $\mathbb{R}$ defined as follows:

(i) b is linear on each of the intervals $[-1/2, -1/4], [1/4, 1/2];$ (4.15)

$$\text{(ii) } b(t) = \begin{cases} 1, & \text{if } -1/4 \leq t \leq 1/4; \\ 0, & \text{if } |t| \geq 1/2. \end{cases}$$

Since $\mathrm{BV}(\mathbb{R}) \subseteq M_{p,\omega}(\mathbb{R})$, we obviously have $b \in M_{p,\omega}(\mathbb{R})$. For $u \in \mathbb{R}$, let $\beta_u = \gamma_u b$. From the relationship of b to the de la Vallée Poussin kernel, we find by Fourier inversion that

$$(\beta_u)^\vee (t) = \frac{2}{\pi^2(t+u)^2} \left\{ \cos \frac{\pi(t+u)}{2} - \cos \left((t+u)\pi \right) \right\}, \quad \text{for } u \in \mathbb{R}, \ t \in \mathbb{R}. \tag{4.16}$$

Next observe that for $u \in \mathbb{R}$, the restriction $\beta_u\big|_{[-1/2,1/2]}$ can be regarded as an element of $\mathrm{BV}(\mathbb{T}) \subseteq M_{p,w}(\mathbb{T})$. It follows by Theorem 4.5 that for $u \in \mathbb{R}$ we have for each $f \in L^p(\omega)$,

$$\left(\mathrm{T}^{(p,\omega)}_{\beta_u^\#} f \right)(x) = \sum_{m=-\infty}^{\infty} (\beta_u)^\vee (m) f(x-m), \quad \text{for almost all } x \in \mathbb{R}. \tag{4.17}$$

We now proceed to use (4.16) and (4.17) in order to show that (in the notation of Theorem 4.16) $\left\{ \mathrm{T}^{(p,\omega)}_{b^{(z)}} \right\}_{z \in \mathbb{T}}$ is a subpositive family of operators on $L^p(\omega)$. Let $u \in [0,1)$. It follows from (4.16) that there is a positive absolute constant δ such that

$$\left| (\beta_u)^\vee (m) \right| \leq \frac{\delta}{m^2}, \text{ for all } m \in \mathbb{Z} \setminus \{0\}, \text{ and } \left| (\beta_u)^\vee (0) \right| \leq \delta. \tag{4.18}$$

It is an elementary fact that there is a continuous function $\mathfrak{f} \in \mathrm{BV}(\mathbb{T})$ such that $\mathfrak{f}^\vee(0) = \delta$, and $\mathfrak{f}^\vee(m) = \delta/m^2$, for all $m \in \mathbb{Z} \setminus \{0\}$. In fact, $\mathfrak{f}$ has the following explicit description:

$$\mathfrak{f}\left(e^{it} \right) = \delta \left(1 + \frac{\pi^2}{3} - \pi t + \frac{t^2}{2} \right), \quad \text{for } 0 \leq t \leq 2\pi.$$

Specializing Λ in Theorem 4.5 to the function $\mathfrak{f}$, we infer that there is a bounded positive operator $\mathfrak{P} \in \mathfrak{B}\left(L^p(\omega) \right)$ such that for each $f \in L^p(\omega)$, $\mathfrak{P}f$ is specified by

$$\left(\mathfrak{P}f \right)(x) = \delta f(x) + \sum_{m \in \mathbb{Z} \setminus \{0\}} \frac{\delta}{m^2} f(x-m), \quad \text{for almost all } x \in \mathbb{R}. \tag{4.19}$$

Comparing (4.17), (4.18), and (4.19), we find that for each $u \in [0, 1)$,

$$\left| \mathrm{T}_{\beta_u^{\#}}^{(p,\omega)} f \right| \leq \mathfrak{P}(|f|) \text{ a.e. on } \mathbb{R}, \qquad \text{for all } f \in L^p(\omega). \tag{4.20}$$

Now let $z \in \mathbb{T}$, and put $z = e^{2\pi i u}$, where $u \in [0, 1)$. Then $\gamma_{-u} \beta_u^{\#} = b^{(z)}$, and hence by (4.20) we see that

$$\left| \mathrm{T}_{b^{(z)}}^{(p,\omega)} f \right| \leq \tau_{-u} \mathfrak{P}(|f|) \text{ a.e. on } \mathbb{R}, \qquad \text{for all } f \in L^p(\omega). \tag{4.21}$$

With the aid of (4.21), we can now state the following theorem.

Theorem 4.17. *Suppose that $\omega \in W_p$. Let b and $\mathfrak{P}$ be as in (4.15) and (4.19), respectively. For $z \in \mathbb{T}$, define $B_z \in \mathfrak{B}\left(L^p(\omega)\right)$ by setting $B_z = \tau_{-u}\mathfrak{P}$, where $z = e^{2\pi i u}$ and $u \in [0, 1)$. Then $\{B_z\}_{z \in \mathbb{T}}$ is a family of positive operators on $L^p(\omega)$ such that*

$$c_B \equiv \sup_{z \in \mathbb{T}} \|B_z\| < \infty. \tag{4.22}$$

For each $z \in \mathbb{T}$,

$$\left| T_{b^{(z)}}^{(p,\omega)} f \right| \leq B_z(|f|) \text{ a.e. on } \mathbb{R}, \qquad \text{for all } f \in L^p(\omega). \tag{4.23}$$

Consequently $\left\{ T_{b^{(z)}}^{(p,\omega)} \right\}_{z \in \mathbb{T}}$ is a subpositive family in the sense of Definition (4.12).

Remark. Notice that the constant c_B in (4.22) depends only on p and ω.

For each $z \in \mathbb{T}$, let $S_z \in \mathfrak{B}\left(L^p(\omega)\right)$ be defined by

$$S_z f = \gamma_{-1/4} \mathrm{T}_{b^{(z)}}^{(p,\omega)}\left(\gamma_{1/4} f\right), \quad \text{for all } f \in L^p(\omega). \tag{4.24}$$

It is clear that for each $z \in \mathbb{T}$, the operator S_z in (4.24) is the multiplier transform on $L^p(\omega)$ corresponding to the element of $M_{p,\omega}(\mathbb{R})$ specified by $t \in \mathbb{R} \mapsto \sum_{m=-\infty}^{\infty} z^{-m} b\left(t + \frac{1}{4} - m\right)$. It follows with the aid of Theorem 4.16 that $z \in \mathbb{T} \mapsto S_z$ is uniformly bounded and strongly continuous.

Suppose next that $\Delta \in M_{p,\omega}(\mathbb{R})$, and the support of Δ is a subset of $[-1/2, 0]$. Applying Theorem 4.16 to Δ, we define the family $\{T_z\}_{z \in \mathbb{T}} \subseteq \mathfrak{B}\left(L^p(\omega)\right)$ by putting $T_z = \mathrm{T}_{\Delta^{(z)}}^{(p,\omega)}$, for each $z \in \mathbb{T}$. Thus, $z \in \mathbb{T} \mapsto T_z$ is strongly continuous, and

$$\sup_{z \in \mathbb{T}} \|T_z\| \leq K_{p,\omega} \|\Delta\|_{M_{p,\omega}(\mathbb{R})}.$$

In view of our hypothesis that the support of Δ is contained in $[-1/2, 0]$, and the definition of b in (4.15), it is easy to see that for $\xi \in \mathbb{T}$, $z \in \mathbb{T}$, we have

$$\left\{ \sum_{m=-\infty}^{\infty} \xi^{-m} b\left(t + \frac{1}{4} - m\right) \right\} \Delta^{(z)}(t) = \Delta^{(\xi z)}(t), \quad \text{for all } t \in \mathbb{R}.$$

This shows that $S_\xi T_z = T_{\xi z}$, and we have established that the pair of operator families (S, T) just defined is a transference couple defined on $\mathbb{T}$ and acting in $L^p(\omega)$. Comparing (4.23) and (4.24), we also see that $\{S_z\}_{z \in \mathbb{T}}$ is a subpositive family.

At this juncture, it will be convenient to exhibit the following result, a corollary of Theorem 4.16. Here and henceforth σ will denote normalized Haar measure on $\mathbb{T}$.

Theorem 4.18. *Assume the hypotheses of Theorem 4.16. Suppose that $k \in L^1(\sigma)$, and define the function $\mathfrak{k} \in L^\infty(\mathbb{R})$ by putting*

$$\mathfrak{k}(t) = \sum_{m=-\infty}^{\infty} \widehat{k}(m)\Psi(t - m), \quad \text{for all } t \in \mathbb{R}.$$

Then $\mathfrak{k} \in M_{p,\omega}(\mathbb{R})$, and

$$T_{\mathfrak{k}}^{(p,\omega)} f = \int_{\mathbb{T}} k(z) T_{\Psi^{(z)}}^{(p,\omega)} f \, d\sigma(z), \quad \text{for all } f \in L^p(\omega). \tag{4.25}$$

Proof. For $f \in L^p(\omega)$, we can use the results in Theorem 4.16 to define $\mathfrak{H}_k f$ by the Bochner integral on the right-hand side of (4.25). Then $\mathfrak{H}_k \in \mathfrak{B}\left(L^p(\omega)\right)$. Let g belong to the linear manifold $\mathcal{L}$ in $L^p(\omega)$ which was described right after (4.13). Since Ψ has support contained in $[-1/2, 1/2]$, easy calculations using the definition of the multipliers $\Psi^{(z)}$, $z \in \mathbb{T}$, along with the compact support of $\widehat{g}$, show that

$$\mathfrak{H}_k g = \left(\mathfrak{k}\, \widehat{g}\right)^{\vee}. \tag{4.26}$$

For $f \in \mathcal{S}(\mathbb{R})$, we obtain a sequence $\{g_n\}_{n=1}^{\infty} \subseteq \mathcal{L}$ such that $\|g_n - f\|_{L^p(\omega)} \to 0$ and $\|g_n - f\|_{L^2(\mathbb{R})} \to 0$ (for example, we can take $g_n = \left(\chi_{[-n,n]}\, \widehat{f}\,\right)^{\vee}$). Since $\mathfrak{H}_k \in \mathfrak{B}\left(L^p(\omega)\right)$, $\mathfrak{H}_k g_n \to \mathfrak{H}_k f$ in $L^p(\omega)$. Applying (4.26) to the sequence $\{g_n\}_{n=1}^{\infty}$, we see that $\mathfrak{H}_k g_n \to \left(\mathfrak{k}\, \widehat{f}\,\right)^{\vee}$ in $L^2(\mathbb{R})$. Consequently, $\mathfrak{H}_k f = \left(\mathfrak{k}\, \widehat{f}\,\right)^{\vee}$, for all $f \in \mathcal{S}(\mathbb{R})$, and this suffices to complete the proof. $\qquad\square$

In order to facilitate further discussion, we now introduce two more items of notation. Given a sequence $\{Q_j\}_{j \geq 1} \subseteq M_{p,\omega}(\mathbb{R})$, we shall symbolize by $\mathfrak{N}_{p,\omega}\left(\{Q_j\}_{j \geq 1}\right) \in [0, \infty]$ the strong type (p, p) norm of the maximal operator on $L^p(\omega)$ defined by the sequence $\left\{ \mathrm{T}_{Q_j}^{(p,\omega)} \right\}_{j\geq 1}$. If $\phi \in \ell^\infty(\mathbb{Z})$, and $\Theta : \mathbb{R} \to \mathbb{C}$ is bounded, measurable, and compactly supported, we define the bounded measurable function $\mathcal{W}_{\phi,\Theta} : \mathbb{R} \to \mathbb{C}$ by writing

$$\mathcal{W}_{\phi,\Theta}(t) = \sum_{m=-\infty}^{\infty} \phi(m)\Theta(t-m), \quad \text{for all } t \in \mathbb{R}. \tag{4.27}$$

The stage is now set for the application of Theorem 4.13 to the transference couple (S,T) described after the second remark. In view of Theorem 4.18, this immediately furnishes the following maximal theorem.

Theorem 4.19. *Suppose that $\omega \in W_p$, $\Delta \in M_{p,\omega}(\mathbb{R})$, and the support of Δ is contained in $[-1/2, 0]$. Then for each sequence $\{k_j\}_{j\geq 1} \subseteq L^1(\mathbb{T})$, we have (in the notation of (4.27)) $\left\{ \mathcal{W}_{\hat{k}_j,\Delta} \right\}_{j\geq 1} \subseteq M_{p,\omega}(\mathbb{R})$, and*

$$\mathfrak{N}_{p,\omega}\left(\left\{ \mathcal{W}_{\hat{k}_j,\Delta} \right\}_{j\geq 1} \right) \leq K_{p,\omega} \left\| \Delta \right\|_{M_{p,\omega}(\mathbb{R})} N_p\left(\{k_j\}_{j\geq 1} \right),$$

where $\mathfrak{N}_{p,\omega}\left(\left\{ \mathcal{W}_{\hat{k}_j,\Delta} \right\}_{j\geq 1} \right)$ (respectively, $N_p\left(\{k_j\}_{j\geq 1} \right)$) denotes the strong type (p,p) norm of the maximal operator on $L^p(\omega)$ (respectively, $L^p(\mathbb{T})$) defined by the multiplier transforms (respectively, convolution operators) corresponding to $\left\{ \mathcal{W}_{\hat{k}_j,\Delta} \right\}_{j\geq 1}$ (respectively, to the convolution kernels $\{k_j\}_{j\geq 1}$).

Suppose next that $\widetilde{\Delta} \in M_{p,\omega}(\mathbb{R})$, and the support of $\widetilde{\Delta}$ is contained in $[0, 1/2]$. We modify the preceding transference couple (S,T) as follows. For each $z \in \mathbb{T}$, define $\widetilde{S}_z \in \mathfrak{B}\left(L^p(\omega)\right)$ by putting

$$\widetilde{S}_z f = \gamma_{1/4} \mathrm{T}_{b^{(z)}}^{(p,\omega)} \left(\gamma_{-1/4} f\right) \quad \text{for all } f \in L^p(\omega).$$

Define the family $\left\{ \widetilde{T}_z \right\}_{z \in \mathbb{T}} \subseteq \mathfrak{B}\left(L^p(\omega)\right)$, by putting $\widetilde{T}_z = \mathrm{T}_{(\widetilde{\Delta})^{(z)}}^{(p,\omega)} f$, for each $z \in \mathbb{T}$. Considerations analogous to those used in establishing Theorem 4.19, now show that $\left(\widetilde{S}, \widetilde{T}\right)$ is a transference couple defined on $\mathbb{T}$ and acting in $L^p(\omega)$, with $\left\{ \widetilde{S}_z \right\}_{z \in \mathbb{T}}$ a subpositive family, and that the conclusions of Theorem 4.19 remain valid for $\widetilde{\Delta}$ in place of Δ. Combining this fact with Theorem 4.19, we arrive at the following result regarding the transference of maximal estimates.

Theorem 4.20. *Suppose that $\omega \in W_p$, $\Psi \in M_{p,\omega}(\mathbb{R})$, and the support of Ψ is contained in $[-1/2, 1/2]$. Then for each sequence $\{k_j\}_{j\geq 1} \subseteq L^1(\mathbb{T})$, we* *have (in the notation of (4.27))* $\left\{ \mathcal{W}_{\widehat{k}_j,\Psi} \right\}_{j\geq 1} \subseteq M_{p,\omega}(\mathbb{R})$, *and*

$$\mathfrak{N}_{p,\omega}\left(\left\{ \mathcal{W}_{\widehat{k}_j,\Psi} \right\}_{j\geq 1} \right) \leq K_{p,\omega} \left\| \Psi \right\|_{M_{p,\omega}(\mathbb{R})} N_p \left(\{k_j\}_{j\geq 1} \right).$$

Theorem 4.20 can be extended so as to transfer maximal multiplier estimates from $L^p(\mathbb{T})$ to $L^p(\omega)$. Let $M_p(\mathbb{Z})$ be the space of Fourier multipliers for $L^p(\mathbb{T})$. Given a sequence of Fourier multipliers $\{\phi_j\}_{j\geq 1} \subseteq M_p(\mathbb{Z})$, we denote by $\mathfrak{N}_p \left(\{\phi_j\}_{j\geq 1} \right)$ the strong type (p,p) norm of the maximal multiplier transform on $L^p(\mathbb{T})$ corresponding to $\{\phi_j\}_{j\geq 1}$. In particular, for $\{k_j\}_{j\geq 1} \subseteq L^1(\mathbb{T})$, we have $N_p \left(\{k_j\}_{j\geq 1} \right) = \mathfrak{N}_p \left(\left\{ \widehat{k}_j \right\}_{j\geq 1} \right)$. The extended version of Theorem 4.20 takes the following form.

Theorem 4.21. *Suppose that $\omega \in W_p$, $\Psi \in M_{p,\omega}(\mathbb{R})$, and the support of Ψ is contained in $[-1/2, 1/2]$. Then for each sequence $\{\phi_j\}_{j\geq 1} \subseteq M_p(\mathbb{Z})$, we have $\left\{ \mathcal{W}_{\phi_j,\Psi} \right\}_{j\geq 1} \subseteq M_{p,\omega}(\mathbb{R})$, and*

$$\mathfrak{N}_{p,\omega}\left(\left\{ \mathcal{W}_{\phi_j,\Psi} \right\}_{j\geq 1} \right) \leq K_{p,\omega} \left\| \Psi \right\|_{M_{p,\omega}(\mathbb{R})} \mathfrak{N}_p \left(\{\phi_j\}_{j\geq 1} \right). \tag{4.28}$$

Proof. It is enough to establish (4.28) in the case of a finite sequence $\{\phi_j\}_{j=1}^N \subseteq M_p(\mathbb{Z})$. Thus we wish to show that

$$\mathfrak{N}_{p,\omega}\left(\left\{ \mathcal{W}_{\phi_j,\Psi} \right\}_{j=1}^N \right) \leq K_{p,\omega} \left\| \Psi \right\|_{M_{p,\omega}(\mathbb{R})} \mathfrak{N}_p \left(\{\phi_j\}_{j=1}^N \right). \tag{4.29}$$

In the special case where each ϕ_j, $1 \leq j \leq N$, is finitely supported, there is a trigonometric polynomial k_j such that $\widehat{k}_j = \phi_j$, and it follows by Theorem 4.20 that (4.29) holds in the case where each ϕ_j has finite support.

In order to treat the general case of a sequence $\{\phi_j\}_{j=1}^N \subseteq M_p(\mathbb{Z})$, let $\{\kappa_n\}_{n=0}^\infty$ denote the Fejér kernel for $\mathbb{T}$, and for $n \geq 0$, $1 \leq j \leq N$, put $\phi_{n,j} = \widehat{\kappa}_n \phi_j$. Since $\phi_{n,j}$ is finitely supported, we have for each $n \geq 0$:

$$\mathfrak{N}_{p,\omega}\left(\left\{ \mathcal{W}_{\phi_{n,j},\Psi} \right\}_{j=1}^N \right) \leq K_{p,\omega} \left\| \Psi \right\|_{M_{p,\omega}(\mathbb{R})} \mathfrak{N}_p \left(\{\phi_{n,j}\}_{j=1}^N \right),$$

and consequently

$$\mathfrak{N}_{p,\omega}\left(\left\{ \mathcal{W}_{\phi_{n,j},\Psi} \right\}_{j=1}^N \right) \leq K_{p,\omega} \left\| \Psi \right\|_{M_{p,\omega}(\mathbb{R})} \mathfrak{N}_p \left(\{\phi_j\}_{j=1}^N \right). \tag{4.30}$$

Temporarily fix j in the range $1 \leq j \leq N$. It is easy to see that the sequence $\left\{ \mathcal{W}_{\phi_{n,j},\Psi} \right\}_{n=0}^\infty$ is uniformly bounded on $\mathbb{R}$ and tends pointwise on $\mathbb{R}$ to $\mathcal{W}_{\phi_j,\Psi}$ as $n \to \infty$. It follows that for $f \in \mathcal{S}(\mathbb{R})$, we have $\left\| \left(\mathcal{W}_{\phi_{n,j},\Psi} \widehat{f} \right)^\vee - \left(\mathcal{W}_{\phi_j,\Psi} \widehat{f} \right)^\vee \right\|_{L^2(\mathbb{R})} \to 0$, as $n \to \infty$. Hence as $n \to \infty$,

$$\left\| \sup_{1\leq j\leq N} \left| \left(\mathcal{W}_{\phi_{n,j},\Psi} \widehat{f} \right)^\vee \right| - \sup_{1\leq j\leq N} \left| \left(\mathcal{W}_{\phi_j,\Psi} \widehat{f} \right)^\vee \right| \right\|_{L^2(\mathbb{R})} \to 0.$$

This furnishes a subsequence $\left\{ \sup_{1 \leq j \leq N} \left| \left(\mathcal{W}_{\phi_{n_\nu,j}, \Psi} \, \widehat{f} \right)^\vee \right| \right\}_{\nu=1}^\infty$ which conver-
ges a.e. on $\mathbb{R}$ to $\sup_{1 \leq j \leq N} \left| \left(\mathcal{W}_{\phi_j, \Psi} \, \widehat{f} \right)^\vee \right|$. Using this together with (4.30), we
can apply Fatou's Lemma to deduce that for each $f \in \mathcal{S}(\mathbb{R})$,

$$\left\| \sup_{1 \leq j \leq N} \left| \left(\mathcal{W}_{\phi_j, \Psi} \, \widehat{f} \right)^\vee \right| \right\|_{L^p(\omega)} \leq K_{p,\omega} \left\| \Psi \right\|_{M_{p,\omega}(\mathbb{R})} \mathfrak{N}_p \left(\{\phi_j\}_{j=1}^N \right) \left\| f \right\|_{L^p(\omega)}.$$

$$(4.31)$$

It follows, in particular, that $\left\{ \mathcal{W}_{\phi_j, \Psi} \right\}_{j=1}^N \subseteq M_{p,\omega}(\mathbb{R})$. Since $\mathcal{S}(\mathbb{R})$ is dense
in $L^p(\omega)$, we can now use (4.31) to obtain (4.29), and thereby complete the
proof of Theorem 4.21. $\qquad \square$

A straightforward "surgical" procedure will enable us to pass from Theorem
4.21 to the case where Ψ is assumed to be an arbitrary compactly supported
element of $M_{p,\omega}(\mathbb{R})$. The specific outcome is formulated as follows.

Theorem 4.22. *Suppose that* $\omega \in W_p$, $\Psi \in M_{p,\omega}(\mathbb{R})$, *and* Ψ *has compact
support. Let* N *be a nonnegative integer such that the support of* Ψ *is con-
tained in* $[-N - (1/2), N + (1/2)]$. *Then for each sequence* $\{\phi_j\}_{j \geq 1} \subseteq M_p(\mathbb{Z})$,
we have $\left\{ \mathcal{W}_{\phi_j, \Psi} \right\}_{j \geq 1} \subseteq M_{p,\omega}(\mathbb{R})$, *and*

$$\mathfrak{N}_{p,\omega} \left(\left\{ \mathcal{W}_{\phi_j, \Psi} \right\}_{j \geq 1} \right) \leq (2N + 1) K_{p,\omega} \left\| \Psi \right\|_{M_{p,\omega}(\mathbb{R})} \mathfrak{N}_p \left(\{\phi_j\}_{j \geq 1} \right).$$

Proof. We can assume without loss of generality that Ψ vanishes outside
$[-N - (1/2), N + (1/2))$. For $-N \leq k \leq N$, let $\Psi_k = \chi_{[k-(1/2),k+(1/2))} \Psi$,
and let Υ_k be the translate of Ψ_k by k. Thus, $\Psi = \sum_{k=-N}^{N} \Psi_k$, and, for
$-N \leq k \leq N$, Υ_k has support contained in $[-1/2, 1/2]$. Given a sequence
$\{\phi_j\}_{j \geq 1} \subseteq M_p(\mathbb{Z})$, we apply Theorem 4.21 to Υ_k. This gives us:

$$\mathfrak{N}_{p,\omega} \left(\left\{ \mathcal{W}_{\phi_j, \Upsilon_k} \right\}_{j \geq 1} \right) \leq K_{p,\omega} \left\| \Psi \right\|_{M_{p,\omega}(\mathbb{R})} \mathfrak{N}_p \left(\{\phi_j\}_{j \geq 1} \right), \quad \text{for } -N \leq k \leq N.$$

Since $\mathfrak{N}_{p,\omega} \left(\left\{ \mathcal{W}_{\phi_j, \Upsilon_k} \right\}_{j \geq 1} \right) = \mathfrak{N}_{p,\omega} \left(\left\{ \mathcal{W}_{\phi_j, \Psi_k} \right\}_{j \geq 1} \right)$, for $-N \leq k \leq N$, we
have

$$\mathfrak{N}_{p,\omega} \left(\left\{ \mathcal{W}_{\phi_j, \Psi_k} \right\}_{j \geq 1} \right) \leq K_{p,\omega} \left\| \Psi \right\|_{M_{p,\omega}(\mathbb{R})} \mathfrak{N}_p \left(\{\phi_j\}_{j \geq 1} \right), \quad \text{for } -N \leq k \leq N.$$

$$(4.32)$$

The remainder of the proof is evident from (4.32) and the identity

$$\mathcal{W}_{\phi_j \Psi} = \sum_{k=-N}^{N} \mathcal{W}_{\phi_j, \Psi_k}, \quad \text{for all } j \geq 1.$$

$$\square$$

We close the discussion with some observations about the relationship between the existing literature and our results in this section. Theorem 4.22 constitutes a weighted generalization of the unweighted result in [2, Theorem (1.1)]; that is, Theorem 4.22 essentially reduces to [2, Theorem (1.1)] when $\omega \in W_p$ is specialized to be the function identically one on $\mathbb{R}$. In the special case of Theorem 4.20 where $\omega \equiv 1$, $\Psi = \chi_{[-1/2,1/2)}$, and $\{k_j\}_{j \geq 1}$ is the Dirichlet kernel for $\mathbb{T}$, we have the unweighted Carleson-Hunt Theorem for $L^p(\mathbb{R})$, $1 < p < \infty$, which was shown by other methods in [14]. The situation regarding the weighted Carleson-Hunt Theorem in the setting of $\mathbb{R}$ is more complicated. It is known in the literature that the demonstration in [12] of the weighted Carleson-Hunt Theorem for $\mathbb{T}$ can be adapted to provide a weighted Carleson-Hunt Theorem for $L^p(\omega)$, $\omega \in A_p(\mathbb{R})$, $1 < p < \infty$ (see the comments in [10, p. 466]). Theorem 4.20 falls short of this much generality, since it only provides the weighted Carleson-Hunt Theorem for $L^p(\omega)$ when $\omega \in W_p$.

References

[1] N. Asmar, E. Berkson, and T. A. Gillespie, *Transferred bounds for square functions* , Houston J. Math. (memorial issue dedicated to Domingo Herrero), 17 (1991), 525–550.

[2] ————, *On Jodeit's multiplier extension theorems*, Journal d'Analyse Math. 64 (1994), 337–345.

[3] P. Auscher and M. J. Carro, *On relations between operators on $\mathbb{R}^N$, $\mathbb{T}^N$ and $\mathbb{Z}^N$*, Studia Math. 101 (1992), 165–182.

[4] E. Berkson, J. Bourgain, and T. A. Gillespie, *On the almost everywhere convergence of ergodic averages for power-bounded operators on L^p-subspaces*, Integral Equations and Operator Theory 14 (1991), 678–715.

[5] E. Berkson and T. A. Gillespie, *Spectral decompositions and vector-valued transference*, in Analysis at Urbana II, Proceedings of Special Year in Modern Analysis at the Univ. of Ill., 1986-87, London Math. Soc. Lecture Note Series 138, Cambridge Univ. Press, Cambridge (1989), 22–51.

[6] ————, *Mean-boundedness and Littlewood-Paley for separation-preserving operators*, Trans. Amer. Math. Soc. 349 (1997), 1169–1189.

[7] ————, *Multipliers for weighted L^p-spaces, transference, and the q-variation of functions*, Bull. des Sciences Math 122 (1998), 427–454.

[8] E. Berkson, M. Paluszyński, and G. Weiss, *Transference couples and their applications to convolution operators and maximal operators* in "Interaction between Functional Analysis, Harmonic Analysis, and Probability," Proceedings of Conference at Univ. of Missouri-Columbia (May 29–June 3, 1994), Lecture Notes in Pure and Applied Mathematics, vol. 175, Marcel Dekker, Inc., New York (1996), 69–84.

[9] K. de Leeuw, *On L_p multipliers*, Annals of Math. 81 (1965), 364–379.

[10] J. García-Cuerva and J. L. Rubio de Francia, *Weighted Norm Inequalities and Related Topics*, North-Holland Mathematics Studies 116 = Notas de Matemática (104), Elsevier Science Publ., New York (1985).

[11] R. Hunt, B. Muckenhoupt, and R. Wheeden, *Weighted norm inequalities for the conjugate function and Hilbert transform*, Trans. Amer. Math. Soc. 176 (1973), 227–251.

[12] R. Hunt and W.-S. Young, *A weighted norm inequality for Fourier series*, Bull. Amer. Math. Soc. 80 (1974), 274–277.

[13] M. Jodeit, *Restrictions and extensions of Fourier multipliers*, Studia Math. 34 (1970), 215–226.

[14] C. Kenig and P. Tomas, *Maximal operators defined by Fourier multipliers*, Studia Math. 68 (1980), 79–83.

Chapter 5

Periodic Solutions of Nonlinear Wave Equations

Jean Bourgain

We are investigating here the construction by the [5] method of space and time periodic solutions of one-dimensional nonlinear wave equations of the following form

$$\Box y + y^3 + F(x, y) = 0, \qquad F(x, y) = O(y^4).$$

This more resonant case, due to the absence of a linear term $\rho \cdot y$, was left aside in [5]. Second, we consider periodic solutions of certain non-Hamiltonian NLW, with nonlinearities involving derivatives (of first order). The model case discussed here is the equation

$$\Box y + \rho y + (y_t)^2 = 0, \qquad \rho \neq 0,$$

but the method is by no means restricted to that example. It mainly demonstrates the applicability of the Lyapounov-Schmidt decomposition method to construct periodic (or quasiperiodic) solutions independently of Hamiltonian structure. In the last section, we construct smooth families in ε of periodic solutions of NLW

$$\Box y + \rho y + y^3 + \varepsilon f(x, y) = 0$$

with $\rho = \rho(\varepsilon)$ a parameter depending smoothly in ε. This result is in the spirit of [12, 13]. We also comment on issues concerning admissible frequencies for the perturbed invariant tori.

The author is grateful to C. Wayne for many comments and improvements of earlier versions of the paper.

5.1   Periodic Solutions of Nonlinear Wave Equations of the form $\Box y + y^3 + O(y^4) = 0$.

Consider the 1D NLW equation

$$y_{tt} - y_{xx} + y^3 + F(x, y) = 0, \qquad (5.1)$$

where[1]

$$F(x, y) = O(y^4). \qquad (5.2)$$

Thus we have no linear term my, $m \neq 0$, at our disposal as in an equation of the form

$$y_{tt} - y_{xx} + my + O(y^3) = 0 \qquad (5.3)$$

as considered in [5]. We assume $F(x, y)$ is an even periodic function in x and in fact take for simplicity F of the form

$$F(x, y) = \sum_{4 \leq j \leq d} a_j(x) y^j, \qquad (5.4)$$

where the a_j are trigonometric cosine polynomials in x (see comments later on). Our aim is to construct periodic solutions of (5.1) with periods in a set of positive measure, using the method from [5] combined with some ingredients from [11]. In [5], equation (5.3) with $m \neq 0$ was considered and the more resonant case (5.1) left open. In [11], periodic solutions of the equation

$$y_{tt} - y_{xx} + y^3 = 0 \qquad (5.5)$$

are produced for periods of bounded type (this avoids the small divisor problems). As in [11], our starting point is the explicit solution by the $4K$-periodic elliptic function

$$
\begin{aligned}
c(\xi) &= cn\left(\tfrac{1}{\sqrt{2}}, \xi\right) \\
&= \frac{2\sqrt{2}\pi}{K\left(\frac{1}{\sqrt{2}}\right)} \sum_{n=1}^{\infty} \frac{1}{e^{\pi(n-1/2)} + e^{-\pi(n-1/2)}} \cos \tfrac{\pi}{2K} (2n-1)\xi
\end{aligned}
\qquad (5.6)
$$

of the differential equation

$$c'' + c^3 = 0. \qquad (5.7)$$

---

[1] It should be mentioned that, if $F(x, y)$ is x-independent, one may easily construct periodic solutions of the form $y(x, t) = y_1(x + \lambda t)$ since the corresponding ODE satisfied by y_1 is Hamiltonian. This observation is due to C. Wayne.

(See [7] pp. 343, 345).

Rescaling (5.1), letting

$$y \longrightarrow \delta y, \tag{5.8}$$

we get the equation

$$y_{tt} - y_{xx} + \delta^2 y^3 + \delta^3 F_1(x, y) = 0 \tag{5.9}$$

(δ will be restricted to a neighborhood of 0).
Defining

$$y_0(x, t) = 4K \ c(4K(x + \lambda t)) = \tilde{c}(x + \lambda t), \tag{5.10}$$

where

$$\lambda = \sqrt{1 + \delta^2}, \tag{5.11}$$

we get thus a solution of the equation

$$\Box y_0 + \delta^2 y_0^3 = 0, \tag{5.12}$$

which is 1-periodic in x and $T = \lambda^{-1}$ periodic in t. The main idea is to perform a perturbative procedure starting from y_0 in order to obtain a solution of (5.9). The perturbation scheme is based on the Newton iteration procedure as in [5], yielding a quadratic error at each step, hence doubly exponentially fast convergent. It is therefore less sensitive to small divisor phenomena than if one would perform an expansion in a δ-series (as in [11]). During this process, a sequence of restrictions will appear that essentially will restrict δ (and hence λ) to a Cantor-like set of positive measure (as in [5]).

The solution y of (5.9) will be given by a Fourier series of the form

$$y(x, t) = \sum_{n, k \in \mathbb{Z}, n \geq 0} \hat{y}(n, k) \ \cos 2\pi(nx + k\lambda t). \tag{5.13}$$

Observe that by (5.6) and (5.10)

$$y_0(x, t) = 8\sqrt{2}\,\pi \sum_{n=1}^{\infty} \frac{1}{e^{\pi(n-1/2)} + e^{-\pi(n-1/2)}} \ \cos 2\pi(2n - 1)\,(x + \lambda t). \tag{5.14}$$

Thus the assumption on F to be an even and periodic function in x is consistent with a representation of y in the form (5.13), in the sense that $F\big(x, y(x, t)\big)$ remains a Fourier series of the form (5.13). As will be apparent later on, treating the problem in generality would lead to certain resonance problems we do not intend to analyze here.

We replace y_0 by a "truncated" version of (5.14), redefining

$$y_0(x, t) = 8\sqrt{2}\,\pi \sum_{n=1}^{A} \frac{1}{e^{\pi(n-1/2)} + e^{-\pi(n-1/2)}} \, \cos 2\pi(2n-1)\,(x + \lambda t) \quad (5.15)$$

taking $A \sim \log \frac{1}{\delta}$, so that in particular

$$\Box y_0 + \delta^2 y_0^3 = O(\delta^3)$$

and

$$\Box y_0 + \delta^2 y_0^3 + \delta^3 F_1(x, y_0) = O(\delta^3). \quad (5.16)$$

Following the procedure described in [1,2,4], we construct a solution y of (5.9) satisfying

$$|\hat{y}(n, k)| < e^{-(|n|+|k|)^c}, \quad (5.17)$$

where $0 < c < 1$ is some constant that will depend on F (in particular on d).

The estimate (5.17) results from the fact that the correction $\Delta_r y$ to the approximate solution y_{r-1} obtained at step $r-1$ of the iteration will satisfy, say,

$$\operatorname{supp} \widehat{\Delta_r y} \subset B_{\mathbb{Z}^2}(0, (10d)^r \log 1/\delta), \quad (5.18)$$

while

$$\log \|\Delta_r y\| < \left(1 + \frac{1}{100}\right)^r \log \delta. \quad (5.19)$$

Thus with this procedure, $\operatorname{supp} \widehat{y_r}$ is contained in some finite region in $\mathbb{Z}^2$ at each stage. It is also possible to obtain a real-analytic solution y (observe that c given by (5.6) and y_0 are real-analytic) and assuming more generally that $F(x, y)$ is real-analytic, even periodic in x, but this requires keeping track as in [5] of a sequence of real-analytic norms

$$\|y\|_\rho = \sum_{n,k \in \mathbb{Z}^2} \rho^{|n|+|k|} \, |\hat{y}(n, k)|, \qquad \rho > 1, \quad (5.20)$$

along the iteration.

Let y_{r-1} be the approximative solution to (5.9) obtained at stage $r-1$. The main point in obtaining $y_r = y_{r-1} + \Delta_r y$ by Newton's method is to control the inverse of the linearized operator

$$T(\Delta_r y) = \Box(\Delta_r y) + 3\delta^2 \, y_{r-1}^2 (\Delta_r y) + \delta^3 (\partial_y F_1(x, y_{r-1})) \cdot (\Delta_r y) \quad (5.21)$$

with $\mathbb{Z}^2$-lattice representation (expressing T passing to Fourier transform)

$$T = D + 3\delta^2 S_{y_{r-1}^2} + \delta^3 S_{\partial_y F_1(x, y_{r-1})}, \tag{5.22}$$

where D is the diagonal operator

$$D_{n,k} = 4\pi^2 \left(-k^2\lambda^2 + n^2\right) \tag{5.23}$$

and S_ϕ denotes the (Toeplitz) operator with matrix elements

$$S_\phi(x, x') = \hat{\phi}(x - x') \tag{5.24}$$

expressing ϕ-multiplication in Fourier space.

In order to obtain $\Delta_r y$ with the desired properties, we ensure bounds and off-diagonal decay estimates on the inverse of restrictions

$$T_M = T\big|_{|n| < M, |k| < M} \tag{5.25}$$

of the form

$$\|T_M^{-1}\| < \delta^{-2} M^C \quad \text{(for some constant } C\text{)} \tag{5.26}$$

and

$$|T_M^{-1}(x, x')| < \delta^{-2} e^{-\frac{1}{2}|x-x'|^c} \quad \text{for say} \quad |x - x'| > M^{\frac{1}{2}} \tag{5.27}$$

(cf. [1]).

One then defines $\Delta_r y$ satisfying $T_M(\Delta_r y) + P_M[\delta^2 y_{r-1}^3 + \delta^3 F_1(x, y_{r-1})] = 0$, where P_M refers to the projection on Fourier modes $|n| < M$, $|k| < M$.

At the first step, we consider for T the operator

$$D + 3\delta^2 S_{y_0^2} \tag{5.28}$$

to obtain thus

$$\|\Delta_1 y\| = O(\delta^{1-}) \tag{5.29}$$

and

$$\Box y_1 + \delta^2 y_1^3 + \delta^3 F_1(x, y_1) = 0(\delta^{4-}). \tag{5.30}$$

The main point is the discussion of T^{-1}. The structure of "singular sites" when $|n| + |k| \to \infty$ has here a simple pattern of lacunarily separated elements as in [5]. However, there is also the behavior of $T_{M_0}^{-1}$ (M_0 a sufficiently large constant) that has to be taken into account, depending on specific properties of y_0 and thus (5.6).

In view of (5.26) and (5.27), it will be convenient to renormalize T defining

$$T = \delta_0^2 T' \tag{5.31}$$

with, say,

$$\delta_0^2 = 10^6 \delta^2. \tag{5.32}$$

Thus by (5.22), (5.23), and (5.11)

$$T' = 4\pi^2 \delta_0^{-2}[n^2 - k^2(1+\delta^2)]\mathbb{1} + 3\frac{\delta^2}{\delta_0^2} S_{y_{r-1}^2} + \frac{\delta^3}{\delta_0^2} S_{\partial_y F_1(x,y_{r-1})}. \tag{5.33}$$

Hence, the off-diagonal part of T' is $O(10^{-10})$, and we may define the set of singular sites (depending on δ) as

$$I_s = \{(n,k) \in \mathbb{Z}_+ \times \mathbb{Z} \,:\, |n^2 - k^2(1+\delta^2)| \,<\, \delta_0^2\}. \tag{5.34}$$

The inverse of the restriction $P_{I_s^c} T' P_{I_s^c}$ may then clearly be analyzed with a Neumann series expansion.

Deleting in $[0, \delta_0]$ a set of small relative measure, one may ensure δ satisfying for any fixed $\gamma > 1$

$$|k\sqrt{1+\delta^2} + n| \geq \|k\sqrt{1+\delta^2}\| > c_1 \frac{\delta^2}{|k|^\gamma} \qquad (k,n \in \mathbb{Z}, \, k \neq 0). \tag{5.35}$$

Here we use the notation $\|\cdot\|$ for the distance to the nearest integer. Let $(n_1, k_1) \in I_s$, $(n_2, k_2) \in I_s$, $|n_1| < |n_2|$. It follows from (5.34) and (5.35) that

$$c_1 \frac{\delta^2}{|n_1 - n_2|^\gamma} < \frac{\delta_0^2}{|n_1|}, \tag{5.36}$$

and thus one has the separation property

$$|n_1 - n_2| > 10^{-12} c_1 |n_1|^{1/\gamma}. \tag{5.37}$$

It is clear from this separation of singular sites that we may choose a sufficiently large constant M_0 such that

$$I_s \subset \Omega_0 \cup \{x_\alpha\}, \, x_\alpha \in I_s \tag{5.38}$$

with

$$\Omega_0 = [0, M_0] \times [-M_0, M_0] \subset \mathbb{Z}^2, \tag{5.39}$$

$$\text{dist}\,(x_\alpha, \Omega_0) > |x_\alpha|^{1/2}, \tag{5.40}$$

$$|x_\alpha - x_\beta| > |x_\alpha|^{1/2} \quad \text{for} \quad \alpha \neq \beta. \tag{5.41}$$

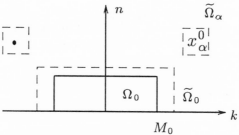

Let $M \gg M_0$. Recall the resolvent identity

$$T_\Lambda^{-1} = (T_{\Lambda_1}^{-1} + T_{\Lambda_2}^{-1}) - (T_{\Lambda_1}^{-1} + T_{\Lambda_2}^{-1})(T_\Lambda - T_{\Lambda_1} - T_{\Lambda_2})T_\Lambda^{-1}, \tag{5.42}$$

where $\Lambda_1 \cap \Lambda_2 = \phi$, $\quad \Lambda = \Lambda_1 \cup \Lambda_2$, $\quad T_\Lambda = P_\Lambda T P_\Lambda$.

Applying (5.42) and arguments going back to the work of J. Fröhlich and T. Spencer on lattice Schrödinger operators (cf. [5] and see [3] for a precise formulation in our context), the estimates (5.26) and (5.29)

$$\| (T'_M)^{-1} \| < M^C \tag{5.43}$$

and

$$| (T'_M)^{-1}(x, x') | < e^{-\frac{1}{2}|x-x'|^c} \quad \text{for} \quad |x - x'| > M^{1/2} \tag{5.44}$$

may be derived from the off-diagonal decay for T'

$$| T'(x, x') | < e^{-|x-x'|^c} \quad \text{for} \quad x \neq x' \tag{5.45}$$

and bounds on the inverse of the restriction of T' to neighborhoods of the "islands" Ω_0 and $\{x_\alpha\}$, say,

$$\| (T'_{\tilde{\Omega}_0})^{-1} \| < C_2 \tag{5.46}$$

and

$$\| (T'_{\tilde{\Omega}_\alpha})^{-1} \| < |x_\alpha|, \tag{5.47}$$

letting for instance

$$\tilde{\Omega}_0 = M_0^{1/10} \quad \text{–neighborhood of} \quad \Omega_0 \tag{5.48}$$

and

$$\tilde{\Omega}_\alpha = |x_\alpha|^{1/10} \quad \text{–neighborhood of} \quad x_\alpha. \tag{5.49}$$

To obtain (5.47), we proceed simply by first-order eigenvalue variation, using δ^2 as parameter. Observe that T' and its restrictions are self-adjoint. Thus if $T'_{\tilde{\Omega}_\alpha} v = \rho v$, $|v|^2 = 1$, (5.33) yields

$$\frac{\partial \rho}{\partial(\delta^2)} = \left\langle \frac{\partial T'_{\tilde{\Omega}_\alpha}}{\partial(\delta^2)} v,\, v \right\rangle \sim -4\pi^2 k_\alpha^2 \delta_0^{-2} + O(1)\delta_0^2 \sim -|x_\alpha|^2 \delta_0^{-2}, \qquad (5.50)$$

where $x_\alpha = (n_\alpha, k_\alpha) \in I_s$. Remark that

$$\frac{\partial y_r}{\partial(\delta^2)} = O(\delta^{-1-}). \qquad (5.51)$$

The main contribution is given by the first increment $\Delta_1 y = \delta^3 T_{\tilde{\Omega}_0}^{-1} F_1(x, y_0)$.

From (5.50) and appropriate restrictions of δ, one may then fulfill (5.47). Considering consecutive size regions $|n| + |k| \sim 2^s$, we first restrict δ to small intervals on which the singular sites set $\{x_\alpha\}$ and hence $\Omega_\alpha, \tilde{\Omega}_\alpha$ may be fixed. A further restriction of δ using (5.50) then allows us also to ensure the properties (5.47). This process is similar to the one in [5].

Next, we turn our attention to condition (5.46), which is the core of the matter. Thus we need to ensure that

$$\left\| \left(T|_{\tilde{\Omega}_0} \right)^{-1} \right\| < C_2 \delta^{-2}. \qquad (5.52)$$

Recall (5.39). We assume

$$M_0^2 \delta^2 < 10^{-6}. \qquad (5.53)$$

Decompose first

$$\tilde{\Omega}_0 = \Omega' \cup \Omega'', \qquad (5.54)$$

where

$$\Omega' = \{(n, k) \in \tilde{\Omega}_0 \mid n^2 \neq k^2\}. \qquad (5.55)$$

Since, for $(n, k) \in \Omega'$, by (5.53)

$$|n^2 - k^2(1 + \delta^2)| > |n^2 - k^2| - M_0^2 \delta^2 > \frac{1}{2}, \qquad (5.56)$$

it follows from (5.33) that

$$\| T_{\Omega'}^{-1} \| < 1. \qquad (5.57)$$

Writing T_{Ω_0} as block matrix

$$T_{\Omega_0} = \begin{pmatrix} T_{\Omega'} & U \\ U^* & T_{\Omega''} \end{pmatrix},$$

where $U = O(\delta^2)$, (5.52) reduces thus to

$$\| (T_{\Omega''} - U^* T_{\Omega'}^{-1} U)^{-1} \| < C\delta^{-2} \tag{5.58}$$

or, since $y_{r-1} = y_0 + O(\delta)$ and (5.33), to

$$\| (-4\pi^2 k^2 \mathbb{1}_{\Omega''} + 3 P_{\Omega''} S_{y_0^2} P_{\Omega''})^{-1} \| < C. \tag{5.59}$$

Write further

$$\Omega'' = \Omega''_+ \cup \Omega''_-, \tag{5.60}$$

where

$$\Omega''_+ = \{(k, k) \mid 0 \le k \le M_0\} \quad \text{and} \quad \Omega''_- = \{(k, -k) \mid 0 < k \le M_0\}. \tag{5.61}$$

Since

$$\operatorname{supp} \widehat{y_0^2} \subset \{(n, k) \in \mathbb{Z}^2 \mid n = k\}, \tag{5.62}$$

it follows from (5.61) and (5.62) that

$$P_{\Omega''_+} S_{y_0^2} P_{\Omega''_-} = 0 = P_{\Omega''_-} S_{y_0^2} P_{\Omega''_+} \quad \text{and} \quad P_{\Omega''_-} S_{y_0^2} P_{\Omega''_-} = \widehat{y_0^2}(0, 0) \, \mathbb{1}_{\Omega''_-}. \tag{5.63}$$

Therefore, (5.59) clearly reduces to the following two statements

$$\| [P_{M_0}(-4\pi^2 k^2 \mathbb{1} + 3 S_{\tilde{c}^2}) P_{M_0}]^{-1} \| < C \tag{5.64}$$

$$\left| -4\pi^2 k^2 + 3 \int_0^1 (\tilde{c})^2 \right| > \frac{1}{C} \quad \text{for all} \quad k \in \mathbb{Z}, \tag{5.65}$$

where $\tilde{c}$ is given by (5.6) and (5.10), i.e.,

$$\tilde{c}(\xi) = 4\sqrt{2}\,\pi \sum_{n=1}^{\infty} \frac{1}{ch(n - \frac{1}{2})\pi} \cos 2\pi\,(2n - 1)\,\xi \tag{5.66}$$

and P_{M_0} denotes the projection on the linear span of *even* functions

$$[\cos 2\pi kx \mid 0 \le k \le M_0]. \tag{5.67}$$

Verification of (5.64).

Assuming

$$f = \sum_{k=0}^{M_0} \hat{f}(k) \, \cos 2\pi k\xi, \tag{5.68}$$

we need to show that

$$\|P_{M_0} \left(f'' + 3\tilde{c}^2 f\right)\|_2 > \frac{1}{C} \, \|f\|_2. \tag{5.69}$$

Consider the Sturm-Liouville operator

$$Q = \frac{d^2}{d\xi^2} + 3(\tilde{c})^2 \tag{5.70}$$

with 1-periodic potential $3(\tilde{c})^2$, acting on 1-periodic functions. Then $L^2(\mathbb{T})$ is the direct sum of invariant spaces of even and odd functions. We are here only interested in the behavior on the space of even functions. It follows from (5.7) and (5.10) that

$$(\tilde{c})'' + (\tilde{c})^3 = 0 \tag{5.71}$$

and hence $(\tilde{c})'$ satisfies

$$Q(\tilde{c})' = 0. \tag{5.72}$$

It follows that 0 is in the periodic (in fact in the Dirichlet) spectrum of Q. Since $\tilde{c}$ is even, $(\tilde{c})'$ is odd. Assume 0 is also in the spectrum of $Q\big|_{\text{epf}}$, where epf denotes even periodic functions. Thus

$$Q\varphi = 0, \quad \varphi \neq 0 \quad \text{even periodic.} \tag{5.73}$$

Since $(\tilde{c})', \varphi$ are linearly independent, the Wronskian

$$W = \begin{vmatrix} (\tilde{c})' & \varphi \\ (\tilde{c})'' & \varphi' \end{vmatrix} = (\tilde{c})'\varphi' - (\tilde{c})''\varphi \tag{5.74}$$

does not vanish identically. On the other hand, (5.72) and (5.73) imply that

$$W' = (\tilde{c})'\varphi'' - (\tilde{c})'''\varphi = 0, \tag{5.75}$$

so that W has to be a nonzero constant. Thus

$$\int_0^1 W \neq 0. \tag{5.76}$$

Now, by (5.74) and (5.71)

$$\int_0^1 W = -2 \int_0^1 (\tilde{c})'' \varphi = 2 \int_0^1 (\tilde{c})^3 \varphi \tag{5.77}$$

while by (5.74) and (5.73) also

$$\int_0^1 W = -2 \int \tilde{c} \varphi'' = 6 \int (\tilde{c})^3 \varphi \tag{5.78}$$

and (5.76), (5.77) and (5.78) are clearly contradictory. Thus 0 is not in the spectrum of $Q|_{\text{ef}}$ (where ef denotes even functions) and there is some constant $a > 0$ so that

$$\|Q f\|_2 \geq a \|f\|_2 \quad \text{for} \quad f \quad \text{even.} \tag{5.79}$$

Coming back to (5.68) and (5.69) write

$$
\begin{aligned}
\| P_{M_0} (f'' + 3(\tilde{c})^2 f) \|_2 &= \| f'' + 3 P_{M_0} (\tilde{c}^2 f) \|_2 \\
&\geq \| Q f \|_2 - 3 \| (\tilde{c})^2 f - P_{M_0} (\tilde{c}^2 f) \|_2 \\
&\geq \| Q f \|_2 - \tfrac{3}{M_0^2} \| ((\tilde{c})^2 f)'' \|_2 \\
&\geq \| Q f \|_2 - \tfrac{A}{M_0^2} (\| f'' \|_2 + \| f \|_2) \\
&\geq \left(1 - \tfrac{A}{M_0^2}\right) \| Q f \|_2 - \tfrac{B}{M_0^2} \| f \|_2 \\
&\geq \left[\left(1 - \tfrac{A}{M_0^2}\right) a - \tfrac{B}{M_0^2}\right] \| f \|_2 \quad \text{(by (5.80)} \\
&\geq \tfrac{1}{C} \| f \|_2
\end{aligned}
$$

if we let M_0 be large enough and for some constants A and B.

Verification of (5.65).

From (5.66), we need to verify that

$$12 \sum_{n=1}^{\infty} \frac{1}{\left[ch \left(n - \tfrac{1}{2}\right) \pi\right]^2} \notin \{k^2 \,|\, k \in \mathbb{Z}\}, \tag{5.80}$$

and this expression equals $1.9\ldots$

Hence (5.64) and (5.65) and thus (5.59), (5.58), (5.52) and (5.46) hold, required to establish (5.43), (5.44) and (5.26), (5.27). This completes the discussion of the linearized operator in the Newton scheme. Thus, following [2], [3], and [5], we obtain the following.

Theorem 5.1. *Consider the equation*

$$\Box y + y^3 + F(x, y) = 0 \qquad (5.81)$$

with

$$F(x, y) = \sum_{j=4}^{d} a_j(x)\, y^j \qquad (5.82)$$

and a_j trigonometric cosine polynomials. Then (5.81) has periodic solutions (1-periodic in space and λ^{-1}-periodic in time) of the form

$$y(x, t) = 4\, K\, \delta\, cn\,(4K(x + \lambda t)) + O(\delta^{2-}) \qquad (5.83)$$

with

$$\lambda = \sqrt{1 + \delta^2} \qquad (5.84)$$

for δ taken in a set of positive measure (the relative measure $\longrightarrow 1$ when $\delta \longrightarrow 0$).

5.2   Periodic Solutions of Certain Non-Hamiltonian Nonlinear Wave Equations

As pointed out in [1] and [5], the construction of time periodic and quasiperiodic solutions using the Lyapounov-Schmidt decomposition and the Newton iteration method is a priori not restricted to Hamiltonian problems. Our purpose here is to treat a model example, suggested to the author by C. Wayne (private communication), a NLW of the form

$$y_{tt} - y_{xx} + \rho y + (y_t)^2 = 0, \quad \rho \neq 0 \qquad (5.85)$$

(for $\rho = 0$, there are only constant solutions).

Recall that our method consists of dividing the problem into an infinite-dimensional piece (the P-equations) containing small divisor difficulties and a remaining finite set of equations (the Q-equations) used to determine the parameters. Essentially speaking, the main ingredient of relevance in solving the P-equations is the geometric structure of "singular sites" of the linearized equation, which technically often constitutes the hardest part of the problem. The methods of [5], as sketched in the example of the previous section and [1] appear as rather general implicit function type results, largely independent of Hamiltonian structure. Most of the structure appears in the context of the Q-equations, deduced in the Hamiltonian case from an appropriate normal form.

They determine the free parameters and are solved by the ordinary implicit function theorem. In example (5.85), the following difficulties appear.

(i) The appearance of derivatives in the nonlinearity. As pointed out in [1], the arguments used there to construct time-periodic or quasiperiodic solutions permit us in the NLW-case to consider also first-order derivatives.

The type of nonlinearity considered in [1] was of the form

$$\left(-\frac{d^2}{dx^2}\right)^{1/2} F(x,y) \tag{5.86}$$

corresponding to a Hamiltonian problem

$$\begin{cases} y_t = Bv \\ v_t = -By + F(x,y), \end{cases} \tag{5.87}$$

where

$$B = \left(-\frac{d^2}{dx^2}\right)^{1/2}.$$

As far as solving the P-equations, one may consider more generally a nonlinear term of the form

$$F(x,y,\partial_x y,\partial_t y) \tag{5.88}$$

and in particular equation (5.85). One observation here is that in the presence of derivatives in the nonlinear perturbation, already in the construction of time-periodic solutions, the structure of the singular sites of the linearized operators presents some of the difficulties of the quasiperiodic case. In particular, one does not have the separation properties at infinity appearing in the [5] work. Indeed, in the case (5.87), the linearized operator T expressed on the $\mathbb{Z}^{1+b}$-lattice has the form

$$\left(-\langle k,\lambda\rangle^2 + n^2\right)\mathbb{1} + |n|\, S_{\partial_y F(x,y)} \tag{5.89}$$

or

$$\frac{-\langle k,\lambda\rangle^2 + n^2}{|n|}\mathbb{1} + S_{\partial_y F(x,y)}. \tag{5.90}$$

In the periodic case, the diagonal part is thus

$$-\frac{k^2\lambda^2 + n^2}{|n|}, \quad n \neq 0, \tag{5.91}$$

and the set of singular sites

$$\left\{ (n,k) \in \mathbb{Z}^2 \middle| \, \big| |n| - |k| \, |\lambda| \big| < \varepsilon \right\} \tag{5.92}$$

has a self-similar structure similar to the quasiperiodic problem.

(ii) The non-Hamiltonian nature of (5.85). This will lead to a linearized operator that is not self-adjoint. We will use the technique from [1] and [3] to control its inverse. This method does not relay on eigenvalue perturbation and self-adjointness. Also, as will be clear below, the Q-equations extracted from (5.85) still permit us to determine the remaining parameters.

Coming back to (5.85) and replacing again y by δy, rewrite the equation as

$$y_{tt} - y_{xx} + \rho y + \delta (y_t)^2 = 0 \tag{5.93}$$

considered as perturbation of the linear equation

$$y_{tt} - y_{xx} + \rho y = 0 \tag{5.94}$$

with spectrum

$$\mu_n = \sqrt{n^2 + \rho}. \tag{5.95}$$

We assume that ρ is such that the sequence $\{\mu_n\}$ given by (5.95) has the expected Diophantine properties.

For instance, let

$$y_0 = p \, \cos \left(x + \mu_1 t \right) + q \, \cos \left(x - \mu_1 t \right) \tag{5.96}$$

be a periodic solution of (5.94). We will construct a perturbed periodic solution of (5.93) of the form

$$y(x,t) = y_\delta(x,t) = \sum_{n,k \in \mathbb{Z}^2, n \geq 0} \hat{y}\,(n,k) \, \cos \left(nx + k\lambda t \right), \tag{5.97}$$

where

$$\hat{y}(1,1) = p, \quad \hat{y}(1,-1) = q, \tag{5.98}$$

$$\sum_{(n,k) \notin \mathcal{S}} e^{(|n|+|k|)^c} \, |\hat{y}(n,k)| = O(\delta), \tag{5.99}$$

with

$$S = \{(1,1),\ (1,-1)\} \tag{5.100}$$

the resonant set and where

$$\lambda^2 = 1 + \rho + O(\delta^2) \tag{5.101}$$

is the perturbed frequency.

Remark: One may construct space and time-periodic solutions of (5.85) of the form $y(x,t) = c(x + \lambda t)$, where c satisfies the ODE

$$c'' + \frac{\lambda^2}{\lambda^2 - 1}(c')^2 + \frac{\rho}{\lambda^2 - 1}c = 0$$

(observation due to C. Wayne). This equation has indeed small amplitude periodic solutions which will in turn give space-time periodic solutions of equation (5.85). This is the reason why we have chosen an unperturbed solution of the form (5.96); cf. [11].

Expressing (5.93) in Fourier yields

$$\left(-k^2\lambda^2 + n^2 + \rho\right)\hat{y}(n,k) + \delta\,\widehat{(y_t)^2}\,(n,k) = 0. \tag{5.102}$$

The P-equations (resp. Q-equations) are obtained restricting $(k,n) \notin S$ (resp. $(k,n) \in S$). Starting from y_0 given by (5.96), we get

$$(y_0)_t = -\mu_1(p\,\sin(x + \mu_1 t) - q\,\sin(x - \mu_1 t)) \tag{5.103}$$

and thus from (5.102)

$$
\begin{aligned}
y \;=\; & p\cos\left(x + \mu_1 t\right) + q\cos\left(x - \mu_1 t\right) \\
& + \frac{\delta\mu_1^2}{2}\left\{-\frac{p^2+q^2}{\rho} + \frac{p^2}{4(1-\lambda^2)+\rho}\cos\left(2x + 2\mu_1 t\right)\right. \\
& + \left.\frac{q^2}{4(1-\lambda^2)+\rho}\cos\left(2x - 2\mu_1 t\right) - \frac{2pq}{4+\rho}\cos 2x - \frac{2pq}{4\lambda^2+\rho}\cos\lambda\mu_1 t\right\} + O(\delta^2).
\end{aligned}
$$

Hence

$$
\begin{aligned}
y_t \;=\; & -\mu_1\left\{p\,\sin\left(x + \mu_1 t\right) - q\,\sin\left(x - \mu_1 t\right)\right\} \\
& - \delta\mu_1^3\left\{\frac{p^2}{4(1-\lambda^2)+\rho}\,\sin\left(2x + 2\mu_1 t\right)\right. \\
& - \left.\frac{q^2}{4(1-\lambda^2)+\rho}\,\sin\left(2x - 2\mu_1 t\right) - \frac{2pq}{4\lambda^2-\rho}\,\sin 2\mu_1 t\right\} + O(\delta^2).
\end{aligned}
$$

Specifying $(n,k) = (1,1)$ and $(n,k) = (1,-1)$, substitution of the previous equation in (5.102) yields for the Q-equations

$$\begin{cases} 1 - \lambda^2 + \rho + \delta^2 \mu_1^4 \left(\frac{p^2}{4(1-\lambda^2)+\rho} - \frac{2q^2}{4\lambda^2-\rho} \right) + O(\delta^3) = 0 \\ 1 - \lambda^2 + \rho + \delta^2 \mu_1^4 \left(\frac{q^2}{4(1-\lambda^2)+\rho} - \frac{2p^2}{4\lambda^2-\rho} \right) + O(\delta^3) = 0. \end{cases} \tag{5.104}$$

Assume $\rho > 0$, $\rho \neq 4/3$ fixed. For $\lambda^2 = 1 + \rho + O(\delta^2)$, one may then solve (5.104) for $p = p(\lambda)$, $q = q(\lambda)$. In order to be able to solve the P-equations by Newton's method, a sequence of conditions on p, q, λ are imposed that eventually will restrict λ down to some Cantor-like set C of positive measure, such that for $\lambda \in C$ one gets a solution of (5.93) of the form (5.97) with $p = p(\lambda)$, $q = q(\lambda)$ in (5.98).

We now turn our attention to the P-equation. The linearization of (5.102) at a given approximative solution y yields the linearized operator

$$T = (-k^2\lambda^2 + n^2 + \rho)\mathbb{1} + 2\delta[S_{iy_t} \circ (k\lambda)\mathbb{1}]. \tag{5.105}$$

What follows is a bit easier in the present context of time-periodic solutions than in the general quasiperiodic case, due to the fact that here, for singular sites, necessarily $|k| \sim |n|$.

Use (5.101) to write

$$-k^2\lambda^2 + n^2 + \rho = (1 - k^2)\rho + n^2 - k^2 + O(k^2\delta^2). \tag{5.106}$$

Since we excluded the resonant cases $(n,k) = (1,1)$ and $(n,k) = (1,-1)$ $(n \geq 0)$ and ρ is assumed to fulfill typical Diophantine conditions, no singular sites appear in the region $|n| + |k| < M_0 = M_0(\delta) = \delta^{-c_1}$. Applying the resolvent identity, the control of T^{-1} may then be achieved from that of T_Λ^{-1}, where Λ is a box of the form

$$\Lambda_+ = \left[L < n < \frac{11}{10}L \right] \times \left[\frac{9}{10\lambda}L < k < \frac{12}{10}\frac{L}{\lambda} \right] \tag{5.107}$$

or

$$\Lambda_- = \left[L < n < \frac{11}{10}L \right] \times \left[\frac{9}{10\lambda}L < -k < \frac{12}{10}\frac{L}{\lambda} \right] \tag{5.108}$$

with $L > \frac{1}{10}M_0(\delta)$, say. Assume (n,k) restricted to a box Λ_+, say. Instead of considering T, one may invert

$$T' = \frac{-k^2\lambda^2 + n^2 + \rho}{k\lambda}\,\mathbb{1} + 2\delta S_{iy_t} \tag{5.109}$$

restricted to Λ_+ (this operator is not self-adjoint since it is the sum of a diagonal and an antisymmetric matrix). Defining the set of singular sites I_s of T' as the pairs $(n, k) \in \Lambda_+$ such that ($c_2 < c_1$ sufficiently small)

$$\left| \frac{-k^2\lambda^2 + n^2 + \rho}{k\lambda} \right| < \delta^{c_2}, \tag{5.110}$$

we get, since $k > M$,

$$|k\lambda - n| < \delta^{c_2}. \tag{5.111}$$

This permits us to conclude a separation of the singular sites by δ^{-c_3}, for some $c_3 > 0$. Indeed, if $(n_1, k_1) \in I_s$, $(n_2, k_2) \in I_s$, and $|k_1 - k_2| < \frac{1}{\delta}$, it follows from (5.101) and (5.111) that

$$\left\| (k_1 - k_2)\sqrt{1 + \rho} \right\| < 3\delta^{c_2}. \tag{5.112}$$

Hence, since ρ is assumed Diophantine,

$$\begin{aligned} &|k_1 - k_2|^{-C} < \left\| (k_1 - k_2)^2\rho \right\| < 10\delta^{c_2}|k_1 - k_2|, \\ &|k_1 - k_2| > \frac{1}{10}\delta^{-c_2/1+C} > \delta^{-c_3} \end{aligned} \tag{5.113}$$

implying the previous statement about separation.

Our aim is to get for suitable restrictions of the parameters p, q, λ again good (polynomial) bounds on the inverse $(T'_{\Lambda_+})^{-1}$ and off-diagonal decay. This is achieved by an inductive process considering restrictions of the operator to boxes $Q \subset \Lambda_+$ of increasing size scale, much in the spirit of [1], [3], and [8]. These boxes will be obtained by restriction of the k-index only, thus

$$Q = \left[L < n < \frac{11}{10}L \right] \times J, \tag{5.114}$$

where J is an interval in $[\frac{9}{10\lambda}L < k < \frac{12}{10\lambda}L]$ of a certain size $o(L)$. The multiscale analysis of the inverse of

$$T'_Q = \frac{-k^2\lambda^2 + n^2 + \rho}{k\lambda} \mathbb{1} + \delta S|_Q \tag{5.115}$$

will relate to size scales of J. Observe that in (5.115) $k, n \sim L$.

Restricting T' to the complement of I_s, the inverse is controlled by a Neumann series.

Next we consider $T'_Q = T'|_Q$ where Q is a box containing at most one singular site and later on restrictions to boxes of increasing size order. This induction process is mainly based on the separation properties of "singular" boxes, meaning essentially that, for Q_1, Q_2 boxes of same size not contained in the doubling of each other, say, large $\|(T'_{Q_1})^{-1}\|$ and $\|(T'_{Q_2})^{-1}\|$ may only occur when Q_1 and Q_2 are sufficiently separated. These separation properties are then used together with the resolvent identity to establish off-diagonal decay of the inverse.

In view of (5.101), we introduce a new parameter τ with variation range $o(1)$, letting

$$\lambda^2 = 1 + \rho + \delta^2 \tau. \tag{5.116}$$

Thus from (5.104) and (5.116) we have

$$\frac{\partial \lambda}{\partial \tau} \sim \delta^2, \quad \left| \frac{\partial p}{\partial \tau} \right| = O(1), \quad \left| \frac{\partial q}{\partial \tau} \right| = O(1) \tag{5.117}$$

(after expressing p, q in λ from (5.104)). Parameter restrictions are eventually expressed in terms of τ, which admissible values will be restricted to a Cantor-type set.

In order to set up the multiscale induction process, we introduce an extra parameter $\sigma, |\sigma| < \frac{L}{3}$ and consider the operators

$$T^\sigma_Q = \frac{-(k\lambda + \sigma)^2 + n^2 + \rho}{k\lambda + \sigma} \mathbb{1} + \delta S|_Q. \tag{5.118}$$

The reason for this is the fact that

$$T^\sigma_{Q+k_0} = T^{\sigma + k_0 \lambda}_Q, \tag{5.119}$$

and hence a discussion of $(T^\sigma_{Q_0})^{-1}$, $Q_0 = [L < n < \frac{11}{10}L] \times J_0$, J_0 fixed, in the full parameter range will also apply to k-translates Q of Q_0. As in [1] and [3], we will control $(T^\sigma_{Q_0})^{-1}$ by a system of monic polynomials in σ with smooth coefficients in λ, p, q. In fact, these polynomials will be here of degree 1, which simplifies matters a bit. In what follows, we briefly point out the main ideas.

From the considerations (5.110)–(5.113), it follows that, if J_0 is sufficiently short, say $< \delta^{-c_3}$, there is at most one singular site (n_0, k_0) for $T^\sigma_{Q_0}$ (fixing the parameters). Thus

$$\left| \frac{-(k\lambda + \sigma)^2 + n^2 + \rho}{k\lambda + \sigma} \right| > \delta^{c_2} \quad \text{for} \quad (n, k) \in Q_0 \backslash \{n_0, k_0\}. \tag{5.120}$$

Denote

$$\sigma_1 = \sigma + k_0 \lambda - n_0 \tag{5.121}$$

assuming thus

$$|\sigma_1| < \delta^{c_2}. \tag{5.122}$$

Rewrite the diagonal of $T_{Q_0}^\sigma$ as

$$\frac{[n_0 + n + (k - k_0)\lambda + \sigma_1][(n - n_0) + (k_0 - k)\lambda - \sigma_1] + \rho}{n_0 + (k - k_0)\lambda + \sigma_1} \tag{5.123}$$

$$= [1 + \frac{n}{n_0}(1 + o(1))][n - n_0) - (k - k_0)\lambda - \sigma_1], \tag{5.124}$$

where the $o(1)$ also refers to σ_1 and λ-derivatives and $\to 0$ for $|n_0| \to \infty$.
    Write

$$T_{Q_0}^\sigma = \begin{pmatrix} T_{Q_0 \backslash \{n_0, k_0\}}^\sigma & -\delta \xi^* \\ \delta \xi & \frac{-(2n_0 + \sigma_1)\sigma_1 + \rho}{n_0 + \sigma_1} \end{pmatrix}, \tag{5.125}$$

where, by (5.120), $(T_{Q_0 \backslash \{n_0, k_0\}}^\sigma)^{-1}$ is well controlled.
    To control the inverse of $(T_{Q_0}^\sigma)^{-1}$, we need thus consider the reciprocal of
the expression

$$\frac{-(2n_0 + \sigma_1)\sigma_1 + \rho}{n_0 + \sigma_1} + \delta^2 \langle (T_{Q_0 \backslash \{n_0, k_0\}}^\sigma)^{-1} \xi, \xi \rangle$$

$$= -2\sigma_1 + o(1) + \delta^2 \langle (T_{Q_0 \backslash \{n_0, k_0\}}^\sigma)^{-1} \xi, \xi \rangle$$

$$= -2\sigma_1 + \varphi(\sigma_1, \lambda, p, q)$$

considered as a function of σ_1, λ, p, q. Thus

$$\partial_{\sigma_1} \varphi = o(1), \tag{5.126}$$

$$|\partial_\lambda \varphi| \lesssim \delta^2, \tag{5.127}$$

$$|\partial_p \varphi|, |\partial_q \varphi| \lesssim \delta^2. \tag{5.128}$$

(In (5.127) and (5.128), an estimate of the form $\delta^{2 - 2c_2}$ is clear from
(5.120); the stated bound results from a slightly more careful analysis of
the last term in (5.126) invoking decay considerations.)

From the implicit function theorem, one may then replace (5.126) by

$$\sigma_1 - \psi(\lambda, p, q), \tag{5.129}$$

where ψ still satisfies (5.127) and (5.128).

Observe that, from (5.121), if we let $\sigma = 0$,

$$\frac{\partial \sigma_1}{\partial \lambda} \sim k_0 \sim L > M_0. \tag{5.130}$$

Hence, considering λ, p, q as a function of τ, we have

$$\left| \frac{\partial}{\partial \tau}[k_0 \lambda - n_0 - \psi(\lambda, p, q)] \right| > M_0 \delta^2 \tag{5.131}$$

by invoking (5.117).

In order to obtain separation of "singular" matrices $T^\sigma_{Q_1}$ and $T^\sigma_{Q_2}$ with

$$Q_i = Q_0 + k_i \quad (i = 1, 2) \tag{5.132}$$

and

$$|k_1 - k_2| \gg |Q_0|, \tag{5.133}$$

one has to avoid simultaneous almost vanishing of two expressions

$$\sigma_1 - \psi_1(\lambda, p, q) \approx 0, \tag{5.134}$$

$$\sigma_2 - \psi_2(\lambda, p, q) \approx 0, \tag{5.135}$$

where

$$\sigma_i = \sigma + k_{0,i}\lambda - n_{0,i}, \qquad |k_i - k_{0,i}| < |Q_0|. \tag{5.136}$$

If (5.134), (5.135), and (5.136) hold, we have (eliminating σ)

$$(k_{0,1} - k_{0,2})\lambda - n_{0,1} + n_{0,2} - \psi_1(\lambda, p, q) + \psi_2(\lambda, p, q) \approx 0. \tag{5.137}$$

Hence

$$\|(k_{0,1} - k_{0,2})\lambda - \psi_1(\lambda, p, q) + \psi_2(\lambda, p, q)\| \approx 0, \tag{5.138}$$

where

$$|k_{0,1} - k_{0,2}| \gg |Q_0|. \tag{5.139}$$

To avoid this, we use the fact that

$$\frac{\partial}{\partial\lambda}[(k_{0,1} - k_{0,2})\lambda - \psi_1(\lambda, p, q) + \psi_2(\lambda, p, q)] \sim k_{0,1} - k_{0,2}. \tag{5.140}$$

Hence

$$\left|\frac{\partial}{\partial\tau}[(k_{0,1} - k_{0,2})\lambda - \psi_1(\lambda, p, q) + \psi_2(\lambda, p, q)]\right| > |Q_0|\delta^2. \tag{5.141}$$

The preceding is the first step in the multiscale analysis. For the continuation of the process, we refer the reader to [1] and [3]. This permits us for the restrictions $T'_Q, Q \subset \Lambda_\pm$, to get estimates on the inverse that are powerlike in $|Q|$ with exponential off-diagonal decay; similarly for the inverse of restrictions of T as needed in the Newton scheme. For details, the reader is referred to [1] and [3].

The result of the preceding is the following.

Theorem 5.2. *Consider a NLW equation*

$$y_{tt} - y_{xx} + \rho y + (y_t)^2 = 0 \tag{5.142}$$

with periodic boundary conditions, where $\rho > 0$ is a typical number. Then for δ sufficiently small taken in a Cantor set of positive measure, (5.139) has a time-periodic solution of the form (5.97)–(5.99)

$$\begin{aligned} y(x, t) = p\cos(x \quad + \quad \lambda t) + q\cos(x - \lambda t) \\ + \quad \sum_{n\geq 0, (n,k)\neq(1,1),(1,-1)} \widehat{y}(n, k)\cos(nx + \lambda t), \end{aligned} \tag{5.143}$$

where

$$p = p(\delta), q = q(\delta) = O(\delta) \quad and \quad \lambda^2 = 1 + \rho + O(\delta^2) \tag{5.144}$$

and the last term in (5.140) is $O(\delta^2)$.

5.3 Some Remarks on Frequencies and Parameters

Consider a perturbed Hamiltonian system

$$i\dot{q} = \frac{\partial H_0}{\partial\bar{q}} + \varepsilon\frac{\partial H_1}{\partial\bar{q}}, \tag{5.145}$$

where $q = \{q_n\}_{n=1}^N$, N finite or $N = \infty$, $q_n \in \mathbb{C}$ are the pairs of conjugate variables. Let $1 \leq b \leq N$, O an interval in $\mathbb{R}^b$, $a \in O$ a b-parameter, and

$$q_0^{(a)} = q_0^{(a)}(t) \tag{5.146}$$

a family of quasiperiodic solutions of

$$i\dot{q} = \frac{\partial H_0}{\partial \bar{q}} \tag{5.147}$$

with frequency vector $\lambda(a) = (\lambda_1(a), \dots, \lambda_b(a))$. Assume the nondegeneracy condition

$$\det\left(\frac{\partial \lambda_j}{\partial a_k}\right)_{1 \leq j,k \leq n} \neq 0 \tag{5.148}$$

is satisfied.

When $b = N$, the problem of persistency of some of these tori for the perturbed equation (5.145) is treated in the classical KAM theory. In this case, invariant tori for (5.145) are obtained for (perturbed) frequency vectors λ' satisfying some Diophantine condition, of the form

$$|\langle \lambda', k \rangle| > c|k|^{-r} \text{ for all } k \in \mathbb{Z}^b \backslash \{0\}, \tag{5.149}$$

where $c > 0$, $r > b - 1$ are some constants.

Our interest goes to the case $N > b$. For $b = 1$, N finite, the discussion below will not apply since there are no small divisors. However, in the PDE context ($N = \infty$) the problems which we recall next appear as well for $b = 1$.

Take for simplicity H_0 of the form

$$H_0(q, \bar{q}) = \sum_{n=1}^b \lambda_n(a)|q_n|^2 + \sum_{n>b} \mu_n(a)|q_n|^2 \tag{5.150}$$

corresponding in (5.147) to a parameter dependent linear equation.

For $b < N$, the persistency problem, which we call "Melnikov problem" has been studied by various authors, including [6], [9], [10] and also in earlier works of J. Moser [12, 13] ($N < \infty$ here). The method used in those works is the standard Hamiltonian procedure. The nonresonance conditions required are of the form

$$|\langle \lambda', k \rangle + \langle \mu', \ell \rangle| > c(|k| + |\ell|)^{-r} \tag{5.151}$$

for $(k, \ell) \in \mathbb{Z}^N \backslash \{0\}$, $|\ell| \leq 2$.

This condition is more restrictive than (5.149) since the normal frequencies $\{\mu'_n\}$ are also involved. Observe that, in the context of (5.145), if we fix H_1 but let ε vary, we have

$$\lambda' = \lambda'(a, \varepsilon) \text{ and } \mu' = \mu'(a, \varepsilon). \tag{5.152}$$

Thus, if we choose $a = (a_1, \ldots, a_b)$ such that

$$\lambda' = \lambda'(a, \varepsilon) = \lambda_0, \qquad (5.153)$$

where λ_0 is fixed, satisfying (5.149), say, (5.151) is not necessarily satisfied when $b < N$. As observed by J. Moser, to fulfill (5.151) requires more parameters. Alternatively, Eliasson [7] weakens (5.153) to the property

$$\lambda' = t\lambda_0 \text{ for some } t \in \mathbb{R} \qquad (5.154)$$

(thus the tangential frequency is parallel to a given vector λ_0).

Essentially speaking, if b parameters are available, (5.154) corresponds to $b-1$ conditions and the remaining degree of freedom is used to ensure (5.151). In conclusion, the set of "admissible" tangential frequencies for invariant tori of (5.145) depends on ε.

In the case $N = \infty$, quasiperiodic solutions of Hamiltonian perturbations of linear or integrable PDE's where studied by S. Kuksin [9, 10] and C. Wayne [14], based again on the KAM methods. In particular, condition (5.151) is needed.

The Lyapounov-Schmidt technique applied in previous sections, originating from [5] and developed further by the author, is less restrictive since multiplicities in the normal frequencies are not excluded. This method allows one to understand in particular how tori with positive Lyapounov exponents appear in perturbations of integrable systems with only elliptic tori. Roughly speaking, the relevant nonresonant expressions here are of the form

$$\langle \lambda', k \rangle + \langle \mu', \ell \rangle \text{ with } (k, \ell) \in \mathbb{Z}^N \backslash \{0\}, |\ell| \leq 1. \qquad (5.155)$$

However (5.155) still involves the normal frequencies and previous comments apply as well.

There are a number of natural questions one may formulate here. For instance, if Eliasson's result [6] with parallel tangential frequency vector may be obtained in the infinite-dimensional (PDE) setting (either following KAM or Lyapounov-Schmidt procedure) or whether (5.153) may be ensured if extra parameters available. Finally, the preceding discussion only comments on the methods and does not claim failure of existence of invariant tori if the nonresonance conditions are violated. One may, however, expect to prove nonexistence of certain families of invariant tori with given frequency vector λ, differentiable in the perturbation parameters, since that statement relates directly to the linearized equation.

In the remainder of this section, we give an example of such smooth families with Diophantine frequency λ in the context of time periodic solutions of a NLW equation with an extra parameter. Consider an equation

$$y_{tt} - y_{xx} + \rho y + y^3 + \varepsilon f(x, y) = 0, \qquad (5.156)$$

where ρ will be a parameter. Assume for simplicity that f is a sum of the form

$$f(x,y) = \sum_{j \geq 1} b_j(x) y^j \qquad (5.157)$$

with $\{b_j\}$ even trigonometric polynomials in x (real-analytic assumptions should work as well). Thus according to (5.145)

$$H_0(y,v) = \int \left[\frac{1}{2}(y_x)^2 + \frac{\rho}{2}y^2 + \frac{1}{4}y^4 + \frac{1}{2}v^2 \right], \qquad (5.158)$$

$$H_1(y,v) = \int F(x,y) \qquad (\partial_y F = f), \qquad (5.159)$$

with $(y, v = \dot{y})$ as canonical variables.

Our parameters a will be extracted from the nonlinearity y^3 by amplitude-frequency modulation, and we let ρ in (5.156) be a second parameter.

For $c > 0$, $\delta > 0$, denote by $\Lambda_{c,\delta}$ the set of frequencies λ satisfying following Diophantine conditions

$$\|k\lambda\| = \min_{n \in \mathbb{Z}} |k\lambda - n| > c|k|^{-1-\delta} \text{ for all } k \in \mathbb{Z}\backslash\{0\} \qquad (5.160)$$

and

$$|-(k^2-1)\lambda^2 + n^2 - 1| > c|k|^{-\delta} \text{ for all } k \in \mathbb{Z}\backslash\{1,-1\}. \qquad (5.161)$$

Clearly $\text{mes}(\mathcal{C}\Lambda_{c,\delta}) \overset{c \to 0}{\longrightarrow} 0$ for all $\delta > 0$.

We prove the following

Theorem 5.3. *Let $\lambda \in \Lambda_{c,\delta}$ (δ small enough). For sufficiently small ε, there are smooth functions of ε*

$$\rho = \rho(\varepsilon) \ , \quad a = a(\varepsilon) \qquad (5.162)$$

such that (5.156) has a λ-periodic solution in time, of the form

$$y_\varepsilon(x,t) = a\cos x \cdot \cos \lambda t + \sum_{n,k \geq 0, (n,k) \neq (1,1)} \hat{y}_\varepsilon(n,k) \cos nx \cdot \cos k\lambda t \qquad (5.163)$$

depending smoothly on ε and such that

$$\sum_{(n,k) \neq (1,1)} e^{-(|n|+|k|)^c} |\hat{y}_\varepsilon(n,k)| = 0(\varepsilon + |a|^3) \qquad (5.164)$$

(a is small).

Note that the representation (5.163) is compatible with the assumption that (5.157) is even in x. This result is in the spirit of some results from [12] and [13].

We construct (5.163) following the Lyapounov-Schmidt scheme from [5], perturbing off from the linear equation

$$y_{tt} - y_{xx} + \rho y = 0 \tag{5.165}$$

with periodic solution

$$y_0(x, t) = a \cos x \cdot \cos \lambda_0 t \tag{5.166}$$

and

$$\lambda_0 = \lambda_0(\rho) = \sqrt{1 + \rho}. \tag{5.167}$$

The Q-equation is obtained by substituting (5.163) in (5.156) and projecting on the $(1, 1)$ mode. Thus we get

$$(-\lambda^2 + 1 + \rho)a + \widehat{y^3}(1, 1) + \varepsilon \widehat{f(x, y)}(1, 1) = 0, \tag{5.168}$$

hence of the form

$$-\lambda^2 + 1 + \rho + \frac{9}{16}a^2 + O(a^3) + O(\varepsilon) = 0. \tag{5.169}$$

The system of P-equations

$$(-\lambda^2 k^2 + n^2 + \rho)\hat{y}(n, k) + \widehat{y^3}(n, k) + \varepsilon \widehat{f(x, y)}(n, k) = 0, \tag{5.170}$$

$(n, k) \in \mathbb{Z}^+ \times \mathbb{Z}^+ \backslash \{1, 1\}$, is solved by the Newton iteration procedure.

At stage r, an approximate solution y_r is obtained and the linearized equation at y_r is given by

$$T = D + S, \tag{5.171}$$

where D is diagonal with diagonal elements

$$D_{n,k} = -\lambda^2 k^2 + n^2 + \rho + 3\widehat{y_r^2}(0, 0) + \varepsilon \widehat{\partial_y f}(0, 0) \tag{5.172}$$

and S with 0-diagonal, given by the Toeplitz operator

$$S = S_{\phi - \hat{\phi}(0,0)} \tag{5.173}$$

and

$$\phi = 3y_r^2 + \varepsilon \partial_y f(x, y_r) \tag{5.174}$$

satisfying

$$\|\phi\| = 0(a^2 + \varepsilon).\tag{5.175}$$

Fix a large number N_0 and let

$$\eta = N_0^{-1} \text{ and } a \sim \eta\,,\ |\varepsilon| < \eta^3\,.\tag{5.176}$$

From (5.169) and (5.176)

$$|\rho - \lambda^2 + 1| \overset{<}{\sim} \eta^2\,.\tag{5.177}$$

Hence, for $\lambda \in \Lambda_{c,\delta}$, it follows from (5.172) and (5.171) that

$$\begin{aligned}
|D_{n,k}| > |-\lambda^2 k^2 + n^2 + \rho| - C\eta^2 &> |-\lambda^2(k^2-1) + n^2 - 1| - C\eta^2 \\
&> c|k|^{-\delta} - C\eta^2
\end{aligned}$$

and thus

$$|D_{n,k}| > cN_0^{-\delta} \text{ for } |n| < N_0.\tag{5.178}$$

Therefore the inverse of $T_{N_0} \equiv T\big|_{|n|<N_0,|k|<N_0}$ may be controlled by a Neumann series and in particular

$$\|(T_{N_0})^{-1}\| \overset{<}{\sim} N_0^\delta\,.\tag{5.179}$$

Since $\operatorname{supp} \hat{y}_r \subset B(0, M^r)$ for some constant M, the preceding allows us to carry out the Newton iteration as long as $M_r \equiv M^r < N_0$, leading to an approximative solution up to $e^{-\left(\frac{3}{2}\right)^r} < e^{-\left(\frac{1}{\eta}\right)^r}$, say. The control of T_N^{-1} for $N > N_0$ requires further conditions. The main point here is an observation made by S. Kuksin.[2]

Assume moreover

$$|D_{n,k}| > \frac{\eta^2}{|n|^{1/2}} \text{ for } |n| \leq N.\tag{5.180}$$

Then (for η small enough)

$$\|D_N^{-1} S_N D_N^{-1}\| < \frac{1}{100}\,.\tag{5.181}$$

The point is that if some site (n, k), $n > N_0$, is "singular" in the sense that $|-\lambda^2 k^2 + n^2| < 1$, hence $\|\lambda k\| < \frac{1}{|k|}$, then for $(n_1, k_1) \neq (n, k)$ and nearby

---

[2]Private communication.

(n, k), necessarily by (5.160) $|-\lambda k_1^2 + n_1^2| \gtrsim |k_1| \|\lambda k_1\| > c|k_1| |k - k_1|^{-2} = c|k|^{1-}$. Hence, if $|D_{n,k}| < 1$, one gets $|D_{n_1, k_1}| > c|n|^{1-}$ for $(n_1, k_1) \neq (n, k)$ nearby (n, k). Property (5.181) is then easily derived from the preceding, (5.180) and since $\|S\| < \eta^2$. It follows that T_N^{-1} may again be controlled by the Neumann series

$$T_N^{-1} = (D_N + S_N)^{-1} = D_N^{-1} \sum_{j \geq 0} (-1)^j (S_N D_N^{-1})^j \tag{5.182}$$

(see [3] for all details). We have in particular

$$\|T_N^{-1}\| \lesssim \frac{N^{1/2}}{\eta^2}. \tag{5.183}$$

In order to fulfill (5.180), we need to restrict the parameters $(\lambda, \varepsilon, a, \rho)$. More specifically, (5.172) yields

$$\frac{\partial D_{n,k}}{\partial \rho} = O(1) \tag{5.184}$$

and hence, for given (n, k), $n > N_0$, (5.180) amounts to removing a set of measure $\lesssim \frac{\eta^2}{|n|^{1/2}}$. Since (5.180) only needs to be verified for $|n^2 - \lambda^2 k^2| < 1$, thus $\|\lambda k\| < \frac{1}{|k|}$, and $\lambda \in \Lambda_{c,\delta}$, there is the following estimate on the total measure to be removed from the parameter set

$$\sum_{\substack{N_0 < \ell < N \\ \ell \text{ dyadic}}} \frac{\eta^2}{\ell^{1/2}} \ell^{1 - \frac{1}{1+\delta}} < \eta^2 N_0^{-1/3} = \eta^{7/3}. \tag{5.185}$$

On the other hand, (5.169) and (5.176) yield for ρ a variation range of size η^2. Observe that when the parameter set excisions first occur, already $\|\Delta y\| < e^{-(1/\eta)^\gamma}$. Thus these excisions are such that the solution to the P-equation may be extended smoothly to the entire parameter set. There is also the issue of varying symbol $\phi = \phi_r$ given by (5.174). Since at stage r, we let $N = M_r$ and $\|\phi_{r-1} - \phi_r\| < e^{-N^c}$, this change of symbol clearly does not lead to problems when fulfilling (5.180). We are reviewing here several issues from [5]. In conclusion, we may obtain a solution $y = y_{\lambda, \varepsilon, a, \rho}$ to the P-equation of the form (5.163), provided that the conditions

$$|D_{n,k}^r| > \frac{\eta^2}{|n|^{1/2}} \text{ for } |n| < M_r \tag{5.186}$$

(only to be verified for $N_0 < |n| < M_r$) are fulfilled and where moreover $y_{\lambda, \varepsilon, a, \rho}$ is smoothly defined on the full parameter set.

Fix $\lambda \in \Lambda_{c,\delta}$. We choose a number $\nu(\lambda)$ such that

$$\nu(\lambda) \sim \eta^2 \tag{5.187}$$

and

$$|-\lambda^2(k^2-1)+n^2-1+\nu(\lambda)| > 10\frac{\eta^2}{|n|^{1/2}} \text{ for } k \neq 1, -1. \qquad (5.188)$$

The same calculation as (5.185) shows that this is possible.

Next we specify (5.162) requiring ρ and a to solve the equations

$$\begin{cases} -\lambda^2 + 1 + \rho + \frac{1}{a}\left[\widehat{y^3}(1,1) + \varepsilon \widehat{f(x,y)}(1,1)\right] = 0 & (Q\text{-equation}), \\ -\lambda^2 + 1 + \rho + 3\widehat{y^2}(0,0) + \varepsilon \widehat{\partial_y f(x,y)}(0,0) = \nu(\lambda), \end{cases} \qquad (5.189)$$

where y is substituted from solving the P-equation. If the latter equation in (5.189) holds, then at stage r, $M_r > N_0$, we get from (5.188)

$$\begin{aligned} |D_{n,k}| &> |-\lambda^2(k^2-1)+n^2-1+\nu(\lambda)| + O(\|y - y_r\|) \\ &> 10\frac{\eta^2}{|n|^{1/2}} - e^{-M_r^c} > \frac{9\eta^2}{|n|^{1/2}} \end{aligned}$$

and (5.186) is fulfilled. Since the equations in (5.189) appear in the form

$$\begin{cases} \tilde{\rho} + \frac{9}{16}a^2 + 0(a^3 + \varepsilon) = 0 \\ \tilde{\rho} + \frac{3}{4}a^2 + 0(a^3 + \varepsilon) = \nu(\lambda) \end{cases} \qquad (5.190)$$

with

$$\rho = \lambda^2 - 1 + \tilde{\rho} \qquad (5.191)$$

one obtains from the implicit function theorem $\rho = \rho(\varepsilon)$, $a = a(\varepsilon) \sim \sqrt{\nu(\lambda)} \sim \eta$ as smooth functions of ε.

References

[1] J. Bourgain, *Construction of quasiperiodic solutions of Hamiltonian perturbations of linear equations and applications to nonlinear PDE*, International Math. Research Notices 11 (1994), 475–497.

[2] ———, *Construction of periodic solutions of nonlinear wave equations in higher dimension*, GAFA 4 (1995), 629–639.

[3] ———, *Nonlinear Schrödinger equations*, IAS Park City Mathematics Series, vol. 5 (1999), 5–157.

[4] ———, *Quasi-periodic solutions of Hamiltonian perturbations of 2D linear Schrödinger equations*, Annals of Math. 148 (1998), 363–439.

[5] W. Craig and C. Wayne, *Newton's method and periodic solutions of nonlinear wave equations*, Comm. Pure and Applied Math. 46 (1993), 1409–1501.

[6] L. Eliasson, *Perturbations of stable invariant tori*, Ann. Sc. Super. Pisa, Cl Sci. IV, Ser. 15 (1988), 115–147.

[7] A. Erdélyi, W. Magnus, F. Oberhettinger, and F. Tricomi, *Higher transcendental functions* (Bateman ms.) (1953).

[8] J. Fröhlich, T. Spencer, and P. Wittwer, *Localization for a class of one dimensional quasi-periodic Schrödinger operators*, Comm. Math. Phys. 132, no. 1 (1990), 5–25.

[9] S. Kuksin, *Perturbation theory for quasi-periodic solutions of infinite dimensional Hamiltonian systems and its applications to the Korteweg – de Vries equation*, Math. USSR Sbornik 64 (1989), 397–413.

[10] ———, *Nearly Integrable Finite-Dimensional Hamiltonian Systems*, Springer LNM 1556 (1993).

[11] B. Lidskij and E. Schulman, *Periodic solutions of the equation $\ddot{u} - u_{xx} + u^3 = 0$*, Funct. Anal. Appl. 22 (1988), 332–333.

[12] J. Moser, *On the theory of quasi-periodic motions*, Siam Review 8 (1966), 145–172.

[13] ———, *Convergent series expansions for quasi-periodic motions*, Math. Annalen 169 (1967), 136–176.

[14] C. Wayne, *Periodic and quasi-periodic solutions of nonlinear wave equations via KAM theory*, CMP 127 (1990), 479–528.

Chapter 6

Some Extremal Problems in Martingale Theory and Harmonic Analysis

Donald L. Burkholder

6.1 Introduction

The main new results are Theorem 6.3 and its corollary for stochastic integrals, Theorem 6.6. However, we begin with a little background with roots in some of the work of Kolmogorov and M. Riesz of more than seventy years ago. Let n be a positive integer, D a domain of $\mathbb{R}^n$, and $\mathbb{H}$ a real or complex Hilbert space. Suppose that u and v are harmonic on D with values in $\mathbb{H}$. Let $|\nabla u| = (\sum_{k=1}^n |\partial u/\partial x_k|^2)^{1/2}$. Then v is *differentially subordinate* to u if, for all $x \in D$,

$$|\nabla v(x)| \leq |\nabla u(x)|.$$

Fix a point $\xi \in D$ and let D_0 be a bounded subdomain satisfying

$$\xi \in D_0 \subset D_0 \cup \partial D_0 \subset D.$$

Denote by μ the harmonic measure on ∂D_0 with respect to ξ. If $1 \leq p < \infty$, let

$$||u||_p = \sup_{D_0} \left[\int_{\partial D_0} |u|^p \, d\mu \right]^{1/p},$$

where the supremum is taken over all such D_0.

Research partially supported by NSF grant DMS-9626398.

Theorem 6.1. [5]. *If $|v(\xi)| \leq |u(\xi)|$ and v is differentially subordinate to u, then*

$$\mu(|v| \geq 1) \leq \mu(|u| + |v| \geq 1) \leq 2\|u\|_1 \tag{6.1}$$

and even for the inequality between the left and right sides of (6.1) the constant 2 is the best possible. Furthermore, if $1 < p < \infty$ and $p^* = \max\{p, q\}$ where $1/p + 1/q = 1$, then

$$\|v\|_p \leq (p^* - 1)\|u\|_p. \tag{6.2}$$

To see that (6.2) holds, define U and V on $\mathbb{H} \times \mathbb{H}$ by

$$U(h, k) = p(1 - 1/p^*)^{p-1}(|k| - (p^* - 1)|h|)(|h| + |k|)^{p-1} \tag{6.3}$$

and

$$V(h, k) = |k|^p - (p^* - 1)^p |h|^p. \tag{6.4}$$

Then U majorizes V, $U(h, k) \leq 0$ if $|k| \leq |h|$, and $U(u, v)$ is superharmonic on D. Therefore,

$$\int_{\partial D_0} V(u, v) \, d\mu \leq \int_{\partial D_0} U(u, v) \, d\mu \leq U(u(\xi), v(\xi)) \leq 0, \tag{6.5}$$

which gives (6.2). See [5] for the details and an application to Riesz systems. The proof of (6.1) has the same pattern. If $|h| + |k| \geq 1$, let

$$U(h, k) = V(h, k) = 1 - 2|h|. \tag{6.6}$$

Otherwise, let

$$U(h, k) = |k|^2 - |h|^2 \text{ and } V(h, k) = -2|h|. \tag{6.7}$$

It is easy to check that, as above, U majorizes V, $U(h, k) \leq 0$ if $|k| \leq |h|$, and $U(u, v)$ is superharmonic on D. So (6.5) holds here also and gives (6.1). That the constant 2 is the best possible even for the inequality between the left and right sides of (6.1) is shown in Remark 13.1 of [7]. In Theorem 6.1, the usual conjugacy condition on v in the classical setting of the open unit disk is replaced by the weaker condition of differential subordination, a condition that makes sense on domains of $\mathbb{R}^n$. Inequalities similar to (6.1) and (6.2) can hold even if u and v are not harmonic. As above, fix $\xi \in D_0$ and denote the harmonic measure on ∂D_0 with respect to ξ by μ.

Theorem 6.2. [7]. *Suppose here that $u : D \to [0, \infty)$ and $v : D \to \mathbb{H}$ are continuous functions with continuous first and second partial derivatives such that*

$$|v(\xi)| \leq |u(\xi)|, \tag{6.8}$$

$$|\nabla v| \leq |\nabla u|, \tag{6.9}$$

$$|\Delta v| \leq |\Delta u|. \tag{6.10}$$

(i) If u is subharmonic, then

$$\mu(|v| \geq 1) \leq \mu(u + |v| \geq 1) \leq 3||u||_1. \tag{6.11}$$

Moreover, if $1 < p < \infty$ *and* $p^{**} = \max\{2p, q\}$ *where* $1/p + 1/q = 1$, *then*

$$||v||_p \leq (p^{**} - 1)||u||_p. \tag{6.12}$$

(ii) If u is superharmonic, then (6.11) holds with the constant 2. This is the best possible constant even for the inequality between the left and right sides.

After appropriate functions U are found the proof follows the same pattern as the proof of Theorem 6.1 (see [7]). These functions U were discovered originally in a martingale setting [4, 7]. In this setting, $p^* - 1$ is the best possible constant for the martingale analogue of (6.2), and the same is true for the other constants mentioned in Theorems 6.1 and 6.2. However, it is an open question whether or not $p^* - 1$ is the best possible constant for (6.2). It is also open whether 3 is the best constant for (6.11) and whether $p^{**} - 1$ is the best for (6.12).

The Beurling-Ahlfors transform (the singular integral operator on $L^p(\mathbb{C})$, $1 < p < \infty$, that maps f to its convolution with $-1/\pi z^2$) is important in the study of quasiconformal mappings and elsewhere; see, for example, the paper of Iwaniec and Martin [13] and references given there. The $L^p(\mathbb{C})$-norm of this operator is not yet known precisely, except for $p = 2$, although this knowledge would be helpful. The function U defined in (6.3) has been used recently (see Bañuelos and Wang [2], and Bañuelos and Lindeman [1]) to obtain sharper information than previously known about the norm of the Beurling-Ahlfors transform. Both papers outline several approaches that might lead to the precise value of this norm. One of these approaches, attributed to the referee of [1], is the following. The function U gives rise to a rank-one convex function on the set of 2×2 matrices with real coefficients [12, 1]. It is an open question [10] whether or not a rank-one convex function on the 2×2 matrices is also quasiconvex. If the answer is positive, then the norm of the Beurling-Ahlfors operator is $p^* - 1$.

6.2 A Martingale Setting

One of the goals of the next few sections is to explain, justify, and augment the following equalities and inequalities, which give some insight into the behavior of martingales and submartingales.

$$\beta_p(\text{posmar}) = \beta_p(\text{mar}) = \beta_p(\text{possub}) \text{ if } 1 < p \leq 3/2, \tag{6.13}$$

$$\beta_p(\text{posmar}) = \beta_p(\text{mar}) < \beta_p(\text{possub}) \text{ if } 3/2 < p \leq 2, \tag{6.14}$$

$$\beta_p(\text{posmar}) < \beta_p(\text{mar}) < \beta_p(\text{possub}) \text{ if } p > 2. \tag{6.15}$$

Suppose that $(\Omega, \mathcal{F}, P)$ is a probability space filtered by $(\mathcal{F}_n)_{n\geq 0}$, a nondecreasing sequence of sub-σ-algebras of $\mathcal{F}$. Let f_n and g_n be strongly integrable, $\mathcal{F}_n$-measurable functions on Ω with values in $\mathbb{H}$, $n \geq 0$. Then $f = (f_n)_{n\geq 0}$ is a martingale under the additional assumption that the conditional expectation $E(f_n|\mathcal{F}_{n-1}) = f_{n-1}$ for all $n > 1$.

Write

$$f_n = \sum_{k=0}^{n} d_k \text{ and } g_n = \sum_{k=0}^{n} e_k$$

and suppose that g is *differentially subordinate* to f:

$$|e_n| \leq |d_n| \text{ for all } n \geq 0. \tag{6.16}$$

Let $||f||_p = \sup_{n\geq 0} ||f_n||_p$ and denote by $\beta_p(\text{mar})$ the least extended real number β such that

$$||g||_p \leq \beta ||f||_p \tag{6.17}$$

for all martingales f and g as above. The probability space and the filtration $(\mathcal{F}_n)_{n\geq 0}$ vary with the pair (f, g). However, it is no loss to assume that the probability space is a fixed nonatomic space such as the Lebesgue unit interval.

Define $\beta_p(\text{posmar})$ similarly: f is a nonnegative real-valued martingale and g is an $\mathbb{H}$-valued martingale that is differentially subordinate to f. Here, of course, d_n in (6.16) is real-valued. The sequence g is *conditionally differentially subordinate* to f if

$$|E(e_n|\mathcal{F}_{n-1})| \leq |E(d_n|\mathcal{F}_{n-1})| \text{ for all } n \geq 1. \tag{6.18}$$

Notice that if f is a martingale then the right side of (6.18) vanishes. Define $\beta_p(\text{possub})$ to be the least extended real number β such that (6.17) holds for all pairs (f, g) where g is $\mathbb{H}$-valued and both differentially subordinate and conditionally differentially subordinate to f where f is a nonnegative real-valued submartingale. Condition (6.16) has the same purpose as the conditions (6.8) and (6.9); condition (6.18) that of (6.10). For example, if $||f||_p$ is finite as we can assume, then

$$EV(f_n, g_n) \leq EU(f_n, g_n) \leq \cdots \leq EU(f_0, g_0) \leq 0 \tag{6.19}$$

replaces (6.5). This is the way (see [4] and [7]) to prove that $\beta(\text{mar}) \leq p^* - 1$ and $\beta_p(\text{possub}) \leq p^{**} - 1$. Equality holds as can be seen by examples or by an alternative approach (see [6], [7]). Note that $p^* = p^{**}$ if $1 < p \leq 3/2$ and $p^* < p^{**}$ if $p > 3/2$. Also, the example on page 671 of [3] shows that if $1 < p \leq 2$, then $\beta_p(\text{posmar}) = \beta_p(\text{mar})$. Therefore, (6.13) and (6.14) hold. The proof of (6.15) will be complete once Theorem 6.3 is established in the next section.

The quantities $\beta_p(\text{posmar})$, $\beta_p(\text{mar})$, and, $\beta_p(\text{possub})$ do not depend on the dimension of the Hilbert space $\mathbb{H}$. They also remain the same if g is restricted to be a ± 1-transform of f ($e_k = \epsilon_k d_k$ where $\epsilon_k \in \{1, -1\}, k \geq 0$) and, as we shall see, in other cases as well.

6.3 On the Size of a Subordinate of a Nonnegative Martingale

Throughout this section, assume that $p > 2$ and $r = (p-1)/2$. Also, assume in the proofs with no loss that $\mathbb{H}$ is the real Lebesgue sequence space ℓ^2.

Theorem 6.3. *If $p > 2$, then*

$$\beta_p^p(\text{posmar}) = pr^{p-1}. \tag{6.20}$$

So if $p > 2$, then $(p-1)/2 < \beta_p(\text{posmar}) < p/2 < p - 1$. Accordingly, (6.15) holds.

Proof. The first part of the proof is to show that the inequality

$$\|g\|_p^p \leq pr^{p-1}\|f\|_p^p \tag{6.21}$$

holds for all martingales f and g relative to the filtration $(\mathcal{F}_n)_{n \geq 0}$ such that g is $\mathbb{H}$-valued and is differentially subordinate to the nonnegative real-valued martingale f. We can and do assume in the proof that $\|f\|_p$ is finite. Then, by differential subordination, g_n is in L^p:

$$|g_n| \leq \sum_{k=0}^{n} |e_k| \leq \sum_{k=0}^{n} |d_k| \leq 2\sum_{k=0}^{n} |f_k|.$$

Here define $V : [0, \infty) \times \mathbb{H} \to \mathbb{R}$ by

$$V(x, y) = |y|^p - pr^{p-1}x^p. \tag{6.22}$$

Then the expectation $EV(f_n, g_n)$ is equal to

$$\|g_n\|_p^p - pr^{p-1}\|f_n\|_p^p$$

and (6.21) follows from the inequality

$$EV(f_n, g_n) \leq 0. \tag{6.23}$$

To prove (6.23), we shall show that there is a function U that majorizes V on $[0, \infty) \times \mathbb{H}$ such that, for all $n \geq 1$,

$$EU(f_n, g_n) \leq EU(f_{n-1}, g_{n-1}) \leq \cdots \leq EU(f_0, g_0) \leq 0. \qquad (6.24)$$

The existence of such a U gives (6.23) as an immediate consequence. Let U be the function on $[0, \infty) \times \mathbb{H}$ defined as follows. If $|y| \leq rx$, then

$$U(x, y) = V(x, y); \qquad (6.25)$$

if $rx < |y| \leq px$, then

$$U(x, y) = \frac{(2p-1)(p-1)^{p-1}}{(p+1)^{p-1}} \left(|y| - \frac{p^2 - p + 1}{2p - 1} x \right) (|y| + x)^{p-1}; \qquad (6.26)$$

and if $0 \leq px < |y|$, then

$$U(x, y) = (|y| + (p^2 - p - 1)x)(|y| - x)^{p-1}. \qquad (6.27)$$

This function, which is continuous, has the desired properties. (The method leading to its discovery is briefly described in §6.6.) By Lemma 6.4 below, it majorizes V. Let $n \geq 1$. Then, by (6.31) in Lemma 6.5 below and the condition of differential subordination,

$$U(f_n, g_n) \leq U(f_{n-1}, g_{n-1}) + \varphi(f_{n-1}, g_{n-1})d_n + \psi(f_{n-1}, g_{n-1}) \cdot e_n. \qquad (6.28)$$

Using (6.32) in Lemma 6.5, we see that the last two terms in this expression are integrable and, by the martingale property, integrate to 0. The other two terms are also integrable so the left side of (6.24) follows. The right side follows from $U(f_0, g_0) \leq 0$: $|g_0| = |e_0| \leq |d_0| = f_0$ and $U(x, y) \leq 0$ if $|y| \leq x$, which is implied by Lemma 6.5 or can be seen directly from the definition of U. Therefore inequality (6.21) holds. It is sharp as we shall prove below, so (6.20) also holds. $\qquad \square$

Lemma 6.4. *If $p > 2$, then $V \leq U$ on $[0, \infty) \times H$.*

It will be clear from the proof that $V(x, y) < U(x, y)$ if and only if $0 < rx < |y|$. Also, notice that, if $p = 2$, then $U = V$.

Proof. The majorization of V by U will follow from the continuity of U and V once we have shown that $V < U$ on each of the domains D_k, $0 \leq k \leq 2$, where

$$
\begin{aligned}
D_0 &= \{(x, y) : 0 < |y| < rx\}, \\
D_1 &= \{(x, y) : rx < |y| < px\}, \\
D_2 &= \{(x, y) : 0 < px < |y|\}.
\end{aligned}
$$

(i) In view of (6.25), the inequality $V \leq U$ holds trivially on D_0. (ii) Let $(x, y) \in D_1$. Then $|y| + x > 0$, so to prove $V(x, y) \leq U(x, y)$ we can assume by homogeneity that $|y| + x = 1$. Thus, $|y| = 1 - x$ and the inequality $V(x, y) \leq U(x, y)$ is equivalent to $F(x) \geq 0$ where

$$F(x) = \frac{(2p-1)(p-1)^{p-1}}{(p+1)^{p-1}} \left(1 - x - \frac{p^2 - p + 1}{2p - 1} x \right) - (1 - x)^p + pr^{p-1}x^p.$$

(6.29)

The assumptions on (x, y) imply that $x \in I$ where $I = (1/(p+1), 2/(p+1))$ but it is convenient to view F as a function on $(0, 1)$, in which case it is easy to see that $F(2/(p+1)) = F'(2/(p+1)) = 0$, $F''(2/(p+1)) > 0$, F''' is positive on $(0, 1)$ so F'' is increasing there, and

$$F''(1/(p+1)) = \frac{p^2(p-1)}{(p+1)^{p-2}} [r^{p-1} - p^{p-3}].$$

Therefore, $F''(1/(p+1)) > 0$ if $2 < p < 3$; $F''(1/4) = 0$ and $F''(1/(p+1)) < 0$ if $p > 3$ as can be seen by checking that the map φ defined on $[2, \infty)$ by

$$\varphi(p) = (p-1)\log(p-1) - (p-1)\log 2 - (p-3)\log p$$

is strictly concave and satisfies $\varphi(2) = \varphi(3) = 0$.

These properties of F imply that if $2 < p \leq 3$, then F is strictly convex and positive on I. Now assume that $p > 3$. Then there is a number $z_p \in I$ such that F is strictly concave on $[1/(p+1), z_p]$ and is strictly convex on $[z_p, 2/(p+1)]$. Therefore, F is positive on $[z_p, 2/(p+1))$. So F is positive at both z_p and $1/(p+1)$ where $F(1/(p+1)) > 0$ follows from $p+1 > (p/(p-1))^p$, which holds for $p > 3$. The concavity implies that F is positive on the whole interval $[1/(p+1), z_p]$. Accordingly, $V \leq U$ on D_1. (iii) Now suppose that (x, y) belongs to D_2. Then $|y| - x > (p-1)x > 0$. Using homogeneity again, we can assume that $|y| - x = 1$. So here $|y| = 1 + x$ and the inequality $V(x, y) \leq U(x, y)$ is equivalent to $G(x) \geq 0$, where

$$G(x) = (1 + x + (p^2 - p - 1)x) - (1 + x)^p + pr^{p-1}x^p.$$

Here the assumptions on (x, y) imply that $x \in J$ where J is the interval $(0, 1/(p-1))$. By (ii), $U \geq V$ on the intersection of the boundary of D_1 with that of D_2. This implies that

$$G(1/(p-1)) \geq 0. \tag{6.30}$$

If $2 < p \leq 3$, then

$$
\begin{aligned}
G'(x) &= p^2 - p - p(1+x)^{p-1} + p^2 r^{p-1} x^{p-1} \\
&> p^2 - p - p2^{p-2}[1 + x^{p-1}] + p^2 r^{p-1} x^{p-1} \\
&> p^2 - p - p2^{p-2} \\
&\geq 0.
\end{aligned}
$$

The first inequality follows from a classical moment inequality; the second is
a consequence of

$$p(p-1)^{p-1} > 2^{2p-3}, \quad p > 2,$$

which follows from $\psi(2) = 0$, $\psi'(2) > 0$, and $\psi''(p) > 0$, $p > 2$, where

$$\psi(p) = \log p + (p-1)\log(p-1) - (2p-3)\log 2;$$

and the third follows from $\eta(2) = \eta(3) = 0$ and the concavity of η on $[2, 3]$,
where $\eta(p) = p - 1 - 2^{p-2}$. Because $G(0) = 0$ and G' is positive on J, we see
that G is positive on J. If $p > 3$, then G is concave on $[0, 1/(p-1)]$, so by
(6.30) and $G(0) = 0$, G is positive on J in this case also. Therefore, $V \leq U$
on D_2. This completes the proof of Lemma 6.4.   □

Lemma 6.5. *There are continuous functions* $\varphi : [0, \infty) \times \mathbb{H} \to \mathbb{R}$ *and* $\psi :$
$[0, \infty) \times \mathbb{H} \to \mathbb{H}$ *such that, if* $x \geq 0$, $x + h \geq 0$, $y \in \mathbb{H}$, $k \in \mathbb{H}$, *and* $|k| \leq |h|$,
then

$$U(x+h, y+k) \leq U(x, y) + \varphi(x, y)h + \psi(x, y) \cdot k \qquad (6.31)$$

and

$$|\varphi(x, y)| + |\psi(x, y)| \leq c_p(|y| + x)^{p-1}. \qquad (6.32)$$

Proof. The functions φ and ψ are defined on $[0, \infty) \times \mathbb{H}$ as follows. If $|y| \leq rx$,
then

$$\varphi(x, y) = -p^2 r^{p-1} x^{p-1} \text{ and } \psi(x, y) = p|y|^{p-2}y;$$

if $rx < |y| \leq px$, then

$$\varphi(x, y) = p\left(\frac{p-1}{p+1}\right)^{p-1}\{(p-2)|y| - (p^2 - p + 1)x\}(|y| + x)^{p-2},$$

$$\psi(x, y) = p\left(\frac{p-1}{p+1}\right)^{p-1}\{(2p-1)|y| - p(p-2)x\}(|y| + x)^{p-2}y',$$

where $y' = y/|y|$; and if $0 \leq px < |y|$, then

$$\varphi(x, y) = p\{(p-2)|y| - (p^2 - p - 1)x\}(|y| - x)^{p-2},$$
$$\psi(x, y) = p\{|y| + p(p-2)x\}(|y| - x)^{p-2}y'.$$

It is easy to check that these functions are continuous and satisfy (6.32) for
some choice of c_p that does not depend on x or y.

We show now that (6.31) holds. By the continuity of the right and left
sides of (6.31), it is enough to show this for $x > 0$, $x + h > 0$, and with
$|k| \notin \{0, r|h|\}$. Fix x, y, h, k and define I and J by

$$I = \{t \in \mathbb{R} : x + ht > 0\},$$

$$J = \{t \in \mathbb{R} : (x + ht, y + kt) \in D_0 \cup D_1 \cup D_2\}.$$

Then $I \backslash J$ is finite: the coefficient of t^2 in $|y + kt|^2 - m^2(x + ht)^2$ is different from zero for $m = 0, r, p$. Define G on I by

$$G(t) = U(x + ht, y + kt).$$

Then G and its first derivative G' are continuous on I and

$$
\begin{aligned}
G'(t) &= U_x(x + ht, y + kt)h + U_y(x + ht, y + kt) \cdot k \\
&= \varphi(x + ht, y + kt)h + \psi(x + ht, y + kt) \cdot k
\end{aligned}
$$

implying that $G'(0) = \varphi(x, y)h + \psi(x, y) \cdot k$. The inequality (6.31) is therefore equivalent to

$$G(1) \leq G(0) + G'(0),$$

which follows from the concavity of G. We shall prove this concavity by showing that if $t \in J$, then $G''(t) \leq 0$ since if the continuous function G' is nonincreasing on each component of J, then it is nonincreasing on I. By translation, it is enough to prove that if $(x, y) \in D_0 \cup D_1 \cup D_2$, then $G''(0) \leq 0$. If $\alpha > 0$, then $U(\alpha x, \alpha y) = \alpha^p U(x, y)$. Therefore $G''(0)$ does not change sign if $(x, y) \in D_j$ is replaced by $(x/|y|, y/|y|) \in D_j$. So it is enough to prove that $G''(0) \leq 0$ under the assumption that $|y| = 1$. If $(x, y) \in D_0$ and $|y| = 1$, then $1 < rx$ and

$$
\begin{aligned}
G''(0) &= p(p - 2)(y \cdot k)^2 + p|k|^2 - p^2(p - 1)r^{p-1}x^{p-2}h^2 \\
&\leq p(p - 2)h^2 + ph^2 - p^2(p - 1)rh^2 \quad (\text{since } rx > 1) \\
&= -p(p - 1)(pr - 1)h^2 \\
&\leq 0.
\end{aligned}
$$

If $(x, y) \in D_1$ and $|y| = 1$, then $rx < 1 < px$ and

$$G''(0) = -p \left(\frac{p - 1}{p + 1} \right)^{p-1} (1 + x)^{p-3}[(p - 2)A_1 + (h^2 - |k|^2)B_1],$$

where, using the positivity of $px - 1$ and $px^2 - p(p - 3)x + 2p - 1$, we see that

$$
\begin{aligned}
A_1 &= h^2[px^2 + p(p + 1)x + 1] + 2h(y \cdot k)(p - 1)(px - 1) \\
&\quad -(y \cdot k)^2[px^2 - p(p - 3)x + 2p - 1] \\
&> h^2[px^2 + p(p + 1)x + 1] - 2h^2(p - 1)(px - 1) \\
&\quad -h^2[px^2 - p(p - 3)x + 2p - 1] \\
&= 0
\end{aligned}
$$

and, since $x < 1/r$, that

$$B_1 = (1+x)[2p - 1 - p(p-2)x]$$

is also nonnegative. Therefore, $G''(0) \leq 0$ on D_1.

If $(x, y) \in D_2$ and $|y| = 1$, then $0 < px < 1$ and

$$G''(0) = -p(1-x)^{p-3}[(p-2)A_2 + (h^2 - |k|^2)B_2].$$

Here

$$
\begin{aligned}
A_2 &= h^2[px^2 - p(p+1)x + 2p - 1] - 2h(y \cdot k)(p-1)(1 - px) \\
&\quad -(y \cdot k)^2(px^2 + p(p-2)x + 1 - px) \\
&\geq h^2[px^2 - p(p+1)x + 2p - 1] - 2h^2(p-1)(1 - px) \\
&\quad -h^2[px^2 + p(p-2)x + 1 - px] \\
&= 0
\end{aligned}
$$

in which we have used the positivity of $1 - px$ and $px^2 + p(p-2)x + 1 - px$. Furthermore,

$$B_2 = (1-x)[1 + p(p-2)x]$$

is also nonnegative, so $G''(0) \leq 0$ on D_2. This completes the proof of (6.31) and (6.21).   □

Sharpness. To complete the proof of Theorem 6.3, we shall show that the inequality (6.21) is sharp by constructing an example. Let β_p be the positive number satisfying

$$\beta_p^p = pr^{p-1}.$$

If $p > 2$, as we continue to assume, then β_p^p is the best constant in the inequality (6.21). In fact, if $0 < \beta < \beta_p$, then there is a ± 1-transform g of a martingale f such that

$$\|g\|_p > \beta\|f\|_p.$$

There are several steps in the construction of such a pair f and g. Throughout, let $0 < 4r\delta < 1$ and $x_n = 1 + n\delta$ for all nonnegative integers n.

(i) As a first step, let $H = (H_n)_{n>0}$ be a Markov chain with values in the closed right half-plane such that $H_0 = (1, p)$ and

$$P[H_{2n+1} = (x_n - r\delta, px_n + r\delta)|H_{2n} = (x_n, px_n)] = \frac{x_n}{x_n + r\delta},$$

$$P[H_{2n+1} = (2x_n, 2rx_n)|H_{2n} = (x_n, px_n)] = \frac{r\delta}{x_n + r\delta},$$

$$P[H_{2n+2} = (x_{n+1}, px_{n+1})|H_{2n+1} = (x_n - r\delta, px_n + r\delta)] = \frac{x_n - r\delta}{x_n + \delta},$$

$$P[H_{2n+2} = (0, (p-1)x_{n+1})|H_{2n+1} = (x_n - r\delta, px_n + r\delta)] = \frac{(r+1)\delta}{x_n + \delta}.$$

Assume further that $(2x_n, 2rx_n)$ and $(0, 2rx_{n+1})$ are absorbing states. Let $\pi_0 = 1$ and, for $n \geq 1$,

$$\pi_n = \pi_n(\delta) = \frac{1}{1+n\delta} \prod_{k=0}^{n-1} \left(1 - \frac{2r\delta}{1+(k+r)\delta}\right).$$

If $n \geq 1$, then

$$P[H_{2n} = (x_n, px_n)] = \pi_n,$$

$$P[H_{2n} = (2x_{n-1}, 2rx_{n-1})] = \pi_{n-1} \frac{r\delta}{1+(n-1+r)\delta},$$

$$P[H_{2n} = (0, 2rx_n)] = \pi_{n-1} \frac{(r+1)\delta}{1+(n-1+r)\delta} \cdot \frac{1+(n-1)\delta}{1+n\delta}.$$

Let F and G be the martingales defined by $H_n = (F_n, G_n)$, and D the difference sequence of F. Then $F_0 = 1$, $G_0 = p$, and, for $n \geq 1$,

$$F_n = 1 + \sum_{k=1}^{n} D_k,$$

$$G_n = p + \sum_{k=1}^{n} (-1)^k D_k.$$

(ii) The second step is to show that, if $n \geq 0$, then

$$\pi_n = \frac{\exp R(n, r, \delta)}{(1+n\delta)^p}, \tag{6.33}$$

where $|R(n, r, \delta)| \leq 2nr\delta^2 + 10nr^2\delta^2$. To see this, note first that, if $0 < t < 1$, then

$$\exp(-t - t^2/(1-t)) = \exp(-t/(1-t)) < 1 - t < \exp(-t).$$

So if $0 < t < 1/2$, then $1 - t = \exp(-t - R_1(t))$ where $0 < R_1(t) < 2t^2$. Also,

$$0 < \frac{2r\delta}{1+k\delta} - \frac{2r\delta}{1+(k+r)\delta} < 2r^2\delta^2.$$

Now use these inequalities and the definition of π_m to see that, for $n \geq 1$,

$$\pi_n = \frac{1}{1+n\delta} \exp\left(-\sum_{k=0}^{n-1} \frac{2r\delta}{1+k\delta} + R_2(n, r, \delta)\right),$$

where $|R_2(n, r, \delta)| \leq 10nr^2\delta^2$. Then use the inequality

$$0 < \sum_{k=0}^{n-1} \frac{2r\delta}{1+k\delta} - 2r\log(1+n\delta) < 2nr\delta^2$$

to obtain (6.33).

(iii) Let $b > 0$. Choose δ_n so that $0 < 4r\delta_n < 1$ and $\lim_{n\to\infty} n\delta_n = b$. Let $(F_{n,k})_{k\geq 0}$ and $(G_{n,k})_{k\geq 0}$ be the martingales F and G as above with δ replaced everywhere by δ_n. Then

$$\lim_{n\to\infty} EF_{n,2n}^p = 1 + 2^p r \log(1+b), \tag{6.34}$$

$$\lim_{n\to\infty} EG_{n,2n}^p = p^p + 2^p r^p p \log(1+b). \tag{6.35}$$

To prove (6.34), observe that if $n \geq 1$, then

$$\begin{aligned}
EF_{n,2n}^p &= (1+n\delta_n)^p \pi_n(\delta_n) \\
&\quad + 2^p \sum_{k=0}^{n-1}(1+k\delta_n)^p \pi_k(\delta_n)\frac{r\delta_n}{1+(k+r)\delta_n} \\
&= \exp(R(n,r,\delta_n)) \\
&\quad + 2^p \sum_{k=0}^{n-1}\exp(R(k,r,\delta_n))\frac{r\delta_n}{1+k\delta_n} \\
&\quad + 2^p \sum_{k=0}^{n-1}\exp(R(k,r,\delta_n))R_3(k,r,\delta_n),
\end{aligned}$$

where $|R_3(k,r,\delta_n)| \leq r^2\delta_n^2$. Consequently, (6.34) holds. Similarly, if $n \geq 1$, then

$$\begin{aligned}
EG_{n,2n}^p &= p^n(1+n\delta_n)^p \pi_n(\delta_n) \\
&\quad + 2^p r^p \sum_{k=0}^{n-1}(1+k\delta_n)^p \pi_k(\delta_n)\frac{r\delta_n}{1+(k+r)\delta_n} \\
&\quad + 2^p r^p \sum_{k=0}^{n-1}(1+(k+1)\delta_n)^p \pi_k(\delta_n)\frac{(r+1)\delta_n}{1+(k+r)\delta_n} \cdot \frac{1+k\delta_n}{1+(k+1)\delta_n},
\end{aligned}$$

which converges to $p^p + 2^p r^p[r + (r+1)]\log(1+b)$. Therefore, (6.35) also holds.

(iv) Now choose b so that

$$\frac{\frac{1}{2} + \frac{1}{2}[p^p + 2^p r^p p \log(1+b)]}{\frac{1}{2}p^p + \frac{1}{2}[1 + 2^p r \log(1+b)]} > \beta^p. \tag{6.36}$$

This is possible because $\beta_p > \beta$ and the limit of the left side of (6.36) as $b \to \infty$ is β_p^p. After this choice of b, choose n so that

$$\frac{\frac{1}{2} + \frac{1}{2}EG_{n,2n}^p}{\frac{1}{2}p^p + \frac{1}{2}EF_{n,2n}^p} > \beta^p. \tag{6.37}$$

This is possible by (6.34), (6.35), and (6.36).

(v) Suppose that $A \in \mathcal{F}$ is independent of the Markov chain H and satisfies $P(A) = 1/2$. (If necessary, the original probability space can be replaced by a slightly larger one to make such a choice possible.) With b and n as in (iv), let $f_0 = g_0 = (p+1)/2$. For $0 \le k \le 2n$, let

$$
\begin{aligned}
(f_{k+1}, g_{k+1}) &= (F_{n,k}, G_{n,k}) \text{ on } A, \\
&= (p, 1) \text{ off } A.
\end{aligned}
$$

For $k > 2n$, let $(f_{k+1}, g_{k+1}) = (f_{2n+1}, g_{2n+1})$. From (i), it follows that f is a martingale and g is a ± 1-transform of f. Furthermore, the ratio

$$\|g\|_p^p / \|f\|_p^p$$

is given by the left side of (6.37). Therefore, $\|g\|_p > \beta \|f\|_p$. This shows that β_p gives the best constant and completes the proof of Theorem 6.3.

6.4 A Sharp Norm Inequality for an Integral with Respect to a Nonnegative Martingale

Suppose that $X = (X_t)_{t \ge 0}$ is a nonnegative martingale on a complete probability space $(\Omega, \mathcal{F}, P)$ filtered by $(\mathcal{F}_t)_{t \ge 0}$, a nondecreasing right-continuous family of sub-σ-algebras of $\mathcal{F}$ where $\mathcal{F}_0$ contains all $A \in \mathcal{F}$ with $P(A) = 0$. So each X_t is integrable and the conditional expectation $E(X_t|\mathcal{F}_s) = X_s$ if $0 \le s \le t$. Let Y be the Itô integral of H with respect to X where H is a predictable process with values in the closed unit ball of the Hilbert space $\mathbb{H}$:

$$Y_t = H_0 X_0 + \int_{(0,t]} H_s \, dX_s.$$

Both X and Y are adapted to the filtration $(\mathcal{F}_t)_{t \ge 0}$ and are right-continuous on $[0, \infty)$ with limits from the left on $(0, \infty)$. Set $\|X\|_p = \sup_{t \ge 0} \|X_t\|_p$.

Theorem 6.6. *Let* $2 < p < \infty$ *and* $r = (p-1)/2$. *Then*

$$\|Y\|_p^p \le p r^{p-1} \|X\|_p^p \tag{6.38}$$

and $p r^{p-1}$ *is the best possible constant.*

The inequality (6.38) follows from (6.21) in the same way that the analogous inequality for nonnegative submartingales in [7] follows from the corresponding discrete-parameter version. The fact that $p r^{p-1}$ is the best possible constant in (6.38) follows from the fact that it is the best possible constant in (6.21) even in the special case that g is a ± 1-transform of f as proved in §6.3.

Let us now modify the definition of β_p(posmar) that is given in §6.2. Here let β_p(posmar) be the least extended real number β such that $||Y||_p \leq \beta ||X||_p$ for all X and Y as above. By Theorem 6.6 and earlier results in [3] and [7], we obtain the same value as in the discrete-parameter case:

$$\begin{aligned} \beta_p(\text{posmar}) &= p^* - 1 = 1/(p-1) \text{ if } 1 < p \leq 2, \\ &= prp^{-1} \text{ if } 2 < p < \infty. \end{aligned}$$

If the definitions of β_p(mar) and β_p(possub) are modified in the same way, then (6.13), (6.14), and (6.15) hold also in this case.

We note here that the sentence near the top of page 996 of [7] beginning with "In fact, the constant . . . " is not clear in its context. It should read "In fact, if $1 < p \leq 2$, then the constant $p^* - 1$ is already the best possible for the smaller class of nonnegative martingales."

6.5   Some Open Questions and Remarks

(i) The following question from [4] is still open even for $\mathbb{H} = \mathbb{R}$. Let M and N be right-continuous martingales on $[0, \infty)$ with limits from the left on $(0, \infty)$. Suppose they are martingales relative to the same filtration, are $\mathbb{H}$-valued, and have respective quadratic variation processes $[M, M]$ and $[N, N]$. If

$$[N, N]_t \leq [M, M]_t \text{ for all } t \geq 0, \tag{6.39}$$

then does $||N||_p \leq (p^* - 1)||M||_p$ for all $p \in (1, \infty)$? If so, is this inequality true under the less restrictive condition

$$[M, M]_\infty \leq [N, N]_\infty? \tag{6.40}$$

There are analogous questions for weak-type, exponential, and other martingale inequalities. Wang [14] and Bañuelos and Wang [2] consider the following condition:

$$[M, M]_t - [N, N]_t \text{ is nonnegative and nondecreasing in } t. \tag{6.41}$$

Under this more restrictive condition, Wang [14] proves the inequality $||N||_p \leq (p^* - 1)||M||_p$ for all $p \in (1, \infty)$. He also proves similar extensions of some of the other inequalities in [4] and [7]. In [2], where the proof of $||N||_p \leq (p^* - 1)||M||_p$ is given under the condition (6.41) in the special case that M and N are continuous, Bañuelos and Wang give applications of this inequality to the Beurling-Ahlfors transform.

(ii) Let $\alpha > 0$. What is the effect of replacing condition (6.10) by

$$|\Delta v| \leq \alpha |\Delta u|, \tag{6.42}$$

and condition (6.18) by

$$|E(e_n|\mathcal{F}_{n-1})| \leq \alpha |E(d_n|\mathcal{F}_{n-1})| \text{ for all } n \geq 1? \tag{6.43}$$

Changsun Choi [9] has proved that if $0 < \alpha < 1$, then (6.12) holds if p^{**} is replaced by $\max\{(\alpha+1)p, q\}$, and that the same replacement in the analogous inequality for nonnegative submartingales (inequality (2.1) of [7]) gives the best constant there. Earlier, Hammack [11] studied the effect of dropping the condition that the submartingale be nonnegative while keeping the original condition (6.18). In this case, the weak-type inequality (inequality (4.1) of [7]) continues to hold but with 6, rather than 3, being the best constant. On the other hand, the L^p-inequalities fail to hold for $1 < p < \infty$.

6.6   A Note on Method

Here are the steps that can lead in a natural way to the function U of §6.3. Start with $\mathbb{H} = \mathbb{R}$ and $\beta > 0$, and as in §6.3, let $p > 2$. Let $S = [0, \infty) \times \mathbb{R}$ and define $V_\beta : S \to \mathbb{R}$ by $V_\beta(x, y) = |y|^p - \beta^p x^p$. Then attempt to find the least majorant U of V_β on S such that the mapping

$$x \mapsto U(x, y + \varepsilon x)$$

is concave on $[0, \infty)$ for all real y and for both $\varepsilon = 1$ and $\varepsilon = -1$. There is no such U if β is too small. However, it seems reasonable to expect that the least β for which there is such a majorant of V_β will lead to a U that has some smoothness, perhaps continuous derivatives of the first order on the interior of S. Furthermore, if this U is least possible and the second-order derivatives exist in a neighborhood of $(x, y) \in S$, then either

$$U_{xx} + 2U_{xy} + U_{yy} = 0 \text{ or } U_{xx} - 2U_{xy} + U_{yy} = 0$$

at (x, y). Otherwise, U would not be extremal. With the use of all of this, the right value of β and the right function U are not hard to find for the case $\mathbb{H} = \mathbb{R}$. For a general Hilbert space $\mathbb{H}$, the right function U has the same formula (interpret the absolute value $|y|$ as the norm of $y \in \mathbb{H}$) as can be seen from the work in §6.3. Perhaps further insight may be gained from Theorem 1.1 of [3] and §2 of [6]. The extremal problems for martingales that we have considered here focus mostly on the comparison of the sizes of two martingales related in some way, such as one being differentially subordinate to the other. These problems lead to interesting (upper) boundary value problems. If instead, the focus is on the comparison of the sizes of their maximal functions, then the boundary value problems become even more interesting; see [8].

References

[1] R. Bañuelos and A. Lindeman, *A martingale study of the Beurling-Ahlfors transform in* $\mathbb{R}^n$, J. Funct. Anal. 145 (1997), 224–265.

[2] R. Bañuelos and G. Wang, *Sharp inequalities for martingales with applications to the Beurling-Ahlfors and Riesz transforms*, Duke Math. J. 30 (1995), 575–600.

[3] D. L. Burkholder, *Boundary value problems and sharp inequalities for martingale transforms*, Ann. Probab. 12 (1984), 647–702.

[4] ———, *Sharp inequalities for martingales and stochastic integrals*, Colloque Paul Lévy (Palaiseau, 1987), Astérisque 157–158 (1988), 75–94.

[5] ———, *Differential subordination of harmonic functions and martingales*, Harmonic Analysis and Partial Differential Equations (El Escorial, 1987), Lecture Notes in Mathematics 1384 (1989), 1–23.

[6] ———, *Explorations in martingale theory and its applications*, Ecole d'Eté de Probabilités de Saint-Flour XIX-1989, Lecture Notes in Mathematics 1464 (1991), 1–66.

[7] ———, *Strong differential subordination and stochastic integration*, Ann. Probab. 22 (1994), 995–1025.

[8] ———, *Sharp norm comparison of martingale maximal functions and stochastic integrals*, Proceedings of the Norbert Wiener Centenary Congress, 1994 (edited by V. Mandrekar and P. R. Masani), Proceedings of Symposia in Applied Mathematics 52 (1997), 343–358.

[9] C. Choi, *A submartingale inequality*, Proc. Amer. Math. Soc. 124 (1996), 2549–2553.

[10] B. Dacorogna, *Direct Methods in the Calculus of Variations*, Springer, 1989.

[11] W. Hammack, *Sharp maximal inequalities for stochastic integrals in which the integrator is a submartingale*, Proc. Amer. Math. Soc. 124 (1996), 931–938.

[12] T. Iwaniec, private communication, June 1996.

[13] T. Iwaniec and G. Martin, *Riesz transforms and related singular integrals*, J. reine angew. Math. 473 (1996), 25–57.

[14] G. Wang, *Differential subordination and strong differential subordination for continuous-time martingales and related sharp inequalities*, Ann. Probab. 23 (1995), 522–551.

Chapter 7

The Monge Ampere Equation, Allocation Problems, and Elliptic Systems with Affine Invariance

Luis A. Caffarelli

The Monge Ampere equation, a classical equation in differential geometry and analysis, has been surfacing more and more often in different contexts of applied mathematics in recent times, posing new and challenging problems.

In this paper I will try to present some of these problems, partial answers and possible directions.

7.1 The Monge Ampere Equation as a Fully Nonlinear Equation

The Monge Ampere equation consists of finding a convex function φ that satisfies

$$\det D^2\varphi = f(x, D\varphi).$$

It has a rich history in differential geometry and analysis.

It appears locally when prescribing the density of the Gauss map, i.e., the map that assigns, to each point $(X, \varphi(X))$ on the surface $S = \text{graph}\,(\varphi)$, the unit normal vector ν (convexity of φ guarantees heuristically that ν is a nice oriented "one to one" map).

Research partially supported by an NSF grant.

It also appears in problems of optimal design in optics (see Oliker [11]) in Riemannian geometry, etc. The convexity restriction on φ, that as we will see later appears naturally in several contexts, has the virtue that it almost puts the Monge Ampere equation in the context of solutions to fully nonlinear equations:

$$F(D^2\varphi) = f.$$

Indeed, in this theory, one requires F to be strictly monotone and Lipschitz as a function of symmetric matrices M, i.e.,

$$F(M) + \lambda||N|| \leq F(M + N) \leq F(M) + \Lambda||N||$$

for any symmetric matrices M, N, with N positive.

In the case of $\det M$, both monotonicity and Lipschitz regularity hold as long as M is positive (φ convex) and M, N remain bounded, but it deteriorates as M goes to infinity.

Our interest in fitting the Monge Ampere equation in the framework of fully nonlinear elliptic equations is of course the powerful local regularity theory available.

7.2 Regularity Theory of Fully Nonlinear Equations

We point out at this time that in the study of second-order, nonlinear equations there are usually two approaches to regularity theory: boundary inherited regularity and local regularity.

Boundary inherited regularity looks, for instance, at a Dirichlet problem,

$$\begin{cases} F(D^2u) = f(x)\,\text{in}\,\mathcal{D} \\ u = g(x)\,\text{along the smooth boundary, }\partial\mathcal{D} \end{cases}$$

and tries first to control first and second derivatives of u along $\partial\mathcal{D}$, by appropriate barriers, and next to extend this control to the interior of $\mathcal{D}$ by maximum principle techniques for Du and D^2u. This method is convenient when the geometry of the problem is good to start with ($g(x)$ and $\partial\mathcal{D}$ are smooth and geometrically constrained).

Another type of theory, the one I want to address here, is that of interior a priori estimates, in which we are given a bounded solution u of

$$F(D^2u) = f(x)$$

in B_1, where u is smooth in any bounded subdomain, B_{1-t}.

The advantage of this theory is in its compactness properties and, as a consequence, that it occurs in the natural functional spaces (u is "two derivatives" better than f both in C^α and L^p spaces).

Since this last theory is available for solutions of uniformly elliptic, fully nonlinear equation, one may hope to apply it with some modification to the Monge Ampere equation. That this is not the case, at least in a straightforward way, is due to the invariances of the Monge Ampere equation.

7.3 The Monge Ampere Equation: Invariance and Counterexamples

The Monge Ampere equation, say

$$\det D^2\varphi = 1,$$

is not only invariant under the action of rigid motions $\mathcal{R}$ (i.e., $\overline{\varphi}(X) = \varphi(RX)$ is again a solution of the same equation) and quadratic dilations (i.e., $\overline{\varphi}(X) = \frac{1}{\lambda^2}\varphi(\lambda X)$) but also under any affine transformation T, with $\det T = 1$ (i.e., $\overline{\varphi}(X) = \varphi(TX)$ solves the same equation).

This last property has as a consequence that we may submit the graph of a solution φ to an enormous deformation $\overline{\varphi}(X) = \varphi(\epsilon x_1, \frac{1}{\epsilon}x_2)$ and still obtain a solution of the same equation.

This makes the existence of nonsmooth solutions almost unavoidable. For instance (Pogorelov), for $n > 2$,

$$\varphi(X) = |(x_1, \dots, x_{n-1})|^{2-2/n} f(x_n)$$

is a solution for

$$\left[\left(1 - \frac{2}{n}\right) f f'' - \left(2 - \frac{2}{n}\right) (f')^2\right] f^{n-2} = \text{constant}$$

(an ODE with nice local solutions). This φ is nonnegative, but $D^2\varphi$ degenerates along the line $x_n = 0$, where $\varphi \equiv 0$. Therefore, an absolute regularity theory is out of the question.

On the other hand, the positive effect of the rich renormalization properties of the Monge Ampere equation is that, if we start from a "reasonable" situation, the possibility of renormalizing at every scale should give us a way of controlling the geometry of φ in an iterative fashion.

7.4 Some Local Theorems

In order to describe these ideas let me state three basic facts.

Fact 1 (Pogorelov):
Let φ be a (smooth) solution of

$$\begin{cases} \det D^2\varphi = 1 & \text{in } \Omega, \\ \varphi \equiv 0 & \text{on } \partial\Omega. \end{cases}$$

Assume that $B_1 \subset \Omega \subset B_K$, then

$$D^2\varphi(0) \le C(K).$$

Fact 2 (Elementary comparison):
Let φ_0 be as in Fact 1, and

$$|\det D^2\varphi_1 - 1| < \epsilon, \quad \varphi_1|_{\partial\Omega} = 0.$$

Then

$$|\varphi_0 - \varphi_1| \le C\epsilon \quad \text{in } \Omega.$$

Fact 3 (Global propagation of singularities):
  (a) $0 < \lambda \le \det D^2\varphi \le \Lambda < \infty$ in B_1,
  (b) $\varphi \ge 0, \varphi(0) = 0$.
Then the set

$$\{X : \varphi = 0\}$$

is (i) a point or (ii) a convex set Γ with *no extremal points* in B_1, i.e., generated by convex combinations of points in $\Gamma \cap \partial B_1$.

Note that the three facts are renormalization invariant. Note also that Fact 3 is by compactness a very strong statement of strict convexity.

Indeed, if φ_n is a sequence of bounded nonnegative solutions to the Monge Ampere equation $0 \le \lambda \le \det D^2\varphi_n \le \Lambda$ that slowly deteriorates around the origin (say $\inf(\varphi_n|_{\partial B_\epsilon})(0)$ goes to zero), a limiting φ_0 has a nontrivial set $\{\varphi_0 = 0\}$ and therefore, from Fact 3, a very large singular set $\{\varphi_0 = 0\}$.

The combination of these facts allows us to prove through renormalization of the sections $S(\ell) = \{x : \varphi(x) - \ell(x) < 0\}$, ℓ a linear function, that the following holds:

"Theorem": Unless φ degenerates according to Fact 3, φ is, in any compact subset of B_1, two derivatives better than f, i.e.,
  (a) If f is bounded, φ is $C^{1,\epsilon}$ for some ϵ (Harnack inequality for the Monge Ampere equation).
  (b) If f is continuous, φ is $W^{2,p}$ for every p (Calderón-Zygmund).
  (c) If f is C^α, φ is $C^{2,\alpha}$ (Schauder).

We will now discuss some applications of this theory.

7.5 The Monge Ampere Equation as a Potential of an "Irrotational Map" in Lagrangian Coordinates

Two concepts at the heart of modelization in continuum mechanics are compression and rotation. In Eulerian coordinates, infinitesimal compression is measured through the divergence of the infinitesimal displacement (velocity) and rotation through the curl.

Mathematically, the important fact is that the prescription of div $v =$ Trace Dv and curl $= Dv - (Dv)^t$ constitutes the simplest linear elliptic system, and thus we have the two typical theorems:

Theorem 7.1. *If div, curl belong to a given function space (say $C^{k,\alpha}$ or $W^{k,p}$), then all of Dv belongs to that space (with estimates), or*

Theorem 7.2 (Helmholtz decomposition). *Given $\overline{v}$ in a function space, it can be decomposed as $\overline{v} = \overline{v}_1 + \overline{v}_2$, with $\overline{v}_1$ incompressible and $\overline{v}_2$ irrotational in the same function space (with estimates). In fact $\overline{v}_2$ is found as $\overline{v}_2 = \nabla u$, with*

$$\begin{cases} \Delta u = \ div\,\overline{v} \\ \text{boundary conditions} \end{cases}$$

and $\overline{v}_1 := \overline{v} - \overline{v}_2$.

In Lagrangian coordinates, the adequate tool to describe large deformations, it is clear what the notion of compression and incompressibility should be: instead of just having a velocity field, our dependent variable $Y(X,t)$ is the position at time t of that particle that at time zero was at X, and therefore the compression that a differential of volume given at time zero has suffered at time t is simply $\det D_X Y$ (i.e., conservation of mass becomes $\rho \det D_X Y = \text{constant}$).

Rotation is a more unsettled question, and I assume that, for different purposes, different definitions occur.

We will discuss briefly later the Cauchy-Novozilov notion of rotation, but one may look at Truesdell and Toupin [16].

Let us agree now that irrotational means $D\overline{v}$ symmetric; that is, heuristically, $\overline{v}$ is the gradient of a potential φ, and then we can state Brenier's factorization [1].

Theorem 7.3. *Let Ω_1, Ω_2 be two bounded domains of $\mathbb{R}^n$, $\overline{v}$ a measurable map from Ω_1 into Ω_2 that does not collapse sets of positive measure (i.e., if $E \subset \Omega_2$ has $|E| = 0$ then $|\overline{v}^{-1}(E)| = 0$). Then $\overline{v}$ can be factorized as*

$$\overline{v} = \overline{v}_2 \cdot \overline{v}_1,$$

where

$$\overline{v}_1 : \Omega_1 \to \Omega_1$$

is incompressible, i.e., for any measurable $E \subset \Omega_1$

$$|(\overline{v}_1)^{-1}(E)| = |E|$$

and v_2 *is irrotational; that is,*

$$v_2 = \nabla\varphi,$$

where φ *is convex.*

Given this decomposition, we note that the questions raised by Theorem 7.2 become relevant: Suppose that $\overline{v}$ is in some function space ($W^{k,p}$ or $C^{k,\alpha}$). Is it true that v_1, v_2 are in the same spaces? This is the content of several papers by the author (see [2], [3], [4], [5]) where this is shown to hold under the necessary hypothesis that Ω_1 and Ω_2 are convex. This is achieved through the local regularity theory for convex solutions φ of the Monge Ampere equation

$$\det D^2\varphi = f$$

provided the "compression" f is under control.

7.6   Optimal Allocation Problems

In fact, the map v_2 above depends on v only through the density $\delta(y) = \det D^2 v$, with which Ω_1 is carried into Ω_2, and it has a classic interpretation as an "optimal allocation" problem between Ω_1 and Ω_2 with prescribed densities $\delta_1(X), \delta_2(Y)$:

Let us describe optimal allocations in a discrete setting: Assume we are given k points $X_1, \ldots, X_k$ and $Y_1, \ldots, Y_k$ in $\mathbb{R}^n$ and we want to map the X_i onto the Y_i's, minimizing a "transportation" cost

$$\tau(Y(X)) = \Sigma_i C(Y(X_i) - X_i),$$

where $C(X - Y)$ represents the cost of transporting X into Y.

In the finite case, an optimal map $Y_0(X)$ clearly exists, and it is cyclically monotone in the sense that, for any permutation $\pi(i)$, $\Sigma C(Y(X_i) - X_{\pi(i)}) \geq \Sigma C(Y(X_i) - X_i)$. If $C(Y - X)$ is $|Y - X|^2$ then this condition transforms into

$$\Sigma\langle Y(X_i), X_{\pi(i)}\rangle \leq \Sigma\langle Y(X_i) - X_i\rangle,$$

the classical definition of cyclically monotone maps.

A theorem of Rockafellar [14] then proves that Y is the gradient of a *convex* potential. The continuous situation can then be described by the following theorem: Suppose we are given the cost function

$$C(Y - X) = |Y - X|^2$$

two domains Ω_1, Ω_2 and densities $\delta_1(X), \delta_2(Y)$ with

$$\int \delta_1 dX = \int \delta_2 \, dY.$$

(a) There exists a unique map $Y_0(X)$ that preserves densities: i.e., for every continuous η

$$\int \eta(Y) \, \delta_2(Y) \, dY = \int \eta(Y_0(X)) \, \delta_1(X) \, dY \Omega_2$$

and such that minimizes

$$\int |Y(X) - X| \delta_1(X) \, dX$$

among all such maps.

(b) $Y = \nabla\varphi$, φ convex and if Ω_i are convex, φ is locally "two derivatives better" than δ_1, δ_2.

A very beautiful problem arises when studying general strictly convex cost functions C, with differential $\overline{C}_j$ $\overline{C}_j$. Then $(Y_0(X) - X)$ is an irrotational vector field $\nabla\varphi$. Heuristically φ satisfies then

$$\det D(X + B_j(\nabla\varphi)) = \frac{\delta_2(X + B_j(\nabla\varphi))}{\delta_1(\nabla\varphi)},$$

where B_j, the inverse map to C_j, is the gradient of B, the conjugate convex function to C. Little is known about solving this Monge Ampere type equation (see [8], [10]).

Note the analogy of this family of Monge Ampere problems with the Euler equations coming from convex functionals:

$$\operatorname{div}(B_j(\nabla\varphi)) = 0.$$

These last are sort of a "linearization" of the ones above.

7.7 Optimal Allocations in Continuum Mechanics

This is an area that is evolving, so I will just make a few remarks. We start by noticing the relation between the functional $(\delta_i(X) \equiv 1)$,

$$J = \int (Y(X) - X)^2 \, dX,$$

and its infinitesimal version, the kinetic energy formula: If $Y(X) = X + \epsilon \bar{v}$, we have, at ϵ-level, $\det(I + \epsilon Dv) = 1 + \epsilon \operatorname{div} \bar{v} = 1$ and

$$J = \epsilon^2 \int |\bar{v}|^2 \, dX.$$

That is, v minimizes kinetic energy among incompressible ($\operatorname{div} \bar{v} = 0$) vector fields.

Brenier used this idea to discretize incompressible Euler's equation [1]. Recently Otto [12], [13] found global weak solutions for several classical evolution equations (degenerate diffusions, lubrication approximation, semi-geosthrophic front formation) using different allocation cost functions. The regularity (or partial regularity) of his solutions remains open.

The semigeostrophic equations are worth noting (see [9]). We have two sets of variables, (x_1, x_2, z), where x_1, x_2 are the physical plane variables, and z, a pseudoheight, a function of pressure instead of x_3, and the momentum-entropy variables M_1, M_2, M_3. The domain Ω_2 in (x_1, x_2, z) is a given cylinder.

The unknowns are a density $\delta_1(M, t)$ and the potential of the map $\varphi(M, t)$ that at any instant of time optimally allocates δ_1 to $X(\Omega_2)$ in the (x_1, x_2, z) variables. The remaining equation tells us how $\delta(M, t)$ evolves, transported by the planar vorticity:

$$\delta_t(M, t) = \delta_{M_1}(-\varphi_{M_2}) + \delta_{M_2}\varphi_{M_1}.$$

Regularity or partial regularity remains open.

7.8   Elliptic Systems Invariant under Affine Transformations

In this final section I would like to take up again the issue of what is a notion of rotation and of an elliptic system invariant under an affine transformation. For that, let us go back to discrete allocations.

We have seen that the optimal map $Y(X_j)$ is cyclically monotone,

$$\Sigma\langle Y(X_i), X_{\pi(i)}\rangle \le \Sigma\langle Y(X_i), X_i\rangle,$$

for any permutation π.

But assume that we try to "flow" to such a minimum by exchanging pairs of Y's. Then monotone maps (satisfying only $\langle Y(X_1) - Y(X_2), X_1 - X_2\rangle$) will be stationary.

Simple examples show that incompressible monotone maps are not regular and can develop local singularities.

But if we "flow" to the minimum by exchanging "triplets" of points Y_1, Y_2, Y_3 we get a "3-monotone map," i.e., a "cyclically monotone map" where only permutations of three elements are allowed. This is enough to control in some sense rotations in an affine invariant fashion (any definition that involves inner products $\langle Y, X \rangle$ is affine invariant), and this is also enough to reconstruct (at least partially) the local regularity theory of the Monge Ampere equation.

Infinitesimally, 3-monotonicity means controlling on each plane the ratio between ω (vorticity) and $\det S$, the symmetric part of the differential, i.e., (rotation)^2 over area dilation. These ratios are called the Cauchy-Novozilov measure of rotation [16]. A related notion described by Truesdell [15] looks at ratios between vorticity and maximal line dilation. They seem to define an affine invariant notion of "elliptic system." The issue of "elliptic systems" with affine invariance is still wide open (see [7]).

References

[1] Y. Brenier, *Polar factorization and monotone rearrangement of vector valued functions*, CPAM, vol. XLIV (1991), 375–417.

[2] L. A. Caffarelli, *A localization property of viscosity solutions to the Monge Ampere equation and their strict convexity*, Ann. of Math. (2) 131 (1990), no. 1, 129–134.

[3] ———, *Interior $W^{2,p}$ estimates for solutions of the Monge Ampere equation*, Ann. of Math. (2) 131 (1990), no. 1, 135–150.

[4] ———, *The regularity of mappings with a convex potential*, J. Amer. Math. Soc. 5 (1992), no. 1, 99–104.

[5] ———, *Boundary regularity maps with convex potentials*, Comm. Pure Appl. Math 45 (1992), no. 9, 1141–1151.

[6] ———, *Allocation maps with general cost functions*. In *Partial Differential Equations and Applications*, Lecture Notes in Pure and Applied Math. 177, Dekker, New York, 1996, 29–35.

[7] ———, *The regularity of monotone maps of finite compression*, CPAM, to appear.

[8] L. A. Caffarelli and M. Milman, *The regularity of planar maps with bounded compression and rotation*, in homage to Dr. Rodolfo A. Ricabarra

(Spanish), 29–34, vol. Homenaje 1, Univ. Nac. del Sur, Bahía Blanca, 1995.

[9] S. Chynoweth and M. J. Sewell, *A concise derivation of the semi-geostrophic equations*, Q. J. of the Royal Meteorological Society 117 (1991), 1109–1128.

[10] W. Gangbo and R. J. McCann, *Optimal maps in Monge's mass transport problem*, C.R.A. Sc. Sér. Mat. 321 (1995), no. 12, 1653–1658.

[11] V. Oliker, *The reflector problem for closed surfaces*. In *Partial Differential Equations and Applications*, Lecture Notes in Pure and Applied Math. 177, Dekker, New York, 1996, 265–270.

[12] F. Otto, *Doubly degenerate diffusion equations as steepest descent*, preprint.

[13] ———, *Lubrication approximation with prescribed non-zero contact angle: An existence result*, preprint.

[14] R. T. Rockafellar, *Characterization of the subdifferentials of a convex function*, Pacific J. of Math 17 (1966), 497–510.

[15] C. Truesdell, *Two measures of vorticity*, J. Rat. Mech. Anal. 2 (1953), 173–217.

[16] C. Truesdell and R. Toupin, *The Classical Field Theories*, Handbuch der Physik, vol. III/1, Springer 274 (1960).

Chapter 8

On a Fourth-Order Partial Differential Equation in Conformal Geometry

8.1 Introduction

In this paper, we will survey some of the recent study of a fourth-order partial differential operator—namely, the Paneitz operator. The study of this operator arises naturally from the consideration of problems in conformal geometry. However, the natural partial differential equations associated with the operator, which when restricted to domains in $\mathbb{R}^n$ becomes the bi-Laplacian, also are interesting in themselves. The content of this paper is an expanded version of a talk the author gave in the conference to honor Professor A. P. Calderón on the occasion of his seventy-fifth birthday.

On a Riemannian manifold (M^n, g) of dimension n, a most well-studied differential operator is the Laplace-Beltrami operator $\Delta = \Delta_g$ which in local coordinates is defined as

$$\Delta = \frac{1}{\sqrt{|g|}} \sum \frac{\partial}{\partial x^i} \left(\sqrt{|g|}\, g^{ij} \frac{\partial}{\partial x^i} \right),$$

where $g = (g_{ij})$, $g^{ij} = (g_{ij})^{-1}$, $|g| = \det(g_{ij})$. On the same manifold, if we change the metric g to a new metric h, we say h is conformal to g if there exists some positive function ρ such that $h = \rho g$. Denote $\rho = e^{2\omega}$; then

Research partially supported by NSF grant DMS-94014.

$g_\omega = e^{2\omega}g$ is a metric conformal to g. When the dimension of the manifold M^n is two, Δ_{g_ω} is related to Δ_g by the simple formula:

$$\Delta_{g_\omega}(\varphi) = e^{-2\omega}\Delta_g(\varphi) \quad \text{for all} \quad \varphi \in C^\infty(M^2). \tag{8.1}$$

When the dimension of M^n is greater than two, an operator which enjoys a property similar to (8.1) is the conformal Laplacian operator $L \equiv -c_n\Delta + R$ where $c_n = \frac{4(n-1)}{n-2}$ and R is the scalar curvature of the metric. We have

$$L_{g_\omega}(\varphi) = e^{-\frac{n+2}{2}\omega}L_g\left(e^{\frac{n-2}{2}\omega}\varphi\right) \tag{8.2}$$

for all $\varphi \in C^\infty(M)$.

In general, we call a metrically defined operator A conformally covariant of bidegree (a, b) if, under the conformal change of metric $g_\omega = e^{2\omega}g$, the pair of corresponding operators A_ω and A are related by

$$A_\omega(\varphi) = e^{-b\omega}A(e^{a\omega}\varphi) \quad \text{for all} \quad \varphi \in C^\infty(M^n). \tag{8.3}$$

It turns out that there are many operators besides the Laplacian Δ on compact surfaces and the conformal Laplacian L on general compact manifold of dimension greater than two which have the conformal covariant property. A particularly interesting one is a fourth-order operator on 4-manifolds discovered by Paneitz [50] in 1983:

$$P\varphi \equiv \Delta^2\varphi + \delta\left(\frac{2}{3}RI - 2\,\text{Ric}\right)d\varphi, \tag{8.4}$$

where δ denotes the divergence, d the de Rham differential and Ric the Ricci tensor of the metric. The *Paneitz* operator P (which we will later denote by P_4) is conformal covariant of bidegree $(0, 4)$ on 4-manifolds; i.e.,

$$P_{g_\omega}(\varphi) = e^{-4\omega}P_g(\varphi) \quad \text{for all} \quad \varphi \in C^\infty(M^4). \tag{8.5}$$

For manifolds of general dimension n, when n is even, the existence of an nth order operator P_n conformal covariant of bidegree $(0, n)$ was verified in [34]. However, it is only explicitly known on the standard Euclidean space $\mathbb{R}^n$ and hence on the standard sphere S^n. The explicit formula for P_n on the standard sphere S^n has appeared in Branson [7] and independently in Beckner [5] and will be discussed in §8.2 below.

This paper is organized as follows. In §8.1, we will list some properties of the Laplace operator and compare it, from the point of view of conformal geometry, to some analogous properties of the Paneitz operator P_4. In section 8.2, we will discuss some natural PDE associated with the Paneitz operators P_n on S^n. We will discuss extremal properties of the variational functional associated with the PDE. We will also discuss some uniqueness properties of the solutions of the PDE. The main point here is that, although P_n is an nth order elliptic operator (which in the case when n is odd, is a pseudo-differential operator), the moving plane method of Alexandrov [2] and Gidas-Ni-Nirenberg [33] can still be applied iteratively to establish the spherical symmetry of the solutions of the PDE. In §8.3, we discuss some general existence and uniqueness results of the corresponding functional for the Paneitz operator P_4 on general compact 4-manifolds. In §8.4, we will discuss another natural geometric functional – namely the zeta functional determinant for the conformal Laplacian operator – where the Paneitz operator plays an important role. We will survey some existence and regularity results of the extremal metrics of the zeta functional determinant, and indicate some recent geometric applications by M. Gursky of the extremal metrics to characterize some compact 4-manifolds. Finally in §8.5, we will discuss some existence results for the P_3 operator, which is conformally covariant of bidegree $(0, 3)$, operating on functions defined on the boundary of compact 4-manifolds. The existence of P_3 would allow us to study boundary value problems associated with the P_4 operator. Our point of view here is that the relation of P_3 to P_4 is parallel to the relation of the Neumann operator to the Laplace operator. Thus, for example for domains in $\mathbb{R}^4$, P_3 is a higher-dimensional analogue with respect to the biharmonic functions of the Dirichlet-Neumann operator with respect to the harmonic functions. Some understanding of the P_3 operator also leads to the study of zeta functional determinant on 4-manifolds with boundary.

In this paper we are dealing with an area of research which can be approached from many different directions. Thus many important research works should be cited. But due to the limited space and the very limited knowledge of the author, we will survey here mainly some recent works of the author with colleagues: T. Branson, M. Gursky, J. Qing, L. Wang, and P. Yang; we will mainly cite research papers which have strongly influenced our work.

8.2 Properties of the Paneitz Operator

On a compact Riemannian manifold (M^n, g) without boundary, when the dimension of the manifold $n = 2$, we denote by $P_2 \equiv -\Delta = -\Delta_g$, the Laplacian operator. When the dimension $n = 4$, we denote $P = P_4$ the

Paneitz operator as defined on (8.4). Thus both operators satisfy conformal covariant property $(P_n)_\omega = e^{-n\omega} P_n$, where $(P_n)_\omega$ denotes the operator with respect to (M^n, g_ω), $g_\omega = e^{2\omega} g$. Other considerations in conformal geometry and partial differential equation also identify P_4 as a natural analogue of $-\Delta$. Here we will list several such properties for comparison.

(i) On a compact surface, a natural curvature invariant associated with the Laplace operator is the Gaussian curvature K. Under the conformal change of metric $g_\omega = e^{2\omega} g$, we have

$$\Delta \omega + K_\omega e^{2\omega} = K \quad \text{on} \quad M^2, \tag{8.6}$$

where K_ω denotes the Gaussian curvature of (M^2, g_ω). While on a 4-manifold, we have

$$-P_4\omega + 2Q_\omega e^{4\omega} = 2Q \quad \text{on} \quad M^4, \tag{8.7}$$

where Q is the curvature invariant

$$12Q = -\Delta R + R^2 - 3|\text{Ric}|^2. \tag{8.8}$$

(ii) The analogy between K and Q becomes more apparent if one considers the Gauss-Bonnet formulae:

$$2\pi \chi(M) = \int_M K dv, \quad \text{where} \quad M = M^2, \tag{8.9}$$

$$4\pi^2 \chi(M) = \int_M \left(Q + \frac{|C|^2}{8} \right) dv, \quad \text{where} \quad M = M^4 \tag{8.10}$$

where $\chi(M)$ denotes the Euler characteristic of the manifold M, and $|C|^2 =$ norm squared of the Weyl tensor. Since $|C|^2 dv$ is a pointwise invariant under conformal change of metric, Q is the term which measures the conformal change in formula (8.10).

(iii) When $n \geq 3$, another natural analogue of $-\Delta$ on M^2 is the conformal Laplacian operator L as defined in (8.2). In this case, if we denote the conformal change of metric by $g_u = u^{\frac{4}{n-2}} g$ for some positive function u, then we may rewrite the conformal covariant property (8.2) for L as

$$L_u(\varphi) = u^{-\frac{n+2}{n-2}} L(u\varphi) \quad \text{on} \quad M^n, n \geq 3 \tag{8.11}$$

for all $\varphi \in C^\infty(M^n)$.

A differential equation which is associated with the operator L is the Yamabe equation:

$$Lu = R_u u^{\frac{n+2}{n-2}} \quad \text{on} \quad M^n, \quad n \geq 3. \tag{8.12}$$

Equation (8.12) has been intensively studied in the recent decade. For example the famous Yamabe problem in differential geometry is the study of the equation (8.12) for solutions $R_u \equiv$ constant; the problem has been completely solved by Yamabe [61], Trudinger [58], Aubin [4], and Schoen [55].

(iv) It turns out there is also a natural fourth-order Paneitz operator P_4^n in all dimensions $n \geq 5$, which enjoys the conformal covariance property with respect to conformal changes in metrics also. The relation of this operator to the Paneitz operator in dimension four is completely analogous to the relation of the conformal Laplacian to the Laplacian in dimension two. On (M^n, g) when $n \geq 4$, define

$$P_4^n = (-\Delta)^2 + \delta(a_n R + b_n R_{i,j})d + \frac{n-4}{2} Q_4^n,$$

where

$$Q_4^n = c_n |\rho|^2 + d_n R^2 - \frac{1}{2(n-1)} \Delta R,$$

and $a_n = \frac{(n-2)^2+4}{4(n-1)(n-2)}$, $b_n = -\frac{4}{n-2}$, $c_n = -\frac{2}{(n-2)^2}$, $d_n = \frac{n^3-4n^2+16n-16}{8(n-1)^2(n-2)^2}$ are dimensional constants. Thus $P_4^4 = P_4$, $Q_4^4 = Q$. Then (Branson [7]) we have for $g_u = u^{\frac{4}{n-4}}g$, $n \geq 5$,

$$\left(P_4^n\right)_u (\varphi) = u^{-\frac{n+4}{n-4}} \left(P_4^n\right) (u\varphi) \tag{8.13}$$

for all $\varphi \in C^\infty(M^n)$. We also have the analogue for the Yamabe equation:

$$P_4^n u = Q_4^n u^{\frac{n+4}{n-4}} \quad \text{on} \quad M^n, \quad n \geq 5. \tag{8.14}$$

We would like to remark on $\mathbb{R}^n$ with Euclidean metric, $P_4^n = (-\Delta)^2$ the bi-Laplacian operator. Equation (8.13) takes the form $(-\Delta)^2 u = c_n u^{\frac{n+4}{n-4}}$, an equation which has been studied in literature, e.g., [52].

8.3 Uniqueness Result on S^n

In this section we will consider the behavior of the Paneitz operator on the standard spheres (S^n, g). First we recall the situation when $n = 2$. On (S^2, g), when one makes a conformal change of metric $g_\omega = e^{2\omega}g$, the Gaussian curvature $K_\omega = K(g_\omega)$ satisfies the differential equation

$$\Delta\omega + K_\omega e^{2\omega} = 1 \tag{8.15}$$

on S^2, where Δ denotes the Laplacian operator with respect to the metric g on S^2.

When $K_\omega \equiv 1$ on (8.15), the Cartan-Hadamard theorem asserts that $e^{2\omega}g$ is isometric to the standard metric g by a diffeomorphism φ; and the conformality requirements says that φ is a conformal transformation of S^2. In particular, $\omega = \frac{1}{2}\log|J_\varphi|$, where J_φ denotes the Jacobian of the transformation φ.

In [28], Chen studied the corresponding equation of (8.15) on $\mathbb{R}^2$ with $K_\omega \equiv 1$, and they proved, using the method of moving plane, the stronger result that, when u is a smooth function defined on $\mathbb{R}^2$ satisfying

$$-\Delta u = e^{2u} \qquad \text{on} \quad \mathbb{R}^2 \tag{8.16}$$

with $\int_{\mathbb{R}^2} e^{2u}dx < \infty$, then $u(x)$ is symmetric with respect to some point $x_0 \in \mathbb{R}^2$ and there exists some $\lambda > 0$, so that $u(x) = \log\frac{2\lambda}{\lambda^2+|x-x_0|^2}$ on $\mathbb{R}^2$. There is an alternative argument by Chanillo-Kiessling [26] for this uniqueness result using the isoperimetric inequality.

As we mentioned in the previous section, when $n \geq 3$, a natural generalization of the Gaussian curvature equation (8.15) above is the Yamabe equation under conformal change of metric. On (S^n, g), denote $g_u = u^{\frac{4}{n-2}}g$ the conformal change of metric of g, where u is a positive function, then the scalar curvature $R_u = R(g_u)$ of the metric is determined by the following differential equation

$$c_n\Delta u + R_u u^{\frac{n+2}{n-2}} = Ru, \tag{8.17}$$

where $c_n = \frac{4(n-1)}{n-2}$, $R = n(n-1)$. When $R_u = R$, a uniqueness result established by Obata [45] again states that this happens if the metric g_u is isometric to g or equivalently $u = |J_\varphi|^{\frac{n-2}{2n}}$ for some conformal transformation φ of S^n. In [12], Caffarelli-Gidas-Spruck studied the corresponding equation of (8.17) on $\mathbb{R}^n$:

$$-\Delta u = n(n-2)u^{\frac{n+2}{n-2}}, \qquad u > 0 \quad \text{on} \quad \mathbb{R}^n. \tag{8.18}$$

They classified all solutions of (8.18), via the method of moving plane, as $u(x) = \left(\frac{2\lambda}{\lambda^2+|x-x_0|^2}\right)^{\frac{n-1}{2}}$ for some $x_0 \in \mathbb{R}^n$, $\lambda > 0$.

For all n, on (S^n, g), there also exists a nth order (pseudo) differential operator $\mathbb{P}_n$ which is the pull back via stereographic projection of the operator $(-\Delta)^{n/2}$ from $\mathbb{R}^n$ with Euclidean metric to (S^n, g). $\mathbb{P}_n$ is conformal covariant of bidegree $(0, n)$; i.e., $(\mathbb{P}_n)_\omega = e^{-n\omega}\mathbb{P}_n$. The explicit formulas for $\mathbb{P}_n$ on S^n

have been computed in Branson [7] and Beckner [5]:

$$\begin{cases} \text{For } n \text{ even } \mathbb{P}_n = \prod_{k=0}^{\frac{n-2}{2}}(-\Delta + k(n-k-1)), \\ \text{For } n \text{ odd } \mathbb{P}_n = \left(-\Delta + \left(\frac{n-1}{2}\right)^2\right)^{1/2} \prod_{k=0}^{\frac{n-3}{2}}(-\Delta + k(n-k-1)). \end{cases}$$

(8.19)

On general compact manifolds in the cases when the dimension of the manifold is two or four, there exist some natural curvature invariants $\widetilde{Q}_n$ of order n which, under conformal change of metric $g_\omega = e^{2\omega}g$, is related to $P_n\omega$ through the following differential equation:

$$-P_n\omega + (\widetilde{Q}_n)_\omega e^{n\omega} = \widetilde{Q}_n \quad \text{on} \quad M.$$

(8.20)

In the case when $n = 2$, P_2 is the negative of the Laplacian operator, $\widetilde{Q}_2 = K$, the Gaussian curvature. When $n = 4$, P_4 is the Paneitz operator, $\widetilde{Q}_4 = 2Q_4$ as defined in (8.8). In the special case of (S^2, g), $P_2 = \mathbb{P}_2$, similarly on (S^4, g), $P_4 = \mathbb{P}_4$. In §8.5 below, we will also discuss the existence of P_1, P_3 operators and corresponding curvature invariants Q_1 and Q_3 defined on boundaries of general compact manifolds of dimension 2 and 4, respectively.

On (S^n, g), when the metric g_ω is isometric to the standard metric, then $(\widetilde{Q}_n)_\omega = \widetilde{Q}_n = (n-1)!$. In this case, equation (8.20) becomes

$$-P_n\omega + (n-1)!e^{n\omega} = (n-1)! \quad \text{on} \quad S^n.$$

(8.21)

One can establish the following uniqueness result for solutions of equation (8.21).

Theorem 8.1 ([21]). *On* (S^n, g), *all smooth solutions of the equation (8.21) are of the form* $e^{2\omega}g = \varphi^*(g)$ *for some conformal transformation* φ *of* S^n; *i.e.,* $\omega = \frac{1}{n}\log|J_\varphi|$ *for the transformation* φ.

We now reformulate the equation (8.21) on $\mathbb{R}^n$. For each point $\xi \in S^n$, denote by x its corresponding point under the stereographic projection π from S^n to $\mathbb{R}^n$, sending the north pole on S^n to ∞; i.e., suppose $\xi = (\xi_1, \xi_2, \ldots, \xi_{n+1})$ is a point $\in S^n \subset \mathbb{R}^{n+1}$, $x = (x_1, \ldots, x_n) \in \mathbb{R}^n$, then $\xi_i = \frac{2x_i}{1+|x|^2}$ for $1 \le i \le n$; $\xi_{n+1} = \frac{1-|x|^2}{1+|x|^2}$. Suppose w is a smooth function on S^n, denote by $\varphi(x) = \log\frac{2}{1+|x|^2} = \log|J_{\pi^{-1}}|$, $u(x) = \varphi(x) + w(\xi)$. Since the Paneitz operator P_n is the pullback under π of the operator $(-\Delta)^{n/2}$ on $\mathbb{R}^n$ ([cf. 8, Theorem 3.3]), w satisfies the equation (8.14) on S^n if and only if u satisfies the corresponding equation

$$(-\Delta)^{n/2}u = (n-1)!e^{nu} \quad \text{on} \quad \mathbb{R}^n.$$

(8.22)

Thus Theorem 8.1 above is equivalent to the following result:

Theorem 8.2. *On* $\mathbb{R}^n$, *suppose* u *is a smooth function satisfying the equation (8.22). Suppose in addition that*

$$u(x) = \log \frac{2}{1 + |x|^2} + w(\xi(x))$$

for some smooth function w *defined on* S^n. *Then* $u(x)$ *is symmetric w.r.t. some point* $x_0 \in \mathbb{R}^n$, *and there exists some* $\lambda > 0$ *so that*

$$u(x) = \log \frac{2\lambda}{\lambda^2 + |x - x_0|^2} \quad \text{for all} \quad x \in \mathbb{R}^n . \tag{8.23}$$

We remark that, in the case when w is a minimal solution of the functional with Euler-Lagrange equation (8.21), the result in Theorem 8.2 is a consequence of some sharp Sobolev type inequalities of Milin-Lebedev when $n = 1$, Moser [43] and Onofri [47] when $n = 2$, and Beckner [5] for general n. Sharp inequalities of this type played an important role in a number of geometric PDE problems; see, for example, the article by Beckner [5] and lecture notes by the author [14].

We would like also to mention that during the course of preparation of the paper, the above theorem was independently proved by C. S. Lin [41] and X. Xu [60] when $n = 4$ for functions satisfying equation (8.22) under some less restrictive growth conditions at infinity. In general, it remains open whether there exist some natural geometric conditions under which functions satisfying equation (8.22) are necessarily of the form (8.23).

We now describe briefly the method of moving plane. First we recall a fundamental result of Gidas-Ni-Nirenberg.

Theorem 8.3 ([33]). *Suppose* u *is a positive* C^2 *function satisfying*

$$\begin{cases} -\Delta u & = f(u) \quad \text{on} \quad B \\ u & = 0 \quad \text{on} \quad \partial B \end{cases} \tag{8.24}$$

in the unit ball B *in* $\mathbb{R}^n$ *and* f *is a Lipschitz function. Then* $u(x) = u(|x|)$ *is a radially symmetric decreasing function in* $r = |x|$ *for all* $x \in B$.

To set up the proof in [33] of the above theorem we introduce the following notation. For each point $x \in \mathbb{R}^n$, denote by $x = (x_1, x')$, where $x_1 \in \mathbb{R}$, $x' \in \mathbb{R}^{n-1}$. For each real number λ, denote

$$\begin{aligned} \Sigma_\lambda &= \{x = (x_1, x') \mid x_1 < \lambda\}, \\ T_\lambda &= \{x = (x_1, x') \mid x_1 = \lambda\}, \\ x_\lambda &= (2\lambda - x_1, x') \quad \text{the reflection point of } x \text{ w.r.t. } T_\lambda . \end{aligned}$$

Define

$$w_\lambda(x) = u(x) - u(x_\lambda) \equiv u(x) - u_\lambda(x) .$$

Suppose u satisfies equation (8.24). The idea of moving plane is to prove that $w_\lambda(x) \geq 0$ on Σ_λ for all $0 \leq \lambda \leq 1$ and actually $w_{\lambda=0} \equiv 0$. This is achieved by an application of the maximum principle and Hopf's boundary lemma to the function u. Thus $u(x_1, x') = u(-x_1, x')$ for $x \in B$. Since one can repeat this argument for any hyperplane passing through the origin of the ball B, one establishes that u is radially symmetric with respect to the origin.

If one attempts to generalize above argument to higher-order elliptic equation such as $(-\Delta)^2 u = f(u)$ with suitable boundary conditions, one quickly realizes that, due to a lack of maximum principle for higher-order elliptic equation, such a result in general cannot be expected to hold. Nevertheless, it turns out that for a special class of Lipschitz functions f; namely for functions f satisfying $f(0) \geq 0$ with f monotonically increasing, e.g., $f(u) = e^u$, one can modify the argument in [33]. The key observation is that, for each f, if $(-\Delta)^2 u = f(u)$ then

$$(-\Delta)^2 w_\lambda(x) = f(u) - f(u_\lambda) = c(x)w_\lambda(x), \qquad (8.25)$$

where $c(x)$ is some positive function whose value at x lies between $f(u(x))$ and $f(u_\lambda(x))$. From (8.25) one then concludes that $w_\lambda(x) \geq 0$ on Σ_λ if and only if $(-\Delta)^2 w_\lambda \geq 0$ on $\sum_\lambda$. Since $w_\lambda(x) = (-\Delta)w_\lambda(x) = 0$ for $x \in T_\lambda$, $w_\lambda(x) \geq 0$ on $\sum_\lambda$ also happens if and only if $(-\Delta)w_\lambda(x) \geq 0$ on Σ_λ. This suggests that one should apply the maximum principle and Hopf's lemma to the function $(-\Delta)w_\lambda$ to generalize the result in [33] to higher-order elliptic operators like $(-\Delta)^2$.

In [22], Theorem 8.2 was proved by applying the above argument to the function $(-\Delta)^{m-1}w_\lambda(x)$ where $m = [\frac{n+1}{2}]$. In the case when n is even ($n = 2m$), one can then apply directly some technical lemmas about "harmonic asymptotic" behavior of w at infinity in [12] to the method of moving planes to finish the proof of Theorem 8.2. In the case when n is odd ($n = 2m - 1$), a form of Hopf's lemma for the pseudodifferential operators $(-\Delta)^{1/2}$ was established, and the technical lemmas in [12] were modified to a version adapted to the operator $(-\Delta)^{1/2}$. One can then apply the method of moving planes to finish the proof of Theorem 8.2.

8.4 Existence and Regularity Result on General 4-Manifolds

On (M^2, g) with Gaussian curvature $K = K_g$, consider the functional

$$J[\omega] = \int |\nabla\omega|^2 dv + 2 \int K\omega dv - \left(\int K dv \right) \log \int e^{2\omega} dv, \qquad (8.26)$$

where the gradient, the volume form are taken with respect to the metric g, and $\int \varphi \, dv = \int \varphi \, dv/\text{volume}$ for all φ.

The Euler-Lagrange equation for J is

$$\Delta \omega + c e^{2\omega} = K \qquad \text{on} \quad M^2, \qquad (8.27)$$

where c is a constant. Notice that (8.27) is a special case of equation (8.6) with $K_\omega \equiv c$. For the special manifold (S^2, g), $K \equiv 1$, and (8.27) is a special case of equation (8.15).

In the special case of (S^2, g), the functional $J[\omega]$ has been extensively studied in [44], [47], [20], [19], [27], [38], [25] in connection with the "Nirenberg problem"; that is, the problem to characterize the set of functions K_ω defined on S^2 which are the Gaussian curvature function of a metric $g_\omega = e^{2\omega} g$ conformal to g. For a general compact surface (M^2, g), there is an intrinsic analytic meaning for the functional $J[\omega]$. Namely it is the logarithmic quotient of determinant of the Laplacian operator with respect to g_ω and g respectively. This is generally known as the Ray-Singer-Polyakov formula [53], [51] which we shall briefly describe in §8.4.

A key analytic fact which has been used in the study of the functional $J[\omega]$ is a sharp Sobolev inequality established by Trudinger [59], Moser [43]: Given a bounded smooth domain Ω in $\mathbb{R}^2$, denote by $W_0^{1,2}(\Omega)$ the closure of the Sobolev space of functions with first derivative in L^2 and with compact support contained in Ω, then $W_0^{1,2}(\Omega) \subseteq \exp L^2$, and there exists a best constant $\beta(1,2) = 4\pi$ such that, for all $u \in W_0^{1,2}(\Omega)$ with $\int |\nabla u|^2 dx \leq 1$, there exists some constant c (independent of u) so that $\int_\Omega e^{\beta |u|^2} dx \leq c |\Omega|$ for all $\beta \leq \beta(1,2)$. Moser's inequality has been generalized to the cases of functions satisfying Neumann boundary condition, and to domains with corners in [20], to general domains in $\mathbb{R}^n$ by Adams [1] and to general compact manifolds [32].

We now state some results which generalize the study of the functional $J[\omega]$ to 4-manifolds. On a compact 4-manifold (M^4, g), denote by $k_p = \int Q dv$, and define

$$II[\omega] = \int (P_4 \omega)\omega + 4 \int Q \omega \, dv - \left(\int Q dv \right) \log \left(\fint e^{4\omega} dv \right). \qquad (8.28)$$

Theorem 8.4 ([21]). *Suppose $k_p < 8\pi^2$, and suppose P_4 is a positive operator with* $\ker P = \{\text{constants}\}$*. Then* $\inf II_{\omega \in W^{2,2}}[\omega]$ *is attained. Denote the infimum by ω_p, then the metric $g_p = e^{2\omega_p} g$ satisfies $Q_p \equiv \text{constant} = k_p / \int dv$.*

Remarks. (1) In general, the positivity of P_4 is a necessary condition for the functional II to be bounded from below. But some recent work of M. Gursky [37] indicates that under the additional assumption that $k_p > 0$ and that g is of positive scalar class (i.e. under some conformal change of metric, g admits a metric of positive scalar curvature, or equivalently the conformal Laplacian operator $L = Lg$ admits only positive eigenvalues) P_4 is always positive. Furthermore, under the same assumption, $k_p < 8\pi^2$ is always satisfied unless (M^4, g) is conformally equivalent to (S^4, g); in the latter case then $k_p = 8\pi^2$ and the extremal metric for $II[\omega]$ has been studied in [9].

(2) Notice that the extremal function ω_p in $W^{2,2}$ for II satisfies the equation

$$-P_4\omega_p + 2Q_p e^{4\omega_p} = 2Q \qquad (8.29)$$

with $Q_p \equiv$ constant. Thus standard elliptic theory can be applied to establish the smoothness of ω_p. This is in contrast with the smoothness property of the extremal function ω_d of the log-determinant functional $F[\omega]$, in which $II[\omega]$ is one of the term. We will discuss regularity property of ω_d in §4.

(3) A key analytic fact used in establishing Theorem 8.4 above is the generalized Moser inequality established by Adams [1], which in the special case of domains Ω in $\mathbb{R}^4$ states that $W_0^{2,2}(\Omega) \hookrightarrow \exp L^2$ with $\beta(2, 4) = 32\pi^2$.

We now briefly mention some partial results in another direction which also indicate P_4 is a natural analogue of $-\Delta$ on 4-manifolds.

On two general compact Riemannian manifolds (M^n, g) and (N^k, h), consider the energy functional

$$E[\omega] = -\int (\Delta\omega)\omega dv = \int |\nabla\omega|^2 dv, \qquad (8.30)$$

defined for all system of functions $\omega \in W^{1,2} : (M^n, g) \to (N^k, h)$. Critical points of $E[\omega]$ are defined as harmonic maps. Regularity of harmonic maps has been and still is a subject under intensive study in geometric analysis (see, e.g., articles [54], [57]). We will here mention some results on compact surfaces which are relevant to the statement of Theorem 8.5 below. In the case when the dimension of M^n is 2, a classical result of Morrey [42] states that all minimal solutions of $E[\omega]$ are smooth. Morrey's result has been extended to all solutions of $E[\omega]$ by the recent beautiful work of Helein [39] [40]. In the case when $n \geq 3$, harmonic maps in general are not smooth. The Hausdorff dimension of the singularity sets of the stationary solutions of $E[\omega]$ has also been studied for example in [56], [30], [6].

In some recent joint work [25] with L. Wang and P. Yang, we studied the functional

$$E_4[\omega] = \int (P_4\omega)\omega dv, \qquad (8.31)$$

which is a natural analogue of the functional $E[\omega]$.

A preliminary result we have obtained so far follows.

Theorem 8.5. (i) *For general target manifold* (N^k, h), *weak solutions of the Euler equations of* $E_4[\omega]$ *which minimizes* $E_4[\omega]$ *are smooth.*

(ii) *When the target manifold is* (S^k, g), *then all critical points of* $E_4[\omega]$ *are smooth.*

In the case when the target manifold is (S^k, g), the Euler equation for the functional E_4 takes the following form:

$$\Delta\Delta\omega^\alpha = -\omega^\alpha(\sum_\beta(\Delta\omega^\beta)^2 + 2|\nabla^2\omega^\beta|^2 \ + \ 4\nabla\omega^\beta \cdot \nabla(\Delta\omega^\beta))$$
$$+ \ \text{lower order terms},$$

where $\omega = (\omega^1, \cdots, \omega^k)$ for all $1 \le \alpha \le k$.

A key step in the proof of (ii) in Theorem 8.5 above is to establish that the function f^α which denotes the function in the right-hand side of equation (8.32) is in fact a function in the Hardy space H^1. Thus duality result of H^1- BMO ([31]) may be applied to establish the continuity of the weak solution ω^α. This is completely parallel to the proof by Helein [39] in establishing the smoothness of harmonic maps for compact surfaces. But the reason for the function f^α to be in H^1 is not quite the same as the case in [39]; in particular, compensated compactness results of [29] cannot be applied directly to establish that f^α is in H^1.

In view of the result of Helein [40], it is most plausible that for all general target manifold (N^k, h), all critical solutions of E_4 are smooth. We have also been informed that recently R. Hardt and L. Mou have some regularity results for minimal points of E_4 on a general manifold M^n of dimension $n \ge 5$.

8.5 Zeta Functional Determinant

There is an interesting connection of the functional $J[\omega]$ in (8.26) to a geometric variation problem. On compact surface (M^2, g), let $\{0 < \lambda_1 \le \lambda_2 \le \cdots\}$ be the spectrum of the (negative of) Laplacian $-\Delta_g$. Let $\zeta(s) = \sum \lambda_i^{-s}$ defined for $\Re e\, s > \frac{1}{2}$, then ζ has a meromorphic continuation to the whole plane and is regular at the origin using the heat kernel expansion of Δ_g. Thus $-\zeta'(0)$ is well defined, and one may define $\log \det \Delta_g$ to be $-\zeta'(0)$ (as in Ray-Singer [53]). In [51], Polyakov further computed the logarithm of the ratio of determinant of two conformally related metrics $g_\omega = e^{2\omega}g$ on a compact surface without boundary.

$$F[\omega] = \log \frac{\det \Delta_\omega}{\det \Delta} = \frac{1}{3} \int_M (|\nabla\omega|^2 + 2K\omega)dv_g, \tag{8.32}$$

under the normalization that $\mathrm{vol}(g_\omega) = \mathrm{vol}(g)$. Notice that $F[\omega]$ is essentially the same as the functional $J[\omega]$ in (8.26). In a series of papers, Osgood-Phillips-Sarnak ([48], [49]) have further studied the functional $F[\omega]$, and they have shown among other things that $F[\omega]$ enjoys a certain compactness property on account of the Moser-Trudinger inequality and proved that in each conformal class, the functional $F[\omega]$ attains its extrema at the constant curvature metrics.

When the dimension of a closed manifold is odd, it was shown in Branson [8] that $\log \det L_g$ is a conformal invariant. Thus the next natural dimension to study the generalized Polyakov formula (8.32) is four.

Suppose (M, g) is a compact, closed 4-manifold, and suppose A is a conformally covariant operator satisfying (8.3) with $b - a = 2$. In [11], Branson-Ørsted gave an explicit computation of the normalized form of $\log \frac{\det A_\omega}{\det A}$ which may be expressed as

$$[\omega] = \gamma_1 I[\omega] + \gamma_2 II[\omega] + \gamma_3 III[\omega], \tag{8.33}$$

where $\gamma_1, \gamma_2, \gamma_3$ are constants depending only on A and

$$
\begin{aligned}
I[\omega] &= 4 \int |C|^2 \omega\, dv - \left(\int |C|^2 dv \right) \log \int e^{4\omega} dv \\
II[\omega] &= \langle P\omega, \omega \rangle + 4 \int Q\omega\, dv - \left(\int Q\, dv \right) \log \int e^{4\omega} dv, \\
III[\omega] &= 12 \left(Y(\omega) - \frac{1}{3} \int (\Delta R)\, \omega\, dv \right),
\end{aligned}
$$

where C is the Weyl tensor, and $Y(\omega) = \int \left(\frac{\Delta(e^\omega)}{e^\omega} \right)^2 - \frac{1}{3} \int R |\nabla \omega|^2$. We also remark that the functional $III[\omega]$ may be written as in [9]

$$III[\omega] = \frac{1}{3} \left[\int R_\omega^2\, dv_\omega - \int R^2\, dv \right]$$

so that, when the background metric is assumed to be the Yamabe metric in a positive conformal class, the functional III is nonnegative.

In [9] we made two observations. The first is that on the standard 4-sphere (S^4, g), the functional $F[\omega]$ for the conformal Laplacian L (and Dirac square $\nabla\!\!\!\!/^2$) is extremized in a strong way, that each term $II[\omega]$ and $III[\omega]$ are extremized by the standard metric $g_\omega = g_0$. For $III[\omega]$ this is a consequence of Obata's result ([45]) that the constant scalar curvature conformal metrics on S^4 are standard. For the functional $II[\omega]$, this is a consequence of Beckner's [5] inequality which is valid for all dimensions, (see the discussion in §8.2). The second observation is that the functional $F[\omega]$ enjoys certain compactness properties for the operators L and $\nabla\!\!\!\!/^2$ for most compact locally symmetric Einstein 4-manifolds. The basic analytic inequality required is Adams's inequality [1].

In [21], we continue the study of the log-determinant formula (8.33) on general 4-manifolds. We then extend the compactness criteria to a more general class of 4-manifolds. We define the conformal invariant:

$$
\begin{aligned}
k_d &= -\gamma_1 \int |C|^2 dv - \gamma_2 \int Q dv \\
&= (-\gamma_2)\, 4\pi^2 \chi(M) + \left(\frac{\gamma_2}{8} - \gamma_1\right) \int |C|^2 dv.
\end{aligned}
\tag{8.34}
$$

Theorem 8.6. *If the functional F satisfies $\gamma_2 < 0$, $\gamma_3 < 0$, and $k_d < (-\gamma_2)8\pi^2$, then $\displaystyle\sup_{\omega \in W^{2,2}} F[\omega]$ is attained by some function ω_d and the metric $g_d = e^{2\omega_d} g_0$ satisfies the equation*

$$
\gamma_1 |C_d|^2 + \gamma_2 Q_d - \gamma_3 \Delta_d R_d = -k_d \cdot Vol(g_d)^{-1}.
\tag{8.35}
$$

Further, all functions $\varphi \in W^{2,2}$ satisfy the inequality:

$$
k_d \log \int e^{4(\varphi - \tilde{\varphi})} dv_d \le (-\gamma_2) \langle P\varphi, \varphi \rangle - 12\gamma_3 Y_d(\varphi),
\tag{8.36}
$$

where $\tilde{\varphi}$ denotes the mean value of φ with respect to the metric g_d, and $\fint$ denotes $\frac{1}{vol(M,g_d)} \int_M dv_d$.

In particular, for the operator L and $\slashed{\nabla}^2$, we obtain existence results for extremal metrics of the corresponding log-determinant functional. Thus for a large class of conformal 4-manifolds, we have the existence of several extremal metrics in addition to the Yamabe metric. The study of the relation among these metrics is interesting. For example, we found in [9] that on S^4 all these extremal metrics coincide; while on $S^3 \times S^1_t$ with the standard metric, we found in [21], [23] depending on the parameter t of that of S^1_t, the metric g_d and the Yamabe metric may not agree. In order to identify these extremal metrics in special circumstances, we provide some uniqueness result:

Theorem 8.7. *If $k_d \le 0$, the extremal metric g_d for the functional F corresponding to the conformal Laplacian operator L is unique.*

This uniqueness assertion is obtained as a consequence of the convexity of the corresponding functionals. Applying the uniqueness result, we were able to identify some of the extremal metrics with known metric in special circumstances.

We remark that, the extremal functions ω_d in Theorems 8.6, 8.7, when first established in [21], are functions in $W^{2,2}$ and satisfy the equation (8.35) weakly in $W^{2,2}$. If we rewrite the equation (8.35) expressing $|C_d|^2$, Q_d, $\Delta_d R_d$

all in terms of the background metric g (with $g_d = e^{2\omega d}g$), then $\omega = \omega_d$ satisfies the following equation

$$\Delta\Delta\omega = c_1|\nabla\omega|^4 + c_2(\Delta\omega)^2 + c_3\Delta\omega|\nabla\omega|^2 + \text{lower order terms} \qquad (8.37)$$

for some constants c_1, c_2, c_3 depending only on γ_1, γ_2, γ_3. Equation (8.37) should be compared to equation (8.4) in §8.4.

In a recent joint work [24] with M. Gursky and P. Yang, we established a general regularity result for weak $W^{2,2}$ solution of equation (8.37) on general compact 4-manifolds. A special case of our result is the following theorem:

Theorem 8.8. *Let $F[\omega]$ be as in Theorem 8.6, then* $\sup_{\omega\in\omega^{2,2}} F[\omega]$, *when attained, is a smooth function.*

The main idea in the proof of Theorem 8.8 follows the same line as the regularity result for harmonic maps in Schoen-Uhlenbeck [56]. The property of the maximal solution of $F[\omega]$ is used in some crucial way which enables us to compare $F[\omega]$ with $F[h]$, where h is a biharmonic extension of the boundary value of ω when restricted to a geodesic ball B_r. Some very basic properties for biharmonic functions are established to estimate the growth of $E_4[\omega]$ on B_r. We will mention below a few such properties.

For simplicity, we denote B_r as a ball of radius r in $\mathbb{R}^4$. For a given function $\omega \in W^{2,2}(B_{2r})$ let h be the solution of the linear equation:

$$\begin{cases} \Delta\Delta h = 0 & \text{on} \quad B_r, \\ h = \omega & \text{on} \quad \partial B_r, \\ \frac{\partial h}{\partial n} = \frac{\partial\omega}{\partial n} & \text{on} \quad \partial B_r. \end{cases}$$

Lemma 8.9. (i) $\int_{B_r} |\nabla^2 h|^2 \leq Cr \int_{\partial B_r} |\nabla^2\omega|^2$, *for some constant C.*

(ii) *Suppose $\omega|_{\partial B_r}$ is Hölder of order α, and $\phi = \frac{\partial\omega}{\partial n}$ is in $L^p(\partial B_r)$ for some $p > 3$, then h is Hölder of order $\beta \leq \min(\gamma, 1 - 3/p)$ on B_r.*

Applying Lemma 8.9, one can prove that the extremal function ω of the functional $F[\omega]$ is Hölder continuous. From there, one can apply some iterative arguments to show that all weak $W^{2,2}$ solutions of (8.37) which are Hölder continuous are in fact C^∞ smooth. It remains open whether all critical solution of the functional $F[\omega]$ are smooth. It is also interesting to study the equation (8.37) on domains of dimension ≥ 5.

In another direction, recently M. Gursky [35] gave some beautiful applications of the extremal metric of the log-determinant functional $F[\omega]$ in Theorem 8.6 and Theorem 8.8 above to characterize certain classes of compact 4-manifolds. To state his results, we will first make some definitions. On a compact manifold (M^n, g), define the Yamabe invariant of g as

$$Y(g) = \inf_{g_\omega = e^{2\omega}g} \text{vol}(g_\omega)^{-\frac{n-2}{n}} \int R_{g_\omega} dv_{g_\omega}. \qquad (8.38)$$

By the work of Yamabe, Trudinger, Aubin, and Schoen mentioned in §8.2, every compact manifold M^n admits a metric g_ω conformal to g which achieves $Y(g)$, hence g_ω has constant scalar curvature. We say (M^n, g) is of positive scalar class if $Y(g) > 0$.

On compact 4-manifolds, both $Y(g)$ and $\int Q_g dv_g$ are conformal invariants. The following result of Gursky [35] indicates that these two conformal invariants constrain the topological type of M^4.

Theorem 8.10. *Suppose* (M^4, g) *is a compact manifold with* $Y(g) > 0$.

(i) *If* $\int Q_g dv_g > 0$, *then* M *admits no nonzero harmonic 1-forms. In particular, the first Betti number of* M *vanishes.*

(ii) *If* $\int Q_g dv_g = 0$, *and if* M *admits a nonzero harmonic 1-form, then* (M, g) *is conformal equivalent to a quotient of the product space* $S^3 \times \mathbb{R}$. *In particular,* (M, g) *is locally conformally flat.*

As a corollary of part (ii) of Theorem 8.10, one can characterize the quotient of the product space $S^3 \times \mathbb{R}$ as compact, locally conformally flat 4-manifold with $Y(g) > 0$ and $\chi(M) = 0$.

A crucial step in the proof of the theorem above is to prove that for suitable choice of γ_1, γ_2, γ_3, the extremal metric g_d for the log-determinant functional $F[\omega]$ exists and is unique. Furthermore, under the assumption $Y(g) > 0$, one has $R_{g_d} > 0$; if $Y(g) = 0$ then $R_{g_d} \equiv 0$. In the case $\int Q_g dv_g = 0$, the existence of nonzero harmonic 1-form actually indicates that $R_{g_d} \equiv$ positive constant. Gursky's proof is highly ingenious.

Using similar ideas, Gursky [36] has also applied the above line of reasoning to the study of Kähler-Einstein surfaces. Suppose M^4 is a compact, 4-manifold in the positive scalar class which also admits a nonzero self-dual harmonic 2-form. He established a lower bound for the Weyl functional $\int_M |C|^2 dV$ over all nonnegative conformal classes on M^4 and proved that the bound is attained precisely at the conformal classes of Kähler-Einstein metrics.

8.6 P_3, a Boundary Operator

In the previous sections, we have discussed the behavior of the Laplacian and the Paneitz operator P_4 on functions defined on compact manifolds without boundary. It turns out that, associated with these operators, there also exist some natural boundary operators for functions defined on the boundary of compact manifolds. We will now briefly describe such operators on the boundary of M^n for $n = 2$ and $n = 4$. Most of the material described in this section is contained in the joint work of Jie Qing with the author [16], [17], and [18]. The reader is also referred to the lecture notes [15] for a more detailed description of such operators derived in conjunction with the generalized formula of Polyakov-Alvarez ([51], [11], [9], [3], [10], [16], [17], [18], and [46]) of the zeta functional determinant for 4-manifolds with boundary. We start with some terminology. On a compact manifold (M^n, g) with boundary, we say a pair of operators (A, B) satisfies the *conformal assumptions* if:

Conformal Assumptions: Both A and B are conformally covariant of bidegree (a_1, a_2) and (b_1, b_2) in the following sense,

$$\begin{aligned} A_\omega(f) &= e^{-a_1\omega}A(e^{a_2\omega}f), \\ B_\omega(g) &= e^{-b_1\omega}B(e^{b_2\omega}g), \end{aligned}$$

for any $f \in C^\infty(M), g \in C^\infty(\partial M)$. Assume also that

$$B(e^{a_2\omega}g) = 0 \text{ if and only if } B_\omega(g) = 0,$$

for any $\omega \in C^\infty(\bar{M})$, where A_ω, B_ω denote the operator A, B, respectively, with respect to the conformal metric $g_\omega = e^{2\omega}g$.

Examples: The typical examples of pairs (A, B) which satisfy all three assumptions above are:

(i) When $n = 2$, $A = -\Delta$, $B = \frac{\partial}{\partial n}$, (negative of) the Laplacian operator and the Neumann operator, respectively.

(ii) When $n \geq 3$, $A = \mathcal{L} = -\frac{4(n-1)}{n-2}\Delta + R$ the conformal Laplacian of bidegree $(\frac{n+2}{2}, \frac{n-2}{2})$, and R is the scalar curvature, and B is either the Dirichlet boundary condition or $B = \mathcal{R} = \frac{2(n-1)}{n-2}\frac{\partial}{\partial n} + H$, the Robin operator of bidegree $(\frac{n}{2}, \frac{n-2}{2})$, where H is the trace of the second fundamental form (the mean curvature) of the boundary ∂M.

(iii) When $n = 4$, in [17] we have discovered a boundary operator P_3, conformal of bidegree $(0,3)$ on the boundary of a compact 4-manifold. On 4-manifolds, (P_4, P_3) is a pair of operators satisfying the conformal covariant assumptions, which, in the sense we shall describe below, is a natural analogue of the pair of operators $(-\Delta, \frac{\partial}{\partial n})$ defined on compact surfaces.

As we have mentioned before, on compact surfaces, from the point of view of conformal geometry, a natural curvature invariant associated with the Laplacian operator is the Gaussian curvature K. K enters the Gauss-Bonnet formula (8.9). The Laplacian operator and K are related by the differential equation (8.6) through the conformal change of metrics $g_\omega = e^{2\omega}g$.

On a compact surface M with boundary, the Gauss-Bonnet formula takes the form

$$2\pi\chi(M) = \int_M K dv + \oint_{\partial M} k d\sigma, \tag{8.39}$$

where k denotes the geodesic curvature of ∂M and $d\sigma$ the arc length measure on ∂M. Through conformal change of metric $g_\omega = e^{2\omega}g$ for ω defined on $\bar{M}$, the Neumann operator $\frac{\partial}{\partial n}$ is related to the geodesic curvature k via the differential equation

$$-\frac{\partial \omega}{\partial n} + k_\omega e^\omega = k \quad \text{on} \quad \partial M. \tag{8.40}$$

Equations (8.6) and (8.40) suggest that we search for the right pair of curvature functions and their corresponding differential operators through the Gauss-Bonnet formula. Recall that on 4-manifolds, Paneitz operator P_4 and the fourth-order curvature operator Q are related via equation (8.7). It turns out that on 4-manifolds there also exists a boundary local invariant of order 3 and a conformal covariant operator P_3 of bidegree $(0,3)$, the relation of (Q, T) to (P_4, P_3) on 4-manifolds is parallel to that of (K, k) to $(\Delta, \frac{\partial}{\partial n})$ on compact surfaces. The expressions of P_3 and T on general compact 4-manifolds, like that of P_4, are quite complicated but can be explicitly written down in terms of geometric intrinsic quantities as in [16], [17]. In particular, via the conformal change of metrics $g_\omega = e^{2\omega} g$, P_3 and T satisfy the equation:

$$-P_3\omega + T_\omega e^{3\omega} = T \quad \text{on} \quad \partial M, \tag{8.41}$$

and

$$(P_3)_\omega = e^{-3\omega} P_3 \quad \text{on} \quad \partial M. \tag{8.42}$$

Perhaps the best way to understand how T and P_3 were discovered in [17] is via the Chern-Gauss-Bonnet formula for 4-manifolds with boundary:

$$\chi(M) = (32\pi^2)^{-1} \int_M (|C|^2 + 4Q)dx + (4\pi^2)^{-1} \oint_{\partial M} (T - \mathcal{L}_4 - \mathcal{L}_5)dy, \tag{8.43}$$

where $\mathcal{L}_4$ and $\mathcal{L}_5$ are boundary invariant of order 3 that are invariant under conformal change of metrics. Hence for a fixed conformal class of metrics,

$$\frac{1}{2} \int_M Qdv + \oint_{\partial M} Tds$$

is a fixed constant. We would like to remark that in the original Chern-Gauss-Bonnet formula T is not exactly the term as we have defined in [17]; actually it differs from T by $\frac{1}{3}\widetilde{\Delta}H$, which does not affect the integration formula (8.43).

Thus on 4-manifolds with boundary it is natural to study the energy functional

$$E[\omega] = \frac{1}{4} \int \omega P_4 \omega + \frac{1}{2} \int \omega Q + \frac{1}{2} \oint_{\partial M} \omega P_3 \omega + \oint_{\partial M} \omega T. \tag{8.44}$$

In view of the complicated expressions of the operators P_4, P_3, Q, and T, at this moment it is difficulty to study the functional $E[\omega]$ defined as above on general compact manifolds. But in the special case of (B^4, S^3) with the standard metrics, we have

$$P_4 = (\Delta)^2, \ P_3 = \frac{1}{2}N\Delta + \widetilde{\Delta}N + \widetilde{\Delta} \ \text{ and } \ Q = 0, \ \text{ and } \ T = 3, \qquad (8.45)$$

where $\widetilde{\Delta}$ denotes the the Laplacian operator Δ on (S^3, g). Thus the expression in $E[\omega]$ becomes relatively simple. In this special case, we are able to study the functional $E[\omega]$. The main analytic tool is the following sharp inequality of Lebedev-Milin type on (B^4, S^3).

Theorem 8.11. *Suppose $\omega \in C^\infty(\bar{B}^4)$. Then*

$$\begin{aligned}
&\log\left\{\frac{1}{2\pi^2}\oint_{S^3} e^{3(\omega - \bar{\omega})} dy\right\} \\
&\leq \ \frac{3}{4\pi^2}\left\{\frac{1}{4}\int_{B^4}\omega\Delta^2\omega + \oint_{S^3}\frac{1}{2}\omega P_3\omega - \frac{1}{4}\frac{\partial\omega}{\partial n} + \frac{1}{4}\frac{\partial^2\omega}{\partial n^2}\right\},
\end{aligned} \qquad (8.46)$$

under the boundary assumptions $\int_{S^3}\tau[\omega]ds[\omega] = 0$ where τ is the scalar curvature of S^3. Moreover, the equality holds if and only if $e^{2\omega}g$ on B^4 is isometric to the canonical metric g.

The key step in the proof of the theorem above is the following analytic lemma.

Lemma 8.12. *Suppose ω solves*

$$\begin{cases}
\Delta^2\omega = 0 & \text{in } \mathbb{R}^4 \\
\omega\mid_{S^3} = u \\
\frac{\partial\omega}{\partial n}\mid_{S^3} = \phi.
\end{cases}$$

Then

$$\Delta\omega\bigg|_{\partial B^4} = 2\,\Delta u + 2\left\{(-\widetilde{\Delta} + 1)^{\frac{1}{2}} + 1\right\}\phi \qquad (8.47)$$

$$-\frac{\partial}{\partial n}\Delta\omega\bigg|_{\partial B^4} = 2\mathbb{P}_3 u + 2\widetilde{\Delta}u - 2\widetilde{\Delta}\phi \qquad (8.48)$$

where $\mathbb{P}_3 = (-\widetilde{\Delta} + 1)^{\frac{1}{2}}(-\widetilde{\Delta})$ is the same as the $\mathbb{P}_3$ operator defined on S^3 as in §8.3.

We would also like to remark that the second term in (8.47) above, i.e., the term $\left\{(-\widetilde{\Delta} + 1)^{\frac{1}{2}} + 1\right\}$ is the Dirichlet to Neumann operator on (B^4, S^3). We can get the following result as a direct corollary of Lemma 8.12.

Corollary 8.13. *On B^4,*

$$P_3(\omega) = \mathbb{P}_3(\omega) \text{ on } \partial B^4,$$

provided that $\Delta^2\omega = 0$ in B^4.

That is, on (B^4, S^3), P^3 is an extension of the Dirichlet-Neumann with respect to biharmonic functions. Thus the study of the P^3 operator can be viewed as an extension of the study of Dirichlet-Neumann operator.

Further developments in this field have been made since the paper was written. We refer the reader to the survey article by S.-Y. A. Chang and P. Yang, *On a fourth-order curvature invariant*, which will soon appear in *Contemporary Mathematics*.

References

[1] D. Adams, *A sharp inequality of J. Moser for higher order derivatives*, Annals of Math., vol. 128 (1988), 385–398.

[2] A. D. Alexandrov, *Uniqueness theorems for surfaces in the large*, V. Vestnik Leningrad Univ. Mat. Mekh. Astronom. 13 (1958), no. 19, 5–8, Amer. Math. Soc. Transl. ser. 2, 21, 412–416.

[3] O. Alvarez, *Theory of strings with boundary*, Nucl. Phys. B 216 (1983), 125–184.

[4] T. Aubin, *Équations différentielles non-linéaires et problème de Yamabe concernant la courbure scalaire*, J. Math. Pures Appl. 55 (1976), 269–296.

[5] W. Beckner, *Sharp Sobolev inequalities on the sphere and the Moser-Trudinger inequality*, Annals of Math. 138 (1993), 213–242.

[6] F. Bethuel, *On the singular set of stationary harmonic maps*, Manuscripta Math. 78 (1993), 417–443.

[7] T. Branson, *Group representations arising from Lorentz conformal geometry*, JFA 74 (1987), 199–293.

[8] ———, *Sharp inequality, the functional determinant and the complementary series*, TAMS 347 (1995), 3671–3742.

[9] T. Branson, S.-Y. A. Chang, and P. C. Yang, *Estimates and extremal problems for the log-determinant on 4-manifolds*, Comm. Math. Physics 149, no. 2 (1992), 241–262.

[10] T. Branson and P. Gilkey, *The functional determinant of a 4-dimensional boundary value problem*, Trans. AMS 344 (1994), 479–531.

[11] T. Branson and B. Ørsted, *Explicit functional determinants in four dimensions*, Proc. Amer. Math. Soc. 113 (1991), 669–682.

[12] L. Caffarelli, B. Gidas, and J. Spruck, *Asymptotic symmetry and local behavior of semilinear equations with critical Sobolev growth*, Comm. Pure Appl. Math. 42 (1989), 271–289.

[13] K. C. Chang and J. Q. Liu, *On Nirenberg's problem*, International J. of Math. 4 (1993), 35–58.

[14] S.-Y. A. Chang, *Moser-Trudinger inequality and applications to some problems in conformal geometry*, Nonlinear Partial Differential Equations, editors: Hardt and Wolf, AMS IAS/Park City Math series, vol. 2, 67–125.

[15] ———, *On zeta functional determinants*, Centre de Recherche Mathématiques, CRM Proceedings and Lecture Notes, vol. 12 (1997), 25–50.

[16] S.-Y. A. Chang and J. Qing, *Zeta functional determinants on manifolds with boundary*, Research announcement, Math. Research Letters 3 (1996), 1–17.

[17] ———, *The zeta functional determinants on manifolds with boundary I–the formula*, JFA 47, no. 2 (1997), 327–362.

[18] ———, *The zeta functional determinants on manifolds with boundary II–extremum metrics and compactness of isospectral set*, JFA 147, no. 2 (1997), 363–399.

[19] S.-Y. A. Chang and P. C. Yang, *Prescribing Gaussian curvature on S^2*, Acta Math. 159 (1987), 215–259.

[20] ———, *Conformal deformation of metrics on S^2*, J. Diff. Geometry 27 (1988), 259–296.

[21] ———, *Extremal metrics of zeta functional determinants on 4-manifolds*, Annals of Math. 142 (1995), 171–212.

[22] ———, *On uniqueness of solution of an n-th order differential equation in conformal geometry*, Math. Res. Letters 4 (1997), 91–102.

[23] S.-Y. A. Chang, M. Gursky, and P. C. Yang, *Remarks on a fourth-order invariant in conformal geometry*, preprint, 1996.

[24] ———, *On regularity of a fourth order PDE with critical exponent*, to appear in Amer. Jour. of Math.

[25] S.-Y. A. Chang, L. Wang, and P. C. Yang, *A regularity theory of biharmonic maps*, to appear in CPAM.

[26] S. Chanillo and M. K.-H. Kiessling, *Conformally invariant systems of non-linear PDEs of Liouville type*, Geom. Funct. Anal. 5 (1995), 924–947.

[27] W. Chen and W. Y. Ding, *Scalar curvature on S^2*, TAMS 303 (1987), 365–382.

[28] W. Chen and C. Li, *Classification of solutions of some non-linear elliptic equations*, Duke Math. J. 63, no 3 (1991), 615–622.

[29] R. Coifman, P. L. Lions, Y. Meyer, and S. Semmes, *Compacité par commensation et éspaces de Hardy*, C.R. Acad. Sci. Paris 311 (1989), 519–524.

[30] C. L. Evans, *Partial regularity for stationary harmonic maps into spheres*, Arch. Rat. Mech. Anal. 116 (1991), 101–113.

[31] C. Fefferman and E. Stein, *H^p spaces of several variables*, Acta Math. 129 (1972), 137–193.

[32] L. Fontana, *Sharp borderline estimates on spheres and compact Riemannian Manifolds*, Ph.D thesis, Washington University, St. Louis, 1991.

[33] B. Gidas, N. W. Ni, and L. Nirenberg, *Symmetry of positive solutions of non-linear equations in $\mathbb{R}^n$*, Analysis and Applications, Part A, 369–402, Advances in Math. Supp. Study 79, Academic Press, New York, London 1981.

[34] C. R. Graham, R. Jenne, L. Mason, and G. Sparling, *Conformally invariant powers of the Laplacian, I: existence*, J. London. Math. Soc. (2) 46 (1992), 557–565.

[35] M. Gursky, *Some Bochner theorems on 4-dimensional conformal geometry*, preprint, 1996.

[36] ———, *The Weyl functional and de Rham cohomology and Kähler-Einstein metrics*, Annals of Math. 148 (1988), 315–337.

[37] ———, *The principal eigenvalue of a conformally invariant differential operator, with an application to semi-linear elliptic PDS*, preprint, 1998.

[38] Z. C. Han, *Prescribing Gaussian curvature on S^2*, Duke Math. J. 61 (1990), 679–703.

[39] F. Helein, *Regularité des applications faiblement harmoniques entre une surface et une varieté riemannienne*, C. R. Acad. Sci. Paris 312 (1991), 591–596.

[40] ———, *Regularity of weakly harmonic maps from a surface into a manifold with symmetries*, Manuscripta Mathematics 70 (1991), 203–218.

[41] C.-S. Lin, *A classification of solutions of a conformally invariant fourth-order equation in $\mathbb{R}^n$*, Comment. Math. Helv. 73 (1998), 206–231.

[42] C. B. Morrey, Jr., *The problem of Plateau on a Riemannian manifold*, Ann. of Math. 49 (1948), 807–851.

[43] J. Moser, *A sharp form of an inequality by N. Trudinger*, Indian Math. J. 20 (1971), 1077–1091.

[44] ———, *On a Non-linear Problem in Differential Geometry and Dynamical Systems* (M. Peixoto, ed.), Acad. Press, New York, 1973.

[45] M. Obata, *Certain conditions for a Riemannian manifold to be isometric with a sphere*, Jour. Math. Society of Japan 14 (1962), 333–340.

[46] K. Okikiolu, *The Campbell-Hausdorff theorem for elliptic operators and a related trace formula*, Duke Math. J. 79 (1995), 687–722.

[47] E. Onofri, *On the positivity of the effective action in a theory of random surfaces*, Comm. Math. Phys. 86 (1982), 321–326.

[48] B. Osgood, R. Phillips, and P. Sarnak, *Extremals of determinants of Laplacians*, J. Funct. Anal. 80 (1988), 148–211.

[49] ———, *Compact isospectral sets of surfaces*, J. Funct. Anal. 80 (1988), 212–234.

[50] S. Paneitz, *A quartic conformally covariant differential operator for arbitrary pseudo-Riemannian manifolds*, preprint, 1983.

[51] A. Polyakov, *Quantum geometry of bosonic strings*, Phys. Lett. B 103 (1981), 207–210.

[52] P. Pucci and J. Serri, *Critical exponents and critical dimensions for polyharmonic operators*, J. Math. Pures Appl. 69 (1990), 55–83.

[53] D. B. Ray and I. M. Singer, *R-torsion and the Laplacian on Riemannian manifolds*, Advances in Math. 7 (1971), 145–210.

[54] R. Schoen, *Analytic aspects of the harmonic map problem*, Math. Sci. Res. Inst. Publ. 2, Springer, Berlin, 1984.

[55] ———, *Conformal deformation of a Riemannian metric to constant scalar curvature*, J. Differential Geom. 20 (1984), 479–495.

[56] R. Schoen and K. Uhlenbeck, *A regularity theory for harmonic maps*, JDG 17 (1982), 307–335.

[57] L. Simon, *Singularities of Geometric Variational Problems*, Nonlinear Partial Differential Equations, Editors: Hardt and Wolf, AMS IAS/Park City Math series, vol. 2, 185–222.

[58] N. Trudinger, *Remarks concerning the conformal deformation of Riemannian structure on compact manifolds*, Ann. Scuo. Norm Sup. Pisa, 3 (1968), 265–274.

[59] ———, *On imbedding into Orlicz spaces and some applications*, J. Math. Mech. 17 (1967), 473–483.

[60] X. Xu, *Classification of solutions of certain fourth-order nonlinear elliptic equations in $\mathbb{R}^4$*, preprint 1996.

[61] H. Yamabe, *On a deformation of Riemannian structures on compact manifolds*, Osaka Math. J. 12 (1960), 21–37.

Chapter 9

Riesz Transforms, Commutators, and Stochastic Integrals

A. B. Cruzeiro and P. Malliavin

9.1 Renormalization in Probability Theory and Elliptic Estimates in Infinite Dimension

The state at time n of the gambler's fortune F_n in a game of heads or tails is almost surely bounded by $n^{\frac{1}{2}+\epsilon}$; denoting by g_k the gain at step k we have $F_n = \sum_{0<k<n+1} g_k$. Therefore F_n appears as a sum which is small because of a large number of cancellations. This elementary example is quite typical of probabilistic situations associated to the Laplace central limit theorem where semiconvergent series always appear. We shall refer to this fact as an example of the more general concept of *renormalization.*

In some branches of classical analysis, as for instance in singular integrals, cancellations play a fundamental role; a methodology for proving convergence consists in obtaining a priori estimates for a sequence of smooth approximations. In our probabilistic context the same approach can work if a priori elliptic estimates in infinite dimension are available. Indeed, limit theorems concern games where the number of trials is tending to infinity and the corresponding probability space must be of infinite dimension.

The appearance of elliptic operators is a more hidden fact: it is linked to the concept of some kind of "universal chaos," that is, some probability space containing in itself all the random fluctuations which could appear in the universe. The Itô theory of stochastic differential equations gives to the probability space X of the Brownian motion this status of "universal chaos;" it is impossible to justify in few words the paramount importance of this space.

The Brownian motion can be characterized as a Gaussian process which has independent increments on disjoint intervals of time; for the sake of brevity we shall grasp the Brownian motion through its Wiener series:

$$x(\tau) = \tau \xi_0 + \sqrt{2} \sum_{k=1}^{\infty} \frac{\sin k\pi\tau}{k\pi} \xi_k, \qquad 0 \leq \tau \leq 1,$$

where the ξ_k are independent Gaussian normal variables. We denote by X the probability space generated by those Gaussian random variables: it is isomorphic to $\mathbb{R}^\infty$ with the Wiener measure μ being defined as the infinite product of the Gaussian measure μ_1 on $\mathbb{R}$: $\mu_1 = \frac{1}{\sqrt{2\pi}} \exp\left(-\frac{\xi^2}{2}\right) d\xi$. The data of the ξ_k constitute some kind of "heat bath." The next step is to say that this heat bath takes its stochastic regularity from the fact that it represents some kind of "stochastic equilibrium."

On $\mathbb{R}$, the space $L^2(\mathbb{R}, \mu_1)$ has for its orthonormal basis the Hermite polynomials H_n; these polynomials can be characterized as the eigenfunctions of the following Sturm-Liouville operator:

$$\mathcal{L}_1 = \frac{1}{2} \exp\left(\frac{\xi^2}{2}\right) \frac{d}{d\xi} \left[\exp\left(-\frac{\xi^2}{2}\right) \frac{d}{d\xi} \right] = \frac{1}{2} \left[\frac{d^2}{d\xi^2} - \xi \frac{d}{d\xi} \right],$$

which will be called the Ornstein-Uhlenbeck operator. It is a symmetric operator in $L^2(\mathbb{R}, \mu_1)$; therefore, the associated parabolic semigroup ${}_1P_t = e^{t\mathcal{L}_1}$ preserves μ_1. This semigroup can be explicitly written through the Mehler formula:

$$ {}_1P_t(f)(\xi^0) = \int_{\mathbb{R}} f(e^{-t}\xi^0 + (1 - e^{-2t})^{\frac{1}{2}}\eta)\mu_1(d\eta). \tag{9.1}$$

It is clear from this formula that ${}_1P_t$ acts as a contraction on all the L^p.

This Mehler formula stays meaningful in infinite dimensions when we replace μ_1 by the Wiener measure μ; we get in this way on X a semigroup P_t; its infinitesimal generator will be formally denoted by $\mathcal{L}$. The elliptic estimates that we shall discuss will be related to $\mathcal{L}$.

9.2 Pisier Transplantation of the Calderón-Zygmund Rotation Method

The inverse of the Cauchy operator, Λ, is defined by

$$\Lambda = \frac{1}{\sqrt{\pi}} \int_0^\infty P_t \frac{e^{-t}}{t^{\frac{1}{2}}} \, dt,$$

where the semigroup P_t is define by the Mehler formula.

It is possible to say that $\mathbb{R}^\infty = \lim \mathbb{R}^n$; therefore it is sufficient to prove the Riesz transform inequalities for the finite-dimensional case, *with constants independent of the dimension*. The gradient ∇ of a smooth function defined on $\mathbb{R}^n$ is an $\mathbb{R}^n$-valued function. The Riesz transform is defined

$$\mathcal{R}f = \nabla \Lambda f; \quad \text{the r.h.s. is an } \mathbb{R}^n - \text{valued function.}$$

Theorem 9.1. *Given $p \in (1, \infty)$, there is a constant c_p, independent of the dimension n, such that*

$$\int_{\mathbb{R}^n} \|\mathcal{R}f\|_{\mathbb{R}^n}^p \, d\mu_n \leq c_p \int_{\mathbb{R}^n} |f|^p \, d\mu_n,$$

where $\|\cdot\|_{\mathbb{R}^n}$ denotes the Euclidean norm and where μ_n denotes the Gaussian normal measure on $\mathbb{R}^n$.

Proof. We shall present the Pisier approach. We denote $X = Y = \mathbb{R}^n$ and we define on $X \times Y$ the measure $\nu = \mu_{\mathbb{R}^n}^{\otimes 2}$. The rotation matrix on $\mathbb{R}^2$ acts on $X \times Y$; this action preserves ν; the orbit of each point is a circle. Making in the Mehler formula the change of variable $t = |\log \cos(\theta)|$, we get

$$(\Lambda f)(x) = \frac{1}{\sqrt{\pi}} \int_0^{\frac{\pi}{s}} \int_Y f(x \cos\theta + y \sin\theta) |\log(\cos\theta)|^{-\frac{1}{2}} \sin\theta \, d\theta \, d\mu. \tag{9.2}$$

The gradient satisfies the identity

$$\nabla_x f(x \cos\theta + y \sin\theta) = \frac{\cos\theta}{\sin\theta} \nabla_y f(x \cos\theta + y \sin\theta).$$

This identity makes possible to transfer the differentiation relative to x of the left-hand side of (9.2) into a differentiation relative to y, on which an integration by parts can be made; we get

$$\left(\frac{\partial}{\partial x_i} \Lambda f \right)(x) = \frac{1}{2\sqrt{\pi}} \int_Y \int_{-\frac{\pi}{2}}^{\frac{\pi}{2}} f(x \cos\theta + y \sin\theta) \phi(\theta) y_i \, d\theta \mu(dy),$$

where $\phi(\theta) = \text{signum}(\theta)(\cos\theta)|\log \cos\theta|^{-\frac{1}{2}}$.

The variable y_i is a Hermite polynomial of degree 1. Using the Hermite expansion of a function in L^p, Pisier proved that

$$\|(\mathcal{R}f)(x)\|_{\mathbb{R}^n}^p \le c_p \int_Y |J(x,y)|^p \mu(dy),$$

where $J(x,y) = \dfrac{1}{2\sqrt{\pi}} \int_{-\frac{\pi}{2}}^{\frac{\pi}{2}} f(x\cos\theta + y\sin\theta)\phi(\theta)\,d\theta.$

We have $\phi(\theta) = \cot\left(\frac{\theta}{2}\right) + \psi(\theta)$, where $\psi \in C^\infty$; using the boundedness of the Hilbert transform on L^p of the circle we get the theorem. $\square$

In the spirit of Benedeck-Calderón-Panzone [2], the same theorem remains true for B-vector-valued functionals, where B is an UMD Banach space [17]. The previous proof of Pisier depends upon beautiful identities whose extension to a general situation is not reasonable to expect.

9.3   Littlewood-Paley-Calderón-Stein Area Integral for Harmonic Extension

We shall replace identities of the previous section by computation of commutators; again on the Gaussian space commutators have a very remarkable expression. In the general case exact expressions will be replaced by majorizations.

Lemma 9.2. *We have*

$$\nabla P_t f = e^{-t} P_t \nabla. \tag{9.3}$$

Denote $\mathcal{C} = \sqrt{-\mathcal{L}}$ *and* $\mathcal{C}' = \sqrt{-\mathcal{L}+1} = \Lambda^{-1}$; *then*

$$\nabla e^{-\eta\mathcal{C}} = e^{-\eta\mathcal{C}'}\nabla. \tag{9.4}$$

Proof. We can differentiate the Mehler formula relatively to x and get (9.3). The semigroup $e^{-\eta\mathcal{C}}$ can be expressed by the symbolic calculus as follows:

$$e^{-\eta\mathcal{C}} = \int_0^\infty P_s q_\eta(s)\,ds, \quad \text{where } q_\eta(s) = \frac{\eta}{4\sqrt{\pi}} s^{-\frac{3}{2}} \exp\left(-\frac{\eta^2}{4s}\right). \tag{9.5}$$

Differentiating the right-hand side and using the commutator relation (9.3) we get the same integral multiplied by $\exp(-s)$ which gives $\exp(-\eta\mathcal{C}')$. $\square$

Given $g \in L^p(X)$ we consider its harmonic extension h_g to the half-space $Z = X \times \mathbb{R}^+$ constructed by

$$h_g(x,\eta) = (\exp(-\eta\mathcal{C})g)(x); \quad \text{then} \quad \left[\frac{\partial^2}{\partial\eta^2} + \mathcal{L}\right]h_g = 0.$$

The gradient on Z has its horizontal component ∇^X and its vertical component $\frac{\partial}{\partial \eta}$. Then, for $p \in (1, \infty)$, the area integral equivalence ([4],[20]) gives

$$\int_X \left[\int_0^\infty ||\nabla^X h_g(x, \eta)||^2 \eta \, d\eta \right]^{\frac{p}{2}} \mu(dx) \simeq \int_X |g(x)|^p \mu(dx), \qquad (9.6)$$

for all g such that $\int g \, d\mu = 0$.

For the vertical gradient the analogous equivalence holds true; we shall state it for the harmonic extension $h' = e^{-\eta C'} g'$:

$$\int_X \left[\int_0^\infty \left(\frac{\partial}{\partial \eta} h' \right)^2 d\eta \right]^{\frac{p}{2}} \mu(dx) \simeq ||g'||_{L^p}^p. \qquad (9.7)$$

Theorem 9.3. (Meyer [18]). *We have the following:*

$$\int ||\nabla f||_{\mathbb{R}^n}^p \, d\mu \simeq \int |Cf|^p \, d\mu.$$

Proof. By differentiating (9.3) relative to η we get

$$\nabla e^{-\eta C} Cf = \frac{d}{d\eta} e^{-\eta C'} \nabla f. \qquad (9.8)$$

Take $g = Cf$, $g' = \nabla f$. The identity (9.8) implies that the left-hand side of (9.6) and (9.7) are equal; therefore their right-hand sides are equivalent. $\square$

9.4 Calderón Factorization through the Cauchy Operator and Stochastic Integrals

We shall not discuss here the general theory of Gross-Sobolev-Stroock spaces, theory where the Riesz transform plays a paramount role [21]. We limit ourselves to a basic existence result for stochastic integrals.

In his paper on the uniqueness of the Cauchy problem, Calderón [5] writes any differential operator as the product of a power of the Cauchy operator by a pseudodifferential operator of degree zero. This idea leads us 25 years later to derive results in stochastic analysis.

We have a stochastic process $\phi(\tau)$ defined on the probability space of the Brownian motion X; we consider partitions Θ of $[0, 1]$ by a finite number of points; given $\theta \in \Theta$ we denote by $\theta^+ \in \Theta$ the point of the partition which is closest on the right to θ. With Nualart-Pardoux we form the following *renormalized* Riemann sum:

$$J_\Theta(\phi) = \sum_{\theta \in \Theta} [x(\theta^+) - x(\theta)] \frac{1}{\theta^+ - \theta} E^{\delta_\theta} \left[\int_\theta^{\theta^+} \phi(\tau) \, d\tau \right],$$

where E^{δ_θ} is the conditional expectation obtained by averaging on the σ-field generated by $x(\tau) - x(\theta)$, $\tau \in [\theta, \theta^+]$. We say that ϕ is *stochastically integrable* if the limit of J_Θ exists when the mesh of the partition Θ tends to zero.

Theorem 9.4 (Watanabe [22]). *Assume that*

$$|||\phi|||^2 := \int_0^1 ||\mathcal{C}'\phi(\tau)||^2_{L^2(X)}\, d\tau < \infty.$$

Then ϕ is stochastically integrable and its stochastic integral satisfies the following L^2 majoration:

$$E\left[\left(\int_0^1 \phi(\tau)\, dx(\tau)\right)^2\right] \leq |||\phi|||^2.$$

Proof: We shall first proceed to a transfer by duality.

Lemma 9.5 (of integration by parts (Gaveau-Trauber [13])). *Assume that ϕ is stochastically integrable. Then, for every smooth functional f, the directional derivative of f along ϕ denoted by $D_\phi(f)$ satisfies the duality relation*

$$E(D_\phi f) = E\left(f \int_0^1 \phi\, dx\right). \tag{9.9}$$

Remark. This lemma identifies the stochastic integral of ϕ with "the divergence of the vector field ϕ."

Proof. (of the theorem) Using the Riesz transform $\mathcal{R}' = \mathcal{C}^{-1}\nabla f$ we have $\nabla f = \mathcal{C}\mathcal{R}'f$; therefore, we write (9.9) as

$$((\mathcal{C}\mathcal{R}'f|\phi) = (\mathcal{R}'f|\mathcal{C}\phi) = (f|[\mathcal{R}']^*\mathcal{C}\phi),$$

where the fact that $\mathcal{C}$ is a symmetric unbounded operator in $L^2(X)$ implies the first equality; the second one is due to the hypothesis $\mathcal{C}\phi \in L^2(X; L^2([0,1]))$ and, since $\mathcal{R}'$ is a bounded operator from $L^2(X)$ to $L^2(X; L^2([0,1]))$, its transpose is well defined. We get finally the following expression for the stochastic integral, which, quite unexpectedly, establishes a beautiful link between probability theory and singular integrals:

$$\int_0^1 \phi\, dx = [\mathcal{R}']^*\mathcal{C}f.$$

The Riesz transform has a norm as a bounded operator in L^2 which is equal to 1. This implies the announced inequality. □

9.5 Commutators in Renormalized Riemannian Geometry on Path Spaces

9.5.1 Itô Global Chart of Path Space

The Brownian motion sits in the Euclidean space $\mathbb{R}$ or $\mathbb{R}^d$ and has therefore a linear structure reflected for instance in its representation by the Wiener series. We want to proceed in the nonlinear setting of the Brownian on a compact Riemannian manifold M; we denote by 2Δ its Laplace-Beltrami operator and by $\pi_t(m_0, m)$ the fundamental solution of the heat operator associated to Δ, with pole at m_0. Then the Brownian motion on M is characterized by the property that it is a stochastic process $p(\tau)$ having for law of its "increments" $\pi_{\theta + -\theta}(p(\theta), *) \, dm$ for any finite subdivision Θ of $[0, 1]$, with the Markovian independence of those increments. Up to operations of finite-dimensional geometry, we can limit ourselves to the case where the Brownian motion starts at $\tau = 0$ from a fixed point. Then the probability space of the Brownian motion will be the path space $P_{m_0}(M)$ with the Brownian probability measure ν.

The first question is how to construct a canonic global chart of $P_{m_0}(M)$. This ambition is feasible according to the fact that P_{m_0} is a contractible space. We recall that Levi-Cività parallel transport $t^q_{\tau \leftarrow 0}$ along a smooth curve $\tau \to q(\tau)$ is defined by solving the matrix differential equation

$$d_\tau t^q_{\tau \leftarrow 0} = \Gamma(\dot{q} \, d\tau) t^q_{\tau \leftarrow 0}, \tag{9.10}$$

where $\Gamma^j_{k,i}$ denotes the Christofel symbols of the metric $ds^2 = g_{i,j} dm^i dm^j$ computed in local coordinates. By some smoothing procedure we can construct a 1-parameter family p^ϵ of smooth curves such that $\lim_{\epsilon \to 0} p^\epsilon = p$. Then

Theorem 9.6 (K. Itô [14]). *The following limit exists:*

$$\lim_{\epsilon \to 0} t^{p^\epsilon}_{* \leftarrow 0} := t^p_{* \leftarrow 0}.$$

Theorem 9.7 (Malliavin [15], Eells-Elworthy). *We define a map* $\mathcal{J}^{-1}$: $P_{m_0}(M) \to P_0(\mathbb{R}^d)$ *by*

$$x(\tau) = \int_0^\tau t^p_{0 \leftarrow \sigma}(dp(\sigma)),$$

where the above integral is an Itô stochastic integral; then this map is an isomorphism of probability spaces when we take on $P_0(\mathbb{R}^d)$ *the Wiener measure. We call* $\mathcal{J}$ *the Itô map.*

Remark. The Itô map provides the desired global chart. This chart preserves the probability measures, but it does not preserve many other canonic geometric objects which must be directly constructed on $P_{m_0}(M)$.

9.5.2   Structure Equation of the Path Space

Tangent process: Any infinitesimal variation Z of a Brownian path p on M will be canonically parameterized by the vector-valued function $t^p_{0 \leftarrow \tau}(Z_\tau)$. We call a *tangent process* along the Wiener space the data of a semimartingale ζ which has its martingale part constructed by multiplying the Brownian differential by an *antisymmetric* matrix: $d\zeta^\alpha = a^\alpha_\beta \, dx^\beta + c^\alpha \, d\tau$. A tangent process along $P_{m_0}(M)$ is defined as $t^p_{0 \leftarrow \tau}(\zeta(\tau))$, where ζ is a tangent process on X.

Theorem 9.8 (Cruzeiro-Malliavin [9], Fang-Malliavin [12]). *The Itô map can be differentiated; its differential $\mathcal{J}'$ realizes an isomorphism between the spaces of tangent process.*

This theorem is false if we are dealing with *restricted tangent processes*, that is, processes with a vanishing martingale part.

The expression of the derivative of the Itô map incorporates a holonomy factor which is a stochastic integral of the curvature of the underlying Riemannian manifold. Then the derivative of a smooth functional f on the path space and along a tangent process $t^p_{\tau \leftarrow 0}(\zeta(\tau))$, that we denote by $\mathbf{D}_\zeta$, is computed via the transfer to the Wiener space through the Itô map. It corresponds to the derivative $D_{\bar{\zeta}}(f \circ \mathcal{J})$, where

$$d\bar{\zeta} = d\zeta + \left[\int_0^\tau \Omega(\sigma, \zeta) o \, dx(\sigma) \right] o \, dx(\tau)$$

with Ω the curvature tensor of M and where the stochastic integral is of Stratonochich type.

Theorem 9.9 (Bismut [3], Driver [11]). *Tangent processes have a formula of integration by parts:*

$$E_\nu(\mathbf{D}_\zeta f) = E_\nu \left(f \int_0^1 \left[c_\tau + \frac{1}{2} R^M \left(\int_0^\tau c \right) \right] dx(\tau) \right)$$

where $R^M_\tau = t^p_{0 \leftarrow \tau} o \; Ricci \; o \, t^p_{0 \leftarrow \tau}$.

Theorem 9.10 (Structure theorem, Cruzeiro-Malliavin [9]). *The Lie bracket of two restricted tangent processes z_1, z_2 has in the parallelism the following expression:*

$$[z_1, z_2] = Q_{z_1} z_2 - Q_{z_2} z_1 \; where \; Q_z(\tau) = \int_0^\tau \Omega_{p(\sigma)}(z_\sigma, o \, dp(\sigma)),$$

where Ω_p denotes the curvature tensor read in the parallelism $t^p_{0 \leftarrow *}$.

We encounter therefore a new hypoelliptic phenomenon, which leads to a necessary renormalization and is present in all subsequent developments of the analysis on the path space. Nevertheless, we still have a natural Lie algebra structure:

Theorem 9.11 (Cruzeiro-Malliavin [9], Driver [11]). *The Lie bracket of two tangent processes is a tangent process.*

9.5.3 Energy Estimates for Stochastic Integrals

The structure theorem provides tools that are needed to compute commutators, which, in the spirit of Calderón [6], are the basic gap to reach elliptic estimates. The commutator effect will bring a small loss in the exponent of the integrability.

The Levi-Cività covariant derivative ∇^M on M induces a Markovian covariant ∇^P derivative on the path space.

Theorem 9.12 (Cruzeiro-Fang [8]). *Given a process Z_r such that for some $p > 2$*

$$E\left[\left(\int_0^1 \int_0^1 |\nabla_\tau^P Z_\sigma|^2 \, d\tau \, d\sigma\right)^{\frac{p}{2}}\right] < \infty,$$

the divergence of Z, $\delta(Z)$, exists and satisfies $E(\delta(Z)^2) < \infty$.

By Bismut's integration by parts formula we can identify the stochastic integral $\int_0^1 Z_\tau \, dp(\tau)$ with the divergence, modulo the Ricci correction term.

9.6 Elliptic Estimates and Riesz Transforms on $P_{m_0}(M)$

In its essence the Littlewood-Paley-Calderón-Stein area integral approach outlined in §9.3 can be worked out in the nonlinear setting [10]. Technically, though, some major gaps have to be filled.

We do not have a representation of the type of Mehler formula for the semigroup and the simple commutators of the Gaussian case are no longer available. The Ornstein-Uhlenbeck operator $\mathbf{L}f = -\delta \mathbf{D}f$ has an explicit expression which has recently been computed by Kazumi.

We take the point of view of Airault-Malliavin [1] that consists in evaluation the intertwining of the differential with the Laplace-Beltrami operator operating on functions, making in this way the Laplace-Beltrami operator on forms, that is, the de Rham–Hodge operator, appear. We denote by Δ the de Rham–Hodge operator on differential forms ω_z of degree one and, given the underlying Hilbert structure of the restricted tangent space, we identify the form with the restricted tangent process z. Then we write

$$\Delta = \delta d + d\delta$$

with $\langle df, z \rangle = \mathbf{D}_z f$.

The semigroup $e^{-t\mathbf{L}}$ on the path space satisfies

$$\frac{\partial}{\partial t}(de^{-t\mathbf{L}}f) = d\delta d e^{-t\mathbf{L}} f = \Delta(de^{-t\mathbf{L}}f)$$

since $ddu = 0$. The problem of estimating the commutator between $\mathbf{D}e^{-t\mathbf{L}}$ and $e^{-t\mathbf{L}}\mathbf{D}$ reduces therefore to estimating the difference between the operators $\mathbf{L}$ and Δ on differential forms, for which one has to obtain explicit expressions. A Weitzenböck formula for 1-differential forms is then derived in the spirit of the results of [9], where such type of theorem was shown with respect to a particular Markovian covariant derivative on the path space.

Theorem 9.13 (Cruzeiro-Malliavin [10]). *The difference between the Ornstein-Uhlenbeck and the de Rham–Hodge operator defined for 1-differential forms on the path space is of type*

$$\Delta(\omega_z) - \mathbf{L}(\omega_z) = A(z) + B(\mathbf{D}z),$$

where A and B are operators with kernel representations expressed in terms of stochastic integrals.

The stochastic integrals appearing in this Weitzenböck formula come from the structure equations on one side and, on the other, from the integration by parts formula. Then a commutation formula allows us to derive stochastic integrals. Since one integrates geometric bounded quantities, the final estimates follow from Itô's stochastic calculus.

References

[1] H. Airault and P. Malliavin, *Semi-martingales with values in a Euclidean vector bundle and Ocone's formula on a Riemannian manifold*, Proc. Cornell Conf. 1993, Amer. Math. Soc., Providence, 1995.

[2] A. Benedeck, A. Calderón, and R. Panzone, *Banach-valued convolution operators*, Proc. Nat. Acad. 48 (1962), 356–365.

[3] J. M. Bismut, *Large deviations and Malliavin calculus*, Birkhauser, Boston, 1984, 216 pages.

[4] A. Calderón, *A theorem of Marcinkiewicz and Zygmund*, Trans. Amer. Math. Soc. 68 (1950), 54–61.

[5] ———, *Uniqueness in the Cauchy problem*, Amer. J. Mathematics 80 (1958), 16–36.

[6] ———, *Commutators, singular integrals on Lipschitz curves, applications*, ICM Helsinki vol. 1, Helsinki (1978), 84–96.

[7] A. Calderón and A. Zygmund, *On singular integrals*, Acta Mathematica 88 (1952), 85–139.

[8] A. B. Cruzeiro and S. Fang, *Une inégalité L^2 pour les intégrales stochastiques anticipatives sur une variété riemannienne*. C. R. Acad. Sci. Paris 321 (1995), 1245–1250.

[9] A. B. Cruzeiro and P. Malliavin, *Renormalized differential geometry on path space: Structural equation, curvature*, J. Funct. Analysis 139 (1996), 119–181.

[10] ———, *Commutators on path spaces and parabolic, elliptic estimates*, in preparation.

[11] B. Driver, *A Cameron-Martin type quasi-invariance theorem for the Brownian motion on a compact manifold*, J. Funct. Analysis 109 (1962), 272–376.

[12] S. Fang and P. Malliavin, *Stochastic analysis on the path space of a Riemannian manifold*, J. Funct. Analysis 118 (1993), 249–274.

[13] B. Gaveau and P. Trauber, *L'intégrale stochastique comme opérateur divergence*, J. Funct. Analysis 38 (192), 230–238.

[14] K. Itô, *The Brownian motion and tensor fields on Riemannian manifolds*, Proc. Int. Cong. Math. Stockholm, Stockholm (1962), 536–539.

[15] P. Malliavin, *Formule de la moyenne, calcul de perturbations et théorème d'annulation pour les formes harmoniques*, J. Funct. Anal. 17 (1974), 274–291.

[16] ———, *Stochastic Analysis*, Springer, Berlin, 1996, 370 pages.

[17] P. Malliavin and D. Nualart, *Quasi-sure analysis of stochastic flows and Banach space valued smooth functionals*, J. Funct. Analysis 112 (1993), 429–457.

[18] P. A. Meyer, *Transformations de Riesz pour les lois gaussiennes*, Sém. Proba. XVII, Springer Lecture Notes 1059 (1984), 179–193.

[19] G. Pisier, *Riesz transform: A simpler proof of P. A. Meyer inequalities*, Sém. Proba. XXII, Springer Lecture Notes in Math. 1321 (1988), 485–501.

[20] E. Stein, *Abstract Littlewood-Paley theory and semi-group*, Princeton Univ. Press, Princeton, 1970.

[21] D. Stroock, *The Malliavin calculus: A functional analytic approach*, J. Funct. Analysis 44 (1981), 212–257.

[22] S. Watanabe, *Lectures on stochastic differential equations and Malliavin calculus*, Tata Inst. Lect. on Math. and Physics 73 (1984), 110 pages.

Chapter 10

An Application of a Formula of Alberto Calderón to Speaker Identification

Ingrid Daubechies and Stéphane Maes

It was an honor as well as a great pleasure for me to be asked to contribute to Alberto Calderón's seventy-fifth birthday conference. More than at any other similar conference, one could feel in the air the friendship, the sense of community among the participants representing the many different branches of analysis with whom Calderón himself is associated. It was wonderful to have been invited to be a part of this celebration.

As the subject of my presentation at a meeting dedicated to Alberto Calderón's wide ranging interests, I chose a topic that may have reminded him of the interest in engineering of his youth. Although most of my work is in mathematics, I make occasional excursions into engineering, assisted by students or other collaborators who are the real engineers in the project. The paper below reports on such an excursion. There was an additional reason to dedicate this presentation to Alberto Calderón: the whole project had been directly motivated by a different reading of a classical integral formula due to him (see formulas (10.2, 10.3) in the paper).

— Ingrid Daubechies

Research by Ingrid Daubechies partially supported by NSF grant DMS-9401785. The bulk of this paper appeared under the title *A Nonlinear Squeezing of the Continuous Wavelet Transform Based on Auditory Nerve Models*, 527–546, in Wavelets in Medicine and Biology (A. Aldroubi and M. Unser, editors), CRC Press, 1996. This material was reprinted by permission from CRC Press, Boca Raton, Florida.

10.1   Introduction

The project concerned the very concrete problem of speaker identification. This usually concerns a situation where speech fragments of a number of speakers have been stored; when a new speech fragment is presented, the speaker identification system should be able to recognize with a reasonably high degree of accuracy whether or not this is one of the previously sampled speakers, and, if so, who it is. Ideally, this should work even if the specific utterance in the piece of speech under scrutiny is different from any that were encountered before. The problem is thus to identify, and later to detect, reliable parameters that characterize the speaker independently of the utterance. There exist various approaches that perform very well on "clean" speech, that is, when both the previously stored samples and the speech fragment for which the speaker has to be identified have very low noise levels. Most models break down at noise levels far below those where our own auditory recognition system starts to fail. Because of the connection of the wavelet transform with the auditory system, and because there existed other indications that an auditory-system-based approach might be more robust than existing methods, we decided to construct a wavelet-based approach to this problem.

This paper is organized as follows. Sections 10.2 to 10.4 present background material, explaining respectively (1) how the (continuous) wavelet transform, which is essentially the same as a decomposition formula proposed by A. Calderón in the early sixties (see (10.2) below), comes up "naturally" in our auditory system, (2) a heuristic approach (the ensemble interval histogram of O. Ghitza [1]) based on auditory nerve models, which eliminates much of the redundancy in the first-stage transform, and (3) the modulation model, valid for large portions of (voiced) speech, and which is used for speaker identification. (Note that our descriptions of the auditory system are very naive and distorted. They are in no way meant as an accurate description of what is well known to be a very complex system. Rather, they are snapshots that motivated our mathematical construction further on, and they should be taken only as such.) In §10.5 we put all this background material to use in our own synthesis, an approach that we call "squeezing" the wavelet transform; with an extra refinement this becomes "synchrosqueezing." The main idea is that the wavelet transform itself has "smeared" out different harmonic components, and that we need to "refocus" the resulting time-frequency or time-scale picture. How this is done is explained in §10.5. Section 10.6 sketches a few implementation issues. Finally, §10.7 shows some results: the "untreated" wavelet transform of a speech segment, its squeezed and synchrosqueezed versions, and the extraction of the parameters used for speaker identification. We conclude with some pointers to and comparisons with similar work in the literature, and with sketching possible future directions.

10.2 The Wavelet Transform as an Approach to Cochlear Filtering

When a sound wave hits our eardrum, the oscillations are transmitted to the basilar membrane in the cochlea. The cochlea is rolled up like a spiral; imagine unrolling it (and with it the basilar membrane) and putting an axis y onto it, so that points on the basilar membrane are labeled by their distance to one end. (For simplicity, we use a one-dimensional model, neglecting any influence of the transverse direction on the membrane, or its thickness.) If a pure tone; i.e., an excitation of the form $e^{i\omega t}$ (or its real part) hits the eardrum, then the response at the level of the basilar membrane, as observed experimentally or computed via detailed models, is in first approximation given by $e^{i\omega t}F_\omega(y)$—a temporal oscillation with the same frequency as the input, but with an amplitude localized within a specific region in y by the hump-shaped function $F_\omega(y)$. In a first approximation, the dependence of F_ω on ω can be modeled by a logarithmic shift: $F_\omega(y) = F(y - \log \omega)$. (Strictly speaking, this model is only good for frequencies above say, 500 Hz; for low frequencies, the dependence of F_ω on ω is approximately linear.) The response to a more complicated $f(t)$ can then be computed as follows:

$$f(t) = \frac{1}{2\pi}\int_{-\infty}^{\infty} \hat{f}(\omega)e^{i\omega t}\, dw$$
$$\Rightarrow \quad \text{response} \quad B(t,y) = \frac{1}{2\pi}\int_{-\infty}^{\infty} \hat{f}(\omega)e^{i\omega t}F(y - \log \omega)\, dw\ . \tag{10.1}$$

(Note that we are assuming linearity here—a superposition of inputs leading to the same superposition of the respective responses. This is again only a first approximation; richer and more realistic auditory models contain significant nonlinearities [5].) This can be rewritten as a continuous wavelet transform. Let us first recall the definition. For a fixed choice of the "wavelet" ψ, a real function that is reasonably localized in "time" and "frequency" (say $|\psi(x)| \le C(1 + |x|)^{-2}$ and $|\hat{\psi}(\xi)| \le C(1 + |\xi|)^{-2}$) and that has mean zero, $\int \psi(x) = 0$, we define the wavelet transform $W_\psi f$ of a function f by

$$W_\psi f(b,a) = \int_{-\infty}^{\infty} f(x)\frac{1}{\sqrt{a}}\psi\left(\frac{x - b}{a}\right)\, dx, \tag{10.2}$$

where the scale parameter a ranges over $\mathbb{R}_+^*$, and the time or space localization parameter b over all of $\mathbb{R}$. The original function f can then be reconstructed from its wavelet transform $W_\psi f$ by

$$f(x) = C_\psi \int_0^{\infty} \int_{-\infty}^{\infty} W_\psi f(b,a)\frac{1}{\sqrt{a}}\psi\left(\frac{x - b}{a}\right), \tag{10.3}$$

where C_ψ is a constant depending on ψ.

If we relabel in (10.1) the axis along the basilar membrane by defining $y := -\log a$ with $a > 0$ and $B'(t, a) = B(t, -\log a)$, and if we moreover define a function G by putting $F(x) =: \hat{G}(e^{-x})$, then the response can be rewritten as

$$B'(t,a) = \frac{1}{2\pi} \int_{-\infty}^{\infty} \int_{-\infty}^{\infty} f(t')e^{i\omega(t-t')}\hat{G}(a\omega) \, dt' \, dw$$

$$= \int_{-\infty}^{\infty} f(t')\frac{1}{a}G\left(\frac{t-t'}{a}\right) \, dt'.$$

(10.4)

By taking $\psi(t) := G(-t)$, we find that $B'(t,a) = |a|^{-\frac{1}{2}}(W_\psi f)(a,t)$, where W_ψ is the continuous wavelet transform as defined above. In this sense, the cochlea can be seen as a "natural" wavelet transformer; all this is of course a direct consequence (and nothing but a reformulation) of the logarithmic dependence on ω of F_ω.

10.3   A Model for the Information Compression after the Cochlear Filters

The cochlear filtering, or the continuous wavelet transform that approximates it, transforms the one-dimensional signal $f(t)$ into a two-dimensional quantity. If we were to sample this two-dimensional transform like an image, then we would end up with an enormous number of data, far more than can in fact be handled by the auditory nerve. Some compression therefore has to take place immediately. The ensemble interval histogram (EIH) method of Oded Ghitza [1] gives such a compression, inspired by auditory nerve models. We describe it here in a nutshell, with its motivation.

Near the basilar membrane, and over its whole length, one finds series of bristles of different stiffness. As the membrane moves near a particular bristle, it can, if the displacement is sufficiently large, "bend" the bristle. For different degrees of stiffness, this happens for different thresholds of displacement. Every time a bristle is bent, we think of this as an "event"; we also imagine that events only count when the bristle is bent away from its equilibrium position, not when it moves back. Figure 10.1 gives a schematic representation of what this means. The curve represents the movement of the membrane, as a function of time, at one particular location y.

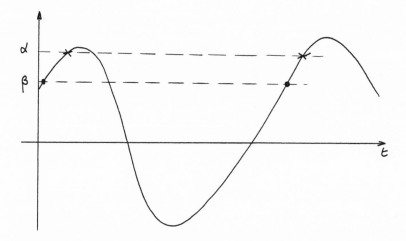

Figure 10.1: Displacement of the basilar membrane, at one fixed point y, as a function of time. The horizontal lines α and β represent the thresholds for bristles of different stiffness.

The two horizontal lines, labeled by α and β, represent two different bristle thresholds, and the dots and crosses mark the corresponding "events" in the timespan represented in the figure. Replacing the information contained in all the curves (for different y) by only the coordinates (level, time, location) of these events would already imply a sizeable compression. The EIH model reduces the information even more, by another transformation. Start by setting a certain resolution level ΔT, and a "window width" t_0. Then, for a given t, look back in time and count within the interval $[t - t_0, t]$, the number $N_{\alpha,y}(T)$ of successive events (for the bristle at position y and with stiffness α) that were spaced apart by an interval between T and $T + \Delta T$. Next, compute $S(t, T)$, the sum over all α and y of these $N_{\alpha,y}(T)$. This new representation $S(t, T)$ of the original signal is still two-dimensional, like the original cochlear or wavelet filtering output; it is, however, often sampled more coarsely than the continuous wavelet transform. More important, from our point of view, than the compression that this represents, is the very non-linear and adaptive transformation represented by $S(t, T)$, which can again be viewed as a time-frequency representation (a second look at the construction of $S(t, T)$ shows that T^{-1} plays the role of an instantaneous frequency). O. Ghitza [1] compared the performance of EIH-based tools for several types of discrimination tests (such as word spotting) with the results obtained from LPC (Linear Predictive Coding, a hidden Markov model for speech); for clean speech, LPC performed better, but the EIH-based schemes were, like the human auditory system itself, much more robust when the noise level was raised, and provided still useful results at noise levels where LPC could no longer be trusted. The nonlinear squeezing of the continuous wavelet transform that we describe in §10.5 is inspired by the EIH-construction.

10.4   The Modulation Model for Speech

The modulation model represents speech signals as a linear combination of amplitude and phase modulated components,

$$f(t) = \sum_{k=1}^{K} A_k(t) \cos[\theta_k(t)] + \eta(t) ,$$

where $A_k(t)$ is the instantaneous amplitude and $\omega_k(t) = \frac{d}{dt}\theta_k(t)$ the instantaneous frequency of component (or formant) k; $\eta(t)$ takes into account the errors of modeling ([6], [7]). In a slightly more sophisticated model, the components are viewed as "ribbons" in the time-frequency plane rather than "curves," and one also associates instantaneous bandwidths $\Delta\omega_k(t)$ to each component. The parameters $A_k(t)$, $\omega_k(t)$, and $\Delta\omega_k(t)$ are all assumed to vary in time (as the notation indicates), but we assume that this variation is slow when compared with the oscillation time of each component, measured by $[\omega_k(t)]^{-1}$. For large parts of speech, the modulation model is very satisfactory, and one can take $\eta(t) \simeq 0$; for other parts (e.g., fricative sounds) it is completely inadequate. The parameters $A_k(t)$, $\omega_k(t)$, and $\Delta\omega_k(t)$ (for those portions of speech where they are meaningful) can be used for speaker recognition. The basic idea is as follows. Imagine that the speech signal can be well represented by, say, $K = 8$ components. For each component, we have 3 parameters that vary in time. The signal can thus be viewed as a path in an $8 \times 3 = 24$-dimensional space. This path depends of course on both the speaker and the utterance. During certain portions (such as within one vowel), the 24 parameters remain in the same neighborhood, after which they make a rapid transition to another neighborhood, where they then dwell for a while, and so on. The order in which these "islands" appear depends on the utterance, but their location in our 24-dimensional space is believed to be independent of the utterance and can be used to characterize the speaker. To use this for a speaker identification project, one must thus do two things: (1) extract the $A_k(t)$, $\omega_k(t)$, $\Delta\omega_k(t)$ (or a subset of these parameters) from the speech signal, and (2) process this information in a classification scheme in order to identify the speaker. When LPC methods are used for this purpose ([8], [9], [10]), one determines in fact only the $\omega_k(t)$ and $\Delta\omega_k(t)$, not the amplitudes $A_k(t)$. They are incorporated into one complex number,

$$z_k(t) = e^{i[\omega_k(t)+i\Delta\omega_k(t)]} ;$$

the $z_k(t)$ are the poles of the vocal tract transfer function

$$\mathcal{H}(z,t) = \sum_{k=1}^{K} \frac{1}{1 - z/z_k(t)} \ .$$

It is not always straightforward to label the $z_k(t)$ correctly with the LPC method, i.e., to decide which of the poles, determined separately, belongs to which component. To circumvent this, one works not with the $z_k(t)$ themselves, but with the so-called LPC-derived cepstrum,

$$c_n(t) = \frac{1}{n} \sum_{k=1}^{K} [z_k(t)]^n \ ,$$

for which the exact attribution of the $z_k(t)$ does not matter; this formula is due to Schroeder [14]. This speaker identification program was developed at CAIP (Center for Aids to Industrial Productivity) at Rutgers University, by K. Assaleh, R. Mammone, and J. Flanagan, [8], [9], [10]. Once the cepstrum is extracted, they use a neural network to do the classification and identification part. They fine-tuned it until it performed so well that it could perfectly distinguish identical twins, when starting from clean speech signals, thus outperforming most humans!

10.5   Squeezing the Continuous Wavelet Transform

Our goal is to use the continuous wavelet transform to extract reliably the different components of the modulation model (when it is applicable) and the parameters characterizing them. Our first problem is that the wavelet transform gives a somewhat "blurred" time-frequency picture. Let us take, for instance, a purely harmonic signal,

$$f(t) = A \cos \Omega t \ .$$

We compute its continuous wavelet transform $(W_\psi f)(a,b)$, using a wavelet ψ that is concentrated on the positive frequency axis (i.e., support $(\hat{\psi}) \subset [0,\infty)$, or $\hat{\psi}(\xi) = 0$ for $\xi < 0$; note that this means that ψ is complex):

$$
\begin{aligned}
(W_\psi f)(a,b) &= \int f(t) \tfrac{1}{\sqrt{a}} \overline{\psi\left(\tfrac{t-b}{a}\right)} \, dt \\
&= \tfrac{1}{2\pi} \int \hat{f}(\xi) \sqrt{a} \, \overline{\hat{\psi}(a\xi)} \, e^{ib\xi} \, d\xi \\
&= \tfrac{1}{2\pi} \int \tfrac{A}{2}[\delta(\xi - \Omega) + \delta(\xi + \Omega)] \sqrt{a} \, \overline{\hat{\psi}(a\xi)} \, e^{ib\xi} \, d\xi \\
&= \tfrac{A}{4\pi} \sqrt{a} \, \overline{\hat{\psi}(a\Omega)} \, e^{ib\Omega} \ .
\end{aligned}
\tag{10.5}
$$

If $\hat{\psi}(\xi)$ is concentrated around $\xi = 1$, then $(W_\psi f)(a, b)$ will be concentrated around $a = \Omega^{-1}$, as expected. But it will be spread out over a region around this value (see figure 10.2), and not give a sharp picture of what was a signal very sharply localized in frequency.

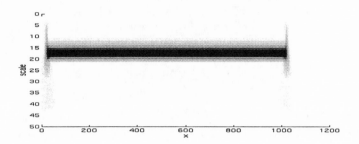

Figure 10.2: Absolute value $|W_\psi f(a, b)|$ of the wavelet transform of a pure tone f.

In order to remedy this blurring, the "Marseilles group" developed the so-called "ridge and skeleton" method [11]. In this method, special curves (the ridges) are singled out in the (a, b)-plane, depending on the wavelet transform $(W_\psi f)(a, b)$ itself (for each b, one finds the values of a where the oscillatory integrand in $(W_\psi f)(a, b)$ has "stationary phase"; for the signals considered here, this amounts to $\partial_b[$ phase of $(W_\psi f)(a, b)] = \frac{\omega_0}{a}$, where ω_0 is the center frequency for ψ). From the restriction of $W_\psi f$ to these ridges (the "skeleton" of the wavelet transform), one can then read off the important parameters, such as the instantaneous frequency. This method has been used with great success for various applications, such as reliably identifying and extracting spectral lines of widely different strengths [11]. In our speech signals, we have many components, some of which can remain very close for a while, to separate later again; components can also die or new components can suddenly appear out of nowhere. For these signals, the ridge and skeleton method does not perform as well. For this reason, we developed a different approach, where we try to squeeze back the defocused information in order to gain a sharper picture; in so doing, we try to use the whole wavelet transform instead of concentrating on special curves.

Let us look back at the wavelet transform (10.5) of a pure tone. Although it is spread out over a region in the a-variable around $a = \Omega^{-1}$, the b-dependence still shows the original harmonic oscillations with the correct frequency, regardless of the value of a. This suggests that we compute, for any (a, b), the instantaneous frequency $\omega(a, b)$ by

$$\omega(a, b) = -i[W_\psi f(a, b)]^{-1} \frac{\partial}{\partial b} W_\psi f(a, b),$$

and that we transfer the information from the (a, b)-plane to a (b, ω)-plane, by taking for instance,

$$S_\psi f(b, \omega_\ell) = \sum_{a_k \text{ such that } |\omega(a_k, b) - \omega_\ell| \leq \Delta\omega/2} |W_\psi f(a_k, b)|. \qquad (10.6)$$

We have assumed here that both the old a-variable and the new ω-variable have been discretized. (A continuous formulation would be to introduce, for every b, a measure $d\mu_b$ in the ω-variable, which assigns to Borel sets A the measure

$$\mu_b(A) = \int |W_\psi f(a, b)| \; \chi_A(\omega(a, b)) \, da,$$

where χ_A is the indicator function of A, $\chi_A(u) = 1$ if $u \in A$, $\chi_A(u) = 0$ if $u \notin A$.) This has *exactly* the same flavor as the EIH transform described in §10.3: we transform to a different time-frequency plane by reassigning contributions with the same instantaneous frequency to the same bin, and we give a larger weight to components with large amplitude $|W_\psi f|$ (just like components with large amplitude in the EIH would give rise to several level crossings, and would therefore contribute more). Our S_ψ is also close to the SBS (in-Synchrony Bands Spectrum, a precursor of the EIH) [12] or to the IFD (Instantaneous Frequency Distribution) [13]. For good measure, one can also sum the $|a_k|^{-\alpha} |W_\psi f(a_k, b)|$ rather than the $|W_\psi f(a_k, b)|$, thus renormalizing the fine-scale regions where often $|W_\psi f(a, b)|$ is much smaller.

When this squeezing operation is performed on the wavelet transform of a pure tone, we find a single horizontal line in the (b, ω)-plane, at $\omega = \Omega$, as expected.

We can, however, refine the operation even further, and define a particular type of squeezing, which we call *synchrosqueezing*, that still allows for reconstruction, even after the (highly nonlinear!) transformation. To see this, we first have to observe that the reconstruction formula of f from $W_\psi f$, given by formula (10.3), is not the only one. We also have, again assuming support $\hat{\psi} \subset [0, \infty)$,

$$
\begin{aligned}
\int_0^\infty W_\psi f(a, b) a^{-3/2} \, da &= \int \int \hat{f}(\xi) e^{ib\xi} \, \overline{\hat{\psi}(a\xi)} \, a^{-1} \, da \, d\xi \\
&= \left[\int_0^\infty \hat{\psi}(\xi) \frac{d\xi}{\xi} \right] \cdot \int \hat{f}(\xi) e^{ib\xi} \, d\xi \qquad (10.7) \\
&= \left[2\pi \int_0^\infty \hat{\psi}(\xi) \frac{d\xi}{\xi} \right] f(b).
\end{aligned}
$$

This suggests that we define

$$(\mathcal{S}_\psi f)(b, \omega_\ell) = \sum_{a_k \text{ such that } |\omega(a_k, b) - \omega_\ell| \leq \Delta\omega/2} W_\psi f(a_k, b) a_k^{-3/2} \qquad (10.8)$$

(without absolute values!); with ω_ℓ spaced apart by $\Delta\omega$, we then still have (in the assumption that the discretizations are sufficiently fine to be good approximations to integrals)

$$\sum_\ell (\mathcal{S}_\psi f)(\omega_\ell, b) = C_\psi^\# f(b) . \qquad (10.9)$$

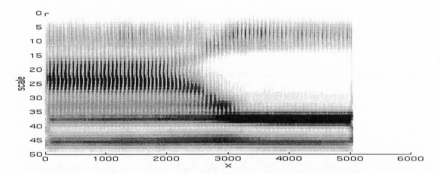

Figure 10.3: Absolute value $|W_\psi f(a, b)|$ of the sound /a–a–i–i/. A colored noise is present with SNR = 15 dB. The horizontal axis is sampled at 8 kHz. The vertical axis represents different subbands (5 octaves, split into 8 equally spaced sub-octaves); low indices are associated to high frequencies.

Having the exact reconstruction (10.9) will be useful to us later on (see the end of this section); note that such an exact reconstruction is not available for the EIH. There is an added bonus to *synchro*squeezing. The process of reassigning components from the (a, b)-plane to the (b, ω)-plane is not perfect, especially when noise is present, and occasionally parts of components that are truly different get assigned to the same ω_ℓ-bin. When this happens, the two pieces from different components are often out of phase with each other, and cancellation takes place in the computation of $\mathcal{S}_\psi$ (but not in S_ψ!). Figures 10.3 and 10.4 show the unprocessed wavelet transform and the synchrosqueezed wavelet transform, respectively, of the speech signal consisting of the two vowels /a–a–i–i/; clearly, the different components can be distinguished much more clearly after the (synchro)squeezing. The extra focusing of the synchrosqueezing over squeezing can be seen in an example in §10.7.

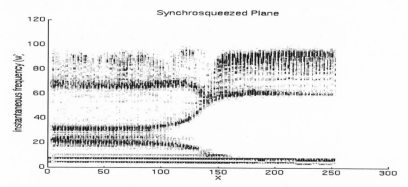

Figure 10.4: Synchrosqueezed representation of /a-a–i-i/ (same signal, same noise level as in figure 10.3). The components can be distinguished much more clearly than in figure 10.3. (Note that because the scale a corresponds to ω^{-1}, there is also a distortion of the vertical axis when compared to figure 10.3.)

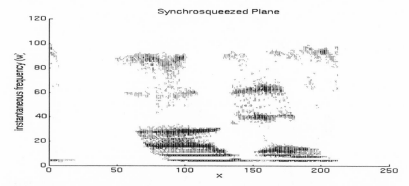

Figure 10.5: Synchrosqueezed representation of /ow-g-λ-s-t/. A colored noise is present with SNR of 15 dB. The "s" part is the cloud in the upper right corner.

One remark is in order here. Both the squeezing and synchrosqueezing operations can be defined with any arbitrary reassigning rule—it does not have to be governed by the instantaneous frequency. In particular, the reconstruction property from $\mathcal{S}_\psi f$ does not depend on the physical interpretation of the reassignment rule. This means that we should not worry about the parts of f where the modulation model does not apply—true, the reassignment will not be as meaningful, because instantaneous frequency does not make much sense there, but we still have not "hurt" the information that was there. In fact, as the synchrosqueezed representation of "august" in figure 10.5 shows, the "s" part is still nicely localized in the upper frequencies, where it belongs, so in practice we do not seem to displace such nonmodulated parts in the time-frequency plane. Of course, the refocusing that we see in the squeezed and synchrosqueezed transform *does* depend on the physical interpretation—an arbitrary reassignment rule would give a messy picture.

After synchrosqueezing, the components are well separated and can be identified. From the synchrosqueezed representation, we can determine the central frequencies $\omega_k(t)$ and the bandwidths $\Delta\omega_k(t)$. How can we find the $A_k(t)$? Remember our exact reconstruction formula (10.9)! If a postprocessing step separates the different components in the synchrosqueezed plane, then we can carve out the component under consideration in the synchrosqueezed plane, delete all the rest, and reconstruct from only this component; this is called the Selective Fusion Algorithm [2]. The direct summation method (10.9) provides fast and relatively accurate results; a slightly slower but even more accurate method uses double integrals (see [2]). This is carried out for speech signals, within the modulation model framework, in [2]. From every reconstructed single component, we can then determine $A_k(t), \theta_k(0)$ so that $A_k(t)\cos(\theta_k(t))$ fits this reconstructed component, within the constraint $\frac{d}{dt}\theta_k(t) = \omega_k(t)$.

This finishes our program of extracting the modulation model parameters from an EIH analog based on the wavelet transform. After a (very summary) discussion of some implementation issues, we shall return to results in §10.7.

10.6   Short Discussion of Some Implementation Issues

First of all, the whole construction is based on a continuous wavelet transform. In practice, this is of course a discrete but very redundant transform, heavily oversampled both in time and in scale. In order to be practical, we need a fast implementation scheme. This was achieved by borrowing a leaf from (nonredundant) wavelet bases, i.e., by using subband filtering schemes. For a given profile $\hat{\psi}(\xi)$ (close to that of a Morlet wavelet), we identified a function $\hat{\phi}$ and trigonometric polynomials $\hat{h}, \hat{g}_\ell$, with $\ell = 1, \dots, L$, so that

$$\hat{\psi}(2^{(\ell-1)/L}\omega) \simeq \hat{g}_\ell(\omega)\hat{\phi}(\omega),$$

$$\hat{\phi}(2\omega) \simeq \hat{h}(\omega)\hat{\phi}(\omega) .$$

This means that the Fourier coefficients of $\hat{h}, \hat{g}_\ell$ can be used for an iterated FIR filtering scheme that gives the redundant wavelet transform in linear time. For details on the algorithm and on the construction of the filters, see [2], [4].

Next we note that the squeezing and synchrosqueezing operations entailed first the determination of the instantaneous frequency $\omega(a, b)$. This was done by a logarithmic differentiation of $W_\psi f(a, b)$. This is, of course, very unstable when $|W_\psi f(a, b)|$ is small; note, however, that these regions will contribute very little to either $S_\psi f$ or $\mathcal{S}_\psi f$ (defined by (10.6) and (10.8), respectively), so that we can safely avoid this problem by putting a lower threshold on $|W_\psi f(a, b)|$. On the other hand, differentiation itself is also a tricky business when the data are noisy; in practice, a standard numerical difference operator was used, involving a weighted differencing operator, spread out over a neighborhood of samples. Again, details can be found in [2]. Alternately, one can also obtain $\omega(a, b)$ by computing the ratio of $|W_{\psi'} f(a, b)|$ and $|W_\psi f(a, b)|$; in a discretized setting, this amounts to a particular weighted differentiation, adapted to ψ.

In the previous section, we glossed over the extraction of the $\omega_k(t)$, $\Delta\omega_k(t)$ from the synchrosqueezed picture. In fact, although we can often clearly see the different components with our eyes, extracting them and their parameters automatically is a different matter. For instance, in "How are you?", an example shown in §10.7, the components are much weaker in some spots than in others, yet we want our "extractor" to bridge those weak gaps. The approach we use, suggested by Trevor Hastie, is to view $|\mathcal{S}_\psi f(b, \omega)|$ as a probability distribution in ω, for every value of b, which can be modeled as a mixture of Gaussians, and which evolves as b changes; moreover, we impose that the centers of the Gaussians follow paths given by splines (cubic or linear). We also allow components to die or to be born. In order to find an evolution law that fits the given $|\mathcal{S}_\psi f(b, \omega)|$, a few steps of an iterative scheme suffice; for details, see [2], [3]. The resulting centers of the Gaussians in the mixture give us the frequencies $\omega_k(t)$; their widths give us the $\Delta\omega_k(t)$.

10.7   Results on Speech Signals

We start by illustrating the enhanced focusing of the synchrosqueezed representation when compared to the squeezed representation of a different example, namely the utterance, "How are you?" or /h-δ-w-a-r-j-u?/; see figures 10.6 and 10.7. Figure 10.8 shows the curves for the corresponding extracted central frequencies $\omega_k(t)$. In this case, the original signal was somewhat noisy; the (pink) noise had an SNR of about 15 dB.

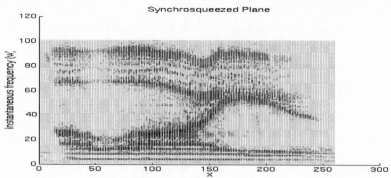

Figure 10.6: Squeezed plane representation for /h-δ-w-a-r-j-u?/. A colored noise is present with SNR = 15 dB.

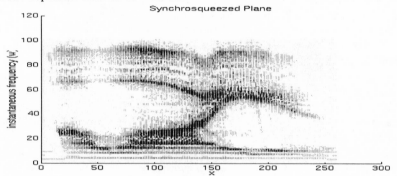

Figure 10.7: Synchrosqueezed plane representation for /h-δ-w-a-r-j-u?/. A colored noise is present with SNR = 15 dB.

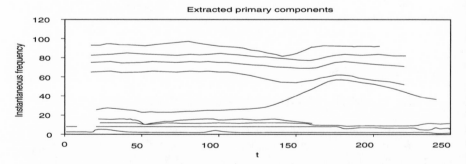

Figure 10.8: Curves for the central frequencies $\omega_k(t)$ for /h-δ-w-a-r-j-u?/. A colored noise is present with SNR = 15 dB.

Next, we illustrate the robustness of our analysis under higher noise levels. We return to the signal /a-a-i-i/, this time with an additional white noise with SNR of 11 dB. Figure 10.9 shows the synchrosqueezed representation of this noisier signal; although the representation is noisier as well, the different components can still be identified clearly, and they have not moved. This is borne out by a comparison of the extracted central frequency curves. Figure 10.10 shows the extracted frequency curves for the slightly noisy original of figure 10.4.

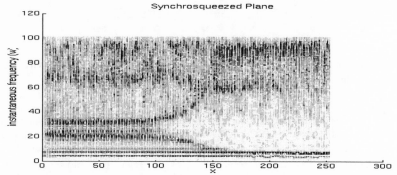

Figure 10.9: Synchrosqueezed plane representation for /····-a-a–i-i-···/. A colored noise is present with SNR = 15 dB. An additional white noise is added with SNR = 11dB.

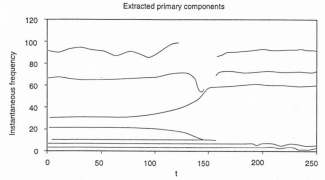

Figure 10.10: Curves for the central frequencies $\omega_k(t)$ for /a-a–i-i/, extracted from Figure 4.

Figure 10.11 shows the extracted frequency curves for the much noisier version given in figure 10.9.

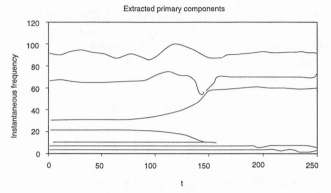

Figure 10.11: Curves for the central frequencies $\omega_k(t)$ for /a-a–i-i/ with additional white noise; see figure 10.9.

Finally, we also show results of a first test of the use of the synchrosqueezed representation for speaker identification. For this first test, we did not use the full strength of the representation, and we did not develop our own classification either. Instead, we took our $\omega_k(t), \Delta\omega_k(t)$ values, and constructed an analog to the LPC-derived cepstrum by defining

$$z_k^w(t) = \exp[i\omega_k(t) - \Delta\omega_k(t)]$$

$$c_n^w(t) = \frac{1}{n}\sum_{k=1}^{K}[z_k^w(t)]^n \; ;$$

we called this the "wastrum." We then used the wastrum as input for the classification scheme that had been developed at CAIP. For the experiment we performed, the input data come from the narrowband part of the *KING* database, released by ITT Aerospace/Communications Division, in April 1992. It is a telephone network database built with 52 American speakers, among whom the first 26 speakers are from the San Diego region. For each speaker, ten sessions have been recorded. The first five sessions were recorded at intervals of one week. Each session is narrowband, with the bandwidth of a telephone channel. Each session consists of roughly 50 to 75 seconds of conversational speech which contains roughly 40% of silences. The sessions are recorded from the interlocutor's side. The first five sessions are within the Great Divide, which means on the West Coast. The SNR is about 15 dB to 20 dB. This noise is introduced by the phone network. The five remaining sessions are recorded across the Great Divide at intervals of one month, and they are much noisier. These last sessions were not used in this experiment. The signal is sampled at 8 kHz and quantized over 12 bits.

For the experiment, the first session of the first 26 speakers is used for training and the following four within divide sessions are used for testing.

The classifier is a vector quantizer. Decisions are made on the basis of the cumulated distances obtained in each frame relatively to the codebooks associated to the different speakers.

Table 10.1 summarizes the results in closed-set speaker identification obtained with the LPC-derived cepstrum and the wastrum. The long-term mean is removed from the features, in agreement with [8]. The silence frames are removed on the basis of energy thresholds for the primary components. The same frames are removed for the LPC approach, in order to compare exactly the same utterances.

Method	Additional SNR	Error Rate
LPC-derived cepstrum	none	~ 0.22
wastrum	none	0.23
LPC-derived cepstrum	15 dB	0.33
wastrum	12 dB	0.30

Table 10.1: Summary of the results obtained on KING database, within the Great Divide, 26 speakers, first section used for training, four other sessions used for testing. Long-term mean removal is used.

The performances of the wastrum are comparable to the LPC-derived cepstrum for the relatively clean speech, which is reassuring: we aim to extract the same cepstral-like information, albeit with very different methods, and so we expect similar performance! The wastrum-method is, however, more robust to noise when the noise cannot be considered as negligible, since we get a lower error rate even though the noise level is significantly higher (12 dB versus 15 dB).

Note that we are comparing here a suboptimal version of our approach (the $A_k(t)$ are not taken into account, and the $\omega_k(t), \Delta\omega_k(t)$ are transformed into the wastrum, which is then put through a classification scheme not specially tailored to our different approach) with a very much optimized version of the LPC-based method. Yet even so, the wastrum method leads to fewer errors for noisy speech than the LPC-derived cepstrum. This indicates that we have indeed inherited (some of) the robustness that characterizes true auditory systems.

The following is a short list of promising future directions to be explored: include the amplitude information $A_k(t)$ (obtained by Selective Fusion [3]) as well; develop a more direct classification scheme, without the detour of the wastrum, and maybe even directly from the synchrosqueezed plane, without extraction of the parameters first; and finally, use of this approach for other tasks in speech analysis.

There is some similarity between our squeezing and synchrosqueezing methods and a technique of "reassignment" developed by Auger and Flandrin [15], with the same goal of "refocusing" in the time-frequency plane; we first heard of their method after the work described here was completed. Auger and Flandrin typically work with Wigner-Ville or similar time-frequency distributions, and their reassignment method is not limited to one direction only (we don't change the b variable in our scheme); on the other hand, their scheme is not linked to an exact reconstruction formula such as our (10.9).

Acknowledgments
Ingrid Daubechies would like to thank CAIP (Rutgers University) and especially Prof. R. Mammone and his group for their hospitality and advice.

References

[1] O. Ghitza, *Auditory models and human performances in tasks related to speech coding and speech recognition*, IEEE Trans. Speech and Audio Proc. 2(1), 115–132, January 1994; see also O. Ghitza, *Advances in speech signal processing*, in S. Furui and M. Sondhi, editors, *Advances in speech signal processing*, Marcel Dekker, New York, NY, 1991.

[2] S. Maes, *The wavelet transform in signal processing, with application to the extraction of the speech modulation model features.* Ph.D. thesis, Université Catholique de Louvain, Louvain-la-Neuve, Belgium, 1994.

[3] ———, *The synchrosqueezed representation yields a new reading of the wavelet transform*, in H. H. Szu, editor, Proc. SPIE 1995 on OE/Aerospace Sensing and Dual Use Photonics—Wavelet Applications for Dual Use— Session on Acoustic and Signal Processing, Wavelet Applications II, vol. 2491, pp. 532–559, Orlando, FL, April 1995, part I.

[4] ———, *Fast quasi-continuous wavelet algorithms for analysis and synthesis of 1-D signals*, SIAM J. Applied Math. 57 (1997), 1763–1801.

[5] J. B. Allen, *Cochlear modeling*, IEEE ASSP Magazine 2 (1) (1985), 3–29.

[6] C. D'Alessandro, *Time-frequency speech transformation based on an elementary waveform representation*, Speech Communication 9 (1990), 419–431.

[7] J. S. Liénard, *Speech analysis and reconstruction using short-time, elementary, waveforms*, in IEEE Proc. ICASSP, 948–951, Dallas, TX, 1987.

[8] K. Assaleh, *Robust features for speaker identification*. Ph.D. thesis, CAIP Center, Rutgers University, The State University of New Jersey, New Brunswick, NJ, 1993.

[9] K. Assaleh, R. J. Mammone, and J. L. Flanagan, *Speech recognition using the modulation model*, in IEEE Proc. ICASSP, vol. 2, 664–667, 1993.

[10] K. T. Assaleh and R. J. Mammone, *New LP-derived features for speaker identification*, IEEE Trans. Speech and Audio Proc. 2 (4) (1994), 630–638.

[11] N. Delprat, B. Escudié, P. Guillemain, R. Kronland-Martinet, Ph. Tchamitchian, and B. Torrésani, *Asymptotic wavelet and Gabor analysis: extraction of instantaneous frequencies*, IEEE Trans. on Information Theory, 38 (vol. 2, part II), 644–664, 1992.

[12] O. Ghitza, *Auditory nerve representation criteria for speech analysis synthesis*, IEEE Trans. ASSP 6 (35), 736–740, June 1987.

[13] D. H. Friedman, *Instantaneous-frequency distribution vs. time: An interpretation of the phase structure of speech*, in IEEE Proc. ICASSP (1985), 1121–1124.

[14] M. Schroeder, *Direct (non-recursive) relations between cepstrum and predictor coefficients*, IEEE Trans. ASSP 29 (1981), 297–301.

[15] F. Auger and P. Flandrin, *Improving the readablilty of time-scale representations by the reassignment method*, IEEE Trans. Signal Processing 43 (5) (1995), 1068–1089.

[16] I. Daubechies and S. Maes, *A nonlinear squeezing of the continuous wavelet transform based on auditory nerve models*, in A. Aldroubi and M. Unser, editors, *Wavelets in Medicine and Biology*, CRC Press, Boca Raton, FL, 1996.

Chapter 11

Analytic Capacity, Cauchy Kernel, Menger Curvature, and Rectifiability

Guy David

I am especially happy and honored to have participated in the conference honoring Professor Calderón, and I am very grateful to the organizers for having made this possible. I decided to talk about analytic capacity and the Cauchy kernel in part because my first contact with research was a very nice graduate course of Yves Meyer on the Cauchy integral on Lipschitz graphs and the accomplishments of Alberto Calderón. These are very pleasant memories, and they influenced me for quite some time. However, I do not know the subject so well, and I hope this text will not contain too many mistakes or omissions.

A more scientific justification for this choice of topic is that there was very impressive progress about a year and a half ago (the characterization by Mattila, Melnikov, and Verdera of the regular measures in the plane for which the Cauchy kernel defines a bounded operator on L^2), and there is some hope of additional progress. I will try to talk a little about both.

11.1 Analytic Capacity

Let us start with standard background material on analytic capacity. For more details, we refer to [3], [12], [20], and [33] (which were the main sources for the preparation of this chapter). For the rest of this text, E will denote a compact set in the complex plane.

Definition 11.1. The analytic capacity of E is the number

$$\gamma(E) = \sup\{|f'(\infty)| : f \text{ analytic on } \mathbb{C}\backslash E, |f(z)| \leq 1 \text{ on } \mathbb{C}\backslash E\}. \quad (11.1)$$

Observe that if f is bounded and analytic on $\mathbb{C} \backslash E$, then it has a removable singularity at ∞ and so $f'(\infty)$ exists. In (11.1) we can restrict our attention to function f such that $f(\infty) = 0$, because a simple computation shows that otherwise $g(z) = \frac{f(z) - f(\infty)}{1 - \overline{f(\infty)} f(z)}$ has the same L^∞-norm as f, but a larger derivative at ∞.

Note that $\gamma(\dot{E}) = 0$ means that all the bounded analytic functions on $\mathbb{C} \backslash E$ are constant. In general, $\gamma(E)$ is a nondecreasing function of E which measures how easy it is to construct analytic functions in the complement of E. It was introduced by L. Ahlfors [1], who proved that the compact set E has zero analytic capacity if and only if it is removable for bounded analytic functions. This means that if U is any open set that contains E and f is a bounded analytic function on $U \backslash E$, then f has an analytic extension to U.

In my opinion, one of the most fascinating things about analytic capacity is that so little is known about it. Here are some of the few things that one can say.

1. If E is a countable union of compact sets of analytic capacity 0, then $\gamma(E) = 0$. (See for instance [12], exercise 1.7 on p. 12.)

2. If E is connected, then $\gamma(E) \geq \text{diam}(E)/4$. To prove this one uses a conformal mapping from the complement of E in the Riemann sphere to the unit disk.

3. If E has Hausdorff dimension > 1, then $\gamma(E) > 0$. This is because there is enough room in E to allow the existence of a positive measure μ with support in E and such that $\mu * \frac{1}{|z|}$ is bounded, and then we can take $f = \mu * \frac{1}{z}$.

4. If $\dim(E) < 1$, or even if the one-dimensional Hausdorff measure $H^1(E)$ is zero, then $\gamma(E) = 0$. This is proved by finding a "curve" Γ that encloses E and has very small length and then computing the derivative at ∞ of any bounded analytic function on $\mathbb{C} \backslash E$ with the help of the Cauchy formula on Γ.

5. On the other hand, there exist compact sets E such that $H^1(E) > 0$ but $\gamma(E) = 0$. This was first observed by Vitushkin; the simplest example is the Cantor set of dimension 1 obtained by taking the Cartesian product of two "middle half" Cantor sets. (See [11] or [13], [14].)

In this chapter we shall focus on the problem of deciding when the compact set E has vanishing analytic capacity, and so we may as well restrict our attention to one-dimensional sets. In fact, we would even be very happy to have a geometric answer to this question when $0 < H^1(E) < +\infty$.

11.2 The Cauchy Kernel; Positive Results

We start with a few notations. If μ is a positive Borel measure on $\mathbb{C}$, we define truncated Cauchy operators $\mathcal{C}_\mu^\varepsilon$ by

$$\mathcal{C}_\mu^\varepsilon f(z) = \int_{|w-z|>\varepsilon} \frac{f(w)\,d\mu(w)}{z-w}, \qquad (11.2)$$

and then we say that $\mathcal{C}_\mu$ is bounded if the $\mathcal{C}_\mu^\varepsilon$ are uniformly bounded on $L^2(d\mu)$. If E is a closed set in the complex plane, we say that $\mathcal{C}_E$ is bounded when $\mathcal{C}_\mu$ is bounded, where μ is the restriction of H^1 to E.

Examples. If E is a line, then $\mathcal{C}_E$ is essentially the Hilbert transform. By an easy comparison argument, $\mathcal{C}_E$ is also bounded when E is a $C^{1+\varepsilon}$ graph. The first really difficult continuity theorem for $\mathcal{C}_E$ was the well-known result of [2], which says that $\mathcal{C}_E$ is bounded when E is the graph of a Lipschitz function with small Lipschitz constant.

The boundedness of $\mathcal{C}_\mu$ when μ is an Ahlfors-regular measure is now quite well understood.

Definition 11.2 (Ahlfors-regularity). A nonnegative Borel measure μ is said to be (Ahlfors)-regular if there is a constant $C > 0$ such that

$$C^{-1}r \leq \mu(B(x,r)) \leq Cr \text{ for all } x \in \text{Supp}(\mu) \text{ and } 0 < r < \text{diam}(\text{Supp}(\mu)). \tag{11.3}$$

An (Ahlfors)-regular set is a closed set E such that the restriction of H^1 to E is an (Ahlfors)-regular measure. [Incidentally, it is easy to prove that if μ is a regular measure, then $E = \text{Supp}(\mu)$ is a regular set and μ is equivalent to the restriction of H^1 to E.]

An (Ahlfors)-regular curve is a set of the form $\Gamma = h(I)$, where I is an interval in $\mathbb{R}$ and h is a Lipschitz mapping such that the Lebesgue measure in $\mathbb{R}$ of $h^{-1}(B(x,r))$ is at most Cr for all balls $B(x,r) \subset \mathbb{C}$.

Regular curves are almost the same thing as connected regular sets. Indeed, every regular curve is a connected regular set, and it is not hard to show that every connected regular set is contained in a regular curve.

Theorem 11.3 ([5]). *The operator $\mathcal{C}_\Gamma$ is bounded when Γ is a regular curve.*

Theorem 11.4 ([22]). *If μ is a regular measure on $\mathbb{C}$ such that $\mathcal{C}_\mu$ is bounded, then there is a regular curve Γ that contains $\text{Supp}(\mu)$.*

We shall return to this very nice result of P. Mattila, M. Melnikov, and J. Verdera later in this lecture. Let us now discuss applications of boundedness results for $\mathcal{C}_\mu$.

Theorem 11.5. *Let E be a regular set such that $\mathcal{C}_E$ is bounded. Then there exists a constant $\eta > 0$ such that $\gamma(K) \geq \eta H^1(K)$ for every compact set $K \subset E$.*

Corollary 11.6. *If K is a compact set such that $H^1(K \cap \Gamma) > 0$ for some rectifiable curve Γ, then $\gamma(K) > 0$.*

By rectifiable curve, we mean a curve with finite length. Corollary 11.6 is known as Denjoy's conjecture. It is an easy consequence of Theorem 11.5 and the result of Calderón cited above ($\mathcal{C}_E$ is bounded when E is a small Lipschitz graph). This is because $\gamma(K)$ is a nondecreasing function of K, and every rectifiable curve can be covered, except for a set of Hausdorff measure zero, by a countable collection of C^1 curves. Thus if K is as in the corollary, then $H^1(K \cap \Gamma') > 0$ for some C^1 curve Γ', and $\gamma(K) \geq \gamma(K \cap \Gamma') > 0$ by Theorem 11.5.

If you use Theorem 11.3 instead of Calderón's result, then you get a slightly more precise lower estimate for $\gamma(K)$. Also see [25], [26] for other estimates of the same type of $\gamma(K)$.

I do not know the precise history of Theorem 11.5. The duality argument which is central in its proof was present in [32] and in [10], and specialists such as Marshall, Havin, or Havinson were aware of something like Theorem 11.5 at the time of Calderón's theorem. See [3] for a proof.

It was conjectured by Vitushkin that for every compact set K of finite (or σ-finite) one-dimensional Hausdorff measure, $\gamma(K) = 0$ if and only if K is purely unrectifiable, i.e., if and only if $H^1(K \cap \Gamma) = 0$ for every rectifiable curve Γ.

This would be a converse to Denjoy's conjecture. Note that Theorem 11.5 goes in the direction of this conjecture, since it says that purely unrectifiable sets are never contained in a regular set E such that $\mathcal{C}_E$ is bounded. Also note that this conjecture is false without the restriction that $H^1(K)$ be σ-finite. P. Jones and T. Murai [16] have an example of a compact set K such that $\gamma(K) > 0$ but whose orthogonal projection on almost every line has zero length.

I will try to describe how far we are from this conjecture. The problem is essentially the following: what sort of geometric information can we derive from the fact that $\gamma(K) > 0$ when K is a compact set with $H^1(K) < +\infty$?

11.3 From Positive Analytic Capacity to the Cauchy Kernel

This is the part of the program that will have to be completed. At this moment, Ahlfors-regular sets and sets on which H^1 is doubling are essentially the only ones to be under complete control in this respect.

Theorem 11.7 ([4]). *Let K be a compact set such that $\gamma(K) > 0$. Suppose in addition that there is an Ahlfors-regular set E that contains K. Then there is a (possibly different) regular set F such that C_F is bounded and $H^1(K \cup F) > 0$. Hence, by Mattila, Melnikov, and Verdera, there is a regular curve Γ such that $H^1(K \cup \Gamma) > 0$.*

Note that, at the qualitative level, we cannot expect more than this: K can always be the union of a nice rectifiable piece with positive analytic capacity and a useless unrectifiable piece.

M. Christ also has more quantitative variants of this theorem. See [3] or [4]. The general idea of the proof is to start with a bounded analytic function on the complement of K, write it down as the Cauchy integral on K of φdH^1 for some bounded complex-valued function φ on K, and then modify E and φ to get a slightly different regular curve F and a para-accretive function b on F whose Cauchy integral lies in BMO. Then apply a $T(b)$-theorem to get that C_F is bounded. The argument is very nice, but it seems to rely strongly on the presence of a regular set E.

Here is an amusing corollary of Theorem 11.7

Theorem 11.8 ([30]). *Let K be a compact set such that $\gamma(K) > 0$ and $H^1(K) < +\infty$, and*

$$0 < \theta_*(x) := \liminf_{r \to 0} r^{-1} H^1(K \cup B(x,r))$$
$$\le \theta^*(x) := \limsup_{r \to 0} r^{-1} H^1(K \cup B(x,r)) < +\infty \tag{11.4}$$

for every point $x \in K$. Then there is a rectifiable curve Γ such that $H^1(K \cup \Gamma) > 0$.

In other words, if the compact set K is purely unrectifiable, $H^1(K) < +\infty$, and (11.4) holds for every $x \in K$, then $\gamma(K) = 0$. Let us sketch the argument. Pajot proves that, if $H^1(K) < +\infty$ and (11.4) holds for every $x \in K$, then K is contained in a countable union of regular sets E_n. His initial goal was to prove that, if in addition we have a suitable control on the P. Jones numbers $\beta_K(x,r)$ on K, then we can construct the set E_n with the same control on the $\beta_{E_n}(x,r)$, and deduce rectifiability results for K from the corresponding results on the E_n. Here, we use the fact that if we start from a purely unrectifiable K, then it is possible to choose the regular sets E_n that cover K so that they are purely unrectifiable as well. Then $\gamma(E_n) = 0$ by Theorems 11.7 and 11.4. Thus K is the countable union of the compact sets of analytic capacity zero $K \cup E_n$, and hence $\gamma(K) = 0$ by the comments after 11.1.

Note that, if the density condition (11.4) is not satisfied at least almost everywhere on K, then E is not contained in a countable union of regular sets and we cannot use Christ's result. It looks very plausible that the hypothesis that (11.4) holds for almost every point of K (instead of every point) is enough for the theorem, but we do not know how to prove this.

When K is a compact set with finite length and positive analytic capacity, but does not necessarily satisfy (11.4), we can still get a measure with some good properties.

Proposition 11.9 ([20]). *Let K be a compact subset of $\mathbb{C}$, with $0 < H^1(K) < +\infty$, and let $f : \mathbb{C}\backslash K \to \mathbb{C}$ be a bounded analytic function with $f(\infty) = 0$ and $\|f\|_\infty \leq 1$. Then there is a (finite) complex Radon measure σ such that $Supp(\sigma) \subset K$, $|\sigma(B(x,r))| \leq r$ for all $x \in \mathbb{C}$ and $r > 0$, and*

$$f(z) = \int_K \frac{d\sigma(w)}{z - w} \quad \text{for all } z \in \mathbb{C}\backslash K. \tag{11.5}$$

The existence of this measure is obtained by writing down the Cauchy formula on curves that surround K, and then taking appropriate weak limits. Note that σ is absolutely continuous with respect to the restriction of H^1 to K, so we may write $\sigma = \varphi 1_k H^1$ for some bounded complex-valued Borel function φ. If $\gamma(K) > 0$, we can take f such that $f'(\infty) = \gamma(K) > 0$, and we get that $\sigma(K) = \gamma(K) > 0$.

If σ were a positive measure, then we would be able to prove that its support is rectifiable, as we shall see in the next section. If $|\sigma|$ were a piece of some nice doubling measure (even if it is not regular), then we would probably be able to use the same sort of technique as in [4] to get a positive measure μ such that $\mathcal{C}_\mu$ is bounded. Unfortunately, we do not know that, or whether it is possible to modify σ and get a better one.

11.4   The Cauchy Kernel and Menger Curvature on Regular Sets

Let us return to Theorem 11.4. The central point of the argument is a computation which relates the integral $\int |\mathcal{C}_\mu(1)|^2 \, d\mu$ to the Menger curvature $c^2(\mu)$.

Definition 11.10 (Menger curvature). For $x, y, z \in \mathbb{C}$ we will denote by $c(x, y, z)$ (the Menger curvature of x, y, z) the inverse of the radius of the circle through x, y, z. [Set $c(x, y, z) = 0$ when x, y, z are on a same line]. Equivalently,

$$c(x, y, z) = 2|x - z|^{-1}|y - z|^{-1} \, \text{dist}(z, L_{x,y}), \tag{11.6}$$

where $L_{x,y}$ is the line through x and y. If μ is a positive measure, set

$$c^2(\mu) = \int \int \int c(x, y, z)^2 \, d\mu(x) \, d\mu(y) \, d\mu(z). \tag{11.7}$$

The numbers $c(x, y, z)$ give some interesting geometric information about the flatness of the support of μ. Note that $c^2(\mu) = 0$ when $\text{Supp}(\mu)$ is contained in a line; an elementary computation shows that $c^2(\mu) = +\infty$ when μ is the restriction of H^1 to the Cantor set of the end of §11.1. Their main advantage, though, is their relation with the Cauchy kernel. Here is the idea. If we write down $\int |\mathcal{C}_\mu(1)|^2 \, d\mu$ brutally (without taking care of truncatures), we get the triple integral

$$\int \mathcal{C}_\mu(1)(z_1)\overline{\mathcal{C}_\mu(1)}(z_1) d\mu(z_1) = \int \int \int \frac{d\mu(z_1) \, d\mu(z_2) \, d\mu(z_3)}{(z_1 - z_2)(\overline{z_1 - z_3})}. \tag{11.8}$$

The value of this integral is not changed if we symmetrize it by replacing $\frac{1}{(z_1 - z_2)(\overline{z_1 - z_3})}$ by the sum $\frac{1}{6} \sum_\sigma \frac{1}{(z_{\sigma(1)} - z_{\sigma(2)})(\overline{z_{\sigma(1)} - z_{\sigma(3)}})}$, where σ runs along the permutations of $\{1, 2, 3\}$. The advantage of doing so is that this last sum is now positive, and is in fact equal to $c^2(z_1, z_2, z_3)/6$. Thus

$$\int |\mathcal{C}_\mu(1)|^2 \, d\mu = \frac{1}{6} \int \int \int c^2(z_1, z_2, z_3) \, d\mu(z_1) \, d\mu(z_2) \, d\mu(z_3). \tag{11.9}$$

This formula is not quite right, because one should also take into account the fact that $\mathcal{C}_\mu(1)$ is not well defined. Thus we have to use truncatures and then the triple integrals are no longer symmetric. However, it is proved in [22] that, if the measure μ satisfies

$$\mu(B(x, r)) \le r \text{ for all } x \in \mathbb{C} \text{ and all } r > 0, \tag{11.10}$$

then the error in (11.9) is at most $C\|\mu\|$. More precisely, denote by E the support of μ and set

$$\Delta(\varepsilon) = \{(z_1, z_2, z_3) \in E^3 : |z_i - z_j| > \varepsilon \text{ for } i, j \in \{1, 2, 3\}, i \ne j\} \tag{11.11}$$

for each $\varepsilon > 0$. Then

$$\left| \int |\mathcal{C}_\mu^\varepsilon(1)|^2 \, d\mu - \frac{1}{6} \int \int \int_{\Delta(\varepsilon)} c^2(z_1, z_2, z_3) \, d\mu(z_1) \, d\mu(z_2) \, d\mu(z_3) \right| \le C\|\mu\| \tag{11.12}$$

for all $\varepsilon > 0$, with a constant C that does not depend on ε.

Now suppose that μ is a regular measure such that $\mathcal{C}_\mu$ is bounded, and set $E = \text{Supp}(\mu)$. Then we may apply (11.12) to $1_{B(x,r)}\mu$ and get the Carleson measure estimate

$$c^2 \left(1_{B(x,r)}\mu \right) \le Cr \text{ for all } x \in E \text{ and } 0 < r \le \text{diam}(E). \tag{11.13}$$

At this stage, we can forget the Cauchy kernel, and Theorem 11.4 is a consequence of the following.

Proposition 11.11. *If μ is a regular measure that satisfies (11.13), then $E = \text{Supp}(\mu)$ is contained in a regular curve.*

The simplest way to prove this lemma is to show that (11.13) implies a similar Carleson measure estimate on the P. Jones numbers $\beta_2(x, r)$. Recall that $\beta_q(x, r)$ is defined for $x \in E$, $0 < r < \text{diam}(E)$, and $1 \leq q < +\infty$ by

$$\beta_q(x, r) = \inf_L \left\{ r^{-1} \int_{E \cap B(x,r)} \left\{ r^{-1} \, \text{dist}(y, L) \right\}^q \, d\mu \right\}^{1/q}, \qquad (11.14)$$

where the infimum is taken over lines. It is not hard to show that, if μ is regular, then

$$\beta_2(x, r)^2 \leq C r^{-1} \kappa(x, r)^2 \text{ for all } x \in E \text{ and } 0 < r < \text{diam}(E), \qquad (11.15)$$

where we set

$$\kappa(x, r)^2 = \int \int \int_{\Delta(x,r)} c^2(y, z, w) \, d\mu(y) \, d\mu(z) \, d\mu(w), \qquad (11.16)$$

$$\Delta(x, r) = \left\{ (y, z, w) \in E \cap B(x, t) : |y - z|, |z - w|, |w - y| > M^{-1} r \right\}, \qquad (11.17)$$

and where the constant M is chosen large enough (depending on the regularity constant for μ). Then (11.13) and Fubini imply the Carleson measure estimate

$$\int_{E \cap B(X,R)} \int_{0 < r < R} \kappa(x, r)^2 \, d\mu(x) \frac{dr}{r^2} \leq CR \qquad (11.18)$$

for all $X \in E$ and $0 < R < \text{diam}(E)$, and then (11.13) yields

$$\int_{E \cap B(X,R)} \int_{0 < r < R} \beta_2(x, r)^2 \, d\mu(x) \frac{dr}{r} \leq CR. \qquad (11.19)$$

There are several ways to see that (11.19) implies the existence of a regular curve Γ that contains E. [See [22] or [8].] The most natural one is perhaps to modify the argument of P. Jones in [15] so that it works with $\beta_2(x, r)$ instead of $\beta_\infty(x, r)$, and construct the curve Γ by hand. (See [31].)

This completes our sketch of Mattila, Melnikov, and Verdera's argument.

Remark. The first use of the estimate (11.12) was to get a new proof of the boundedness of $\mathcal{C}_\Gamma$ for Lipschitz graphs. The idea is that Lipschitz graphs satisfy the curvature estimate (11.13), and hence

$$\left\| \mathcal{C}_\Gamma \left(1_{B(x,r)} \right) \right\|^2 \leq Cr \text{ for all } x \in \Gamma \text{ and } r > 0. \qquad (11.20)$$

The boundedness of $\mathcal{C}_\Gamma$ then follows by a $T(1)$-type argument. See [24] for details. The Menger curvature was first used in connection with analytic capacity by M. Melnikov [23].

11.5 Menger Curvature and Rectifiability

Define the total Menger curvature of a compact set E in the plane by

$$c^2(E) = \int \int \int_{E \times E \times E} c(x, y, z)^2 \, dH^1(x) dH^1(y) dH^1(z).$$

Theorem 11.12. *If $E \subset \mathbb{C}$ is a compact set such that $H^1(E)$ is σ-finite and $c^2(E) < +\infty$, then E is rectifiable.*

This means that E is contained in a countable union of rectifiable (or equivalently, C^1) curves, plus perhaps a set of zero H^1-measure. This result can be seen as a qualitative version of Proposition 11.11. Its interest is that we now allow E to be very thin at places, and the price we pay for this is that we no longer have good quantitative estimates on the rectifiability of E.

Theorem 11.12 is the result of discussions with P. Mattila. We have a manuscript for the proof that is vaguely outlined below, but it is likely that there will be no final version of it because Jean-Christophe Léger has a slightly better proof based on ideas from the "corona construction" [17]. The proof in [17] is actually shorter, and has the advantage of extending to sets of higher dimensions, but the other one is perhaps slightly easier to present when one does not have to check the details.

We shall say a few words about the proof of Theorem 11.12 later, but let us first see its connection with the Cauchy kernel.

Let μ be a positive measure such that (11.10) holds, and suppose that $\mathcal{C}_\mu^\varepsilon(1) \in L^2(d\mu)$ with estimates that do not depend on ε. Note that the measure σ in Proposition 11.9 is almost like this, except that we do not know that it is positive. Then (11.12) tells us that $c^2(\mu) < +\infty$. Because of (11.10), μ is absolutely continuous with respect to the restriction of H^1 to $E = \text{Supp}(\mu)$ and is given by a bounded density function φ. By elementary measure theory, we can cover μ-almost all of E by a countable union of compact subsets F_n of E where $\varphi(x) \geq 1/n$. Then $c^2(F_n) \leq n^3 c^2(\mu) < +\infty$ for all n. Theorem 11.12 says that F_n is rectifiable, and certainly E contains a nontrivial rectifiable piece if $\mu \neq 0$. Thus, as was mentioned before, the main obstruction to our program seems to be getting a *positive* measure whose Cauchy integral is not too large.

Let us now talk about the proof of Theorem 11.12. The reader will be asked to forgive the few lies in the rapid sketch that follows. We start with a few reductions.

Let μ denote the restriction of H^1 on E. Because we are only interested in rectifiability and can take countable unions at leisure, it does not cost us anything to assume that μ is finite, and even that

$$\mu(B(x, r)) \leq Cr \text{ for all } x \in \mathbb{C} \text{ and all } r > 0. \tag{11.21}$$

(This reduction relies on the fact that $\theta^*(x) = \limsup_{r \to 0} r^{-1}\mu(E \cap B(x,r))$ $< +\infty$ for μ-almost every point $x \in E$, by a classical theorem of geometric measure theory.)

Also, it will be enough to show that one-tenth of E, say, is rectifiable, because we may always start the argument again with the rest of E.

Our hypothesis that $c^2(\mu) < +\infty$ implies that

$$\int_E \int_r \kappa(x,r)^2 \, d\mu(x) \frac{dr}{r^2} < +\infty, \tag{11.22}$$

where the constant M in (11.17) is chosen later, depending on the constant in (11.21). We may choose $t_0 > 0$ so small that in (11.22) the part of the integral that comes from $r < t_0$ is smaller than $\varepsilon\mu(E)$, where ε is as small as we want. (In the real argument, ε is chosen at the end.) We can cover E by balls $B(y,t)$ such that $t < t_0$, $\mu(B(y,t)) \geq t/10$, and for which the sum of the radii t is at most $100\mu(E)$. (Use the definition of H^1 and a covering lemma.) Then the above remark and Tchebychev imply that for most $B(y,t)$,

$$\int_{E \cap B(y,t)} \int_{0<r<t} \kappa(x,r)^2 \, d\mu(x) \frac{dr}{r^2} \leq \int_{E \cap B(y,t)} \int_{0<r<t_0} \kappa(x,r)^2 \, d\mu(x) \frac{dr}{r^2} \leq C\varepsilon t. \tag{11.23}$$

Since the $B(y,t)$ that satisfy (11.23) cover most of E, we see that it is enough to prove the following.

Lemma 11.13. *Suppose that E satisfies (11.21) and is contained in a ball $B(y,t)$ such that $\mu(B(y,t)) \geq C^{-1}t$ and*

$$\int_{E \cap B(y,t)} \int_{0<r<t} \kappa(x,r)^2 \, d\mu(x) \frac{dr}{r^2} \leq \varepsilon t. \tag{11.24}$$

Then there is a curve Γ with length $\leq 3t$ such that $\mu(E \backslash \Gamma) \leq \mu(E)/10$. (The precise quantifiers are that for each $C > 0$, we can find M [in (11.17)] and then ε so that the statement holds.)

We are finished with our preliminary reductions. The last one was perhaps a little less innocent as it may seem, because it takes care of the following apparent problem.

Our set E could very well be a collection of about N little circles of the same radius N^{-1}, spread uniformly in the unit square (so that they lie at distances comparable to $N^{-\frac{1}{2}}$ from their neighbors). Then the curvature of E is not very small, but it is less than some constant that does not depend on N. This means that we may force μ to look a lot like the 2-dimensional Lebesgue measure, and yet keep the curvature $c^2(E)$ bounded. This is of course coherent with the fact that the 2-dimensional Lebesgue measure has finite curvature (the $\frac{dr}{r}$ integral in (11.22) converges brutally). It is important to understand that there is no contradiction so far. The Lebesgue measure is not allowed in Theorem 11.12 because $\mathbb{R}^2$ does not have σ-finite H^1-measure, and we see this in the argument above when we say that we can cover the set E by very small balls with a control on the sum of the radii. On the other hand, our example with many little circles should not disturb us, because it is rectifiable. The point of the reduction above is that if we are patient enough, we simply wait long enough and start working in tiny balls where the normalized curvature is very small. We do not know in advance how long we will have to wait, but this is to be expected because our example with little circles suggests that we can only hope to prove that the set E is rectifiable, not that it is rectifiable with uniform estimates.

The point of the main lemma is that once you have been patient enough, the bad guy that tries to construct a set E which is unrectifiable does not have enough curvature left to cause much harm. For instance, replacing each little circle by a bunch of very tiny circles spread in a disk will cost him a definite amount of curvature. The bad guy may have used this trick a few times at larger scales, but now he cannot anymore.

Let us now try to explain why the main lemma may be true. The idea of the proof is to construct the curve Γ by the same sort of algorithm as P. Jones [15] used to embed in a rectifiable curve any set E with appropriate quadratic estimates on its numbers $\beta_\infty(x, r)$. Thus we start from a line at scale t, and then we start modifying it at smaller and smaller scales to make it go through more and more points of E. When we are working at scale 2^{-n}, say, we choose the new points that we have to add to be at distance comparable to 2^{-n} from previously introduced points and from each other.

As long as the density of E near the points that we want to add is not too small, everything works as if the set E was regular, and the length that we have to add to the current curve to go through the new point can be controlled by a local $r\beta_2(x, r)^2$, which in turn can be estimated by a $\kappa(x, r)^2$, just like for (11.15). The argument is similar to the ones in [15] and [31].

If the density near a point becomes too small, we do not have enough control, and so we decide to ignore that point. To be a little more precise, if we are currently working at scale 2^{-n} and the density of E in a ball of radius 2^{-n} centered on the current Γ is lower than a certain threshold, we decide to leave this ball alone for the rest of the construction. We also decide not to try to incorporate new points of E around which the density is too small.

With this sort of strategy, and because we trust the argument of P. Jones in the regular case, it is believable that we shall construct a limit curve Γ with finite length, but of course we have to say why many points of E will lie on Γ.

There are essentially two interesting types of points of E which do not lie on Γ. Let us call E_1 the set of points of $E\backslash\Gamma$ the lie in a ball centered on Γ where we decided to stop the modifications of Γ in the construction because the density of E on that ball was too small. We can cover E_1 by a collection of such balls B_i such that the $B_i/10$ are disjoint, and then $\mu(E_1)$ is much less than the sum of the radii of the B_i, which is controlled by the length of Γ. Thus E_1 is as small as we want (depending on the density threshold).

The second type of points of $E\backslash\Gamma$ are the points around which the construction of Γ may have continued forever, but which are a little off Γ and which were never added as new points because the density near them is too small. These points may exist when E looks like a line plus a thin mist around it. Then the curve Γ will simply go through the line, and we shall miss the mist. We have to control the total amount of E that evaporates into mist.

The most interesting case for this estimate is the case of points $x \in E$ with the following properties. Let d denote the distance from x to Γ. Suppose that the density of E in the ball $B(x, d/10)$ is very small, but the density of E in the ball $B(x, 10d)$ is not too small. Also suppose that $\beta_2(x, 10d)$ is very small, so that there is a line L such that most of the points of $E\cap B(x, 10d)$ lie very close to L. Finally, assume that Γ crosses $B(x, 10d)$ and that $\Gamma\cap(x, 10d)$ stays quite close to the line L. (The other cases can somehow be treated by the way the curve Γ is constructed, and the estimates are not too interesting for this lecture, in the sense that they only use estimates where the density is not too small.)

Because $\mu(E \cap B(x, 10d))$ is not too small (and (11.21) holds), we can find reasonably many pairs (y, z) of points of $E \cap B(x, 10d)$ such that y, z lie very close to L and $|y - z|$ is not too small. Since $\text{dist}(x, L) \geq d/2$ because $\Gamma \cap B(x, 10d)$ stays close to L, we get that $c(x, y, z) \geq (Cd)^{-1}$ for all those pairs (y, z). Then the contribution of the point x to the integral in (11.24) is $\geq C^{-1}$. This tells us that the total mass of the points of E of this type is controlled by (11.24), and thus is as small as we want (depending on ε).

This ends our rough sketch of proof. Of course the details of the construction are slightly more painful than this.

Remark. This last part of the proof is the only one where we have to use curvature (as opposed to the numbers $\beta_1(x, r)$ on balls $B(x, r)$ where the density is not too small, and where the curvature indeed controls $\beta_1(x, r)$). Also, in the main lemma, the triples $(x, y, z) \in E^3$ are not all needed, but only those were $|x - y|$, $|y - z|$ and $|x - z|$ are all comparable (with a constant that depends on the regularity constant in (11.21)).

As we explained before, a variant of Theorem 11.12 holds in higher dimension, with the curvature $c(x, y, z)$ replaced with a function of $d+1$ points of E that measures the distance of one of them to the d-plane that passes through the other ones (d is the dimension of E). See [17].

11.6 Conclusion

This part of the text was added in June 1998. It is with great sorrow that I learned of the recent death of A. Calderón. He was always a great model for me; I am sure many of us will miss him a lot.

Since the time of the conference, the progress that could be hoped for after the breakthrough of [22] indeed took place. It is now known that for a compact set K with $H^1(K) < +\infty$, $\gamma(K) = 0$ if and only if K is purely unrectifiable. This was first proved for the analogue of γ where we consider Lipschitz harmonic functions (instead of bounded analytic functions) in the complement of K in $\mathbb{R}^2$. (See [7].) The proof uses an appropriate generalization of the argument of M. Christ [4]. The case of analytic capacity itself followed a little later [6]; the new part of the argument is a generalization of the $T(b)$-Theorem without doubling. It has been proved (simultaneously, independently, and for slightly different reasons) by F. Nazarov, S. Treil, and S. Volberg [28]. Their method also gives the result on analytic capacity [29], with a more clever and pleasant proof. From their work, it also appears that we have been wrong to assume systematically doubling properties (or Ahlfors-regularity, or spaces of homogeneous type) whenever we dealt with Calderón-Zygmund operators.

References

[1] L. Ahlfors, *Bounded analytic functions*, Duke Math. J. 14 (1947), 1–11.

[2] A. P. Calderón, *Cauchy integrals on Lipschitz curves and related operators*, Proc. Nat. Acad. Sci. USA 74 (1977), 1324–1327.

[3] M. Christ, *Lectures on singular integral operators*, NSF-CBMS Regional Conference series in Mathematics 77, AMS 1990.

[4] ———, *A T(b) theorem with remarks on analytic capacity and the Cauchy integral*, Colloq. Math. 60/61 (1990), 1367–1381.

[5] G. David, *Opérateurs intégraux singuliers sur certaines courbes du plan complexes*, Ann. Sci. Ec. Norm. Sup. 17 (1984), 157–189.

[6] ———, *Unrectifiable 1-sets have vanishing analytic capacity*, Revista Matemática Iberoamericana, vol. 14, no. 2 (1998), 369–479.

[7] G. David and P. Mattila, *Removable sets for Lipschitz harmonic functions in the plane*, to appear, Revista Matemática Iberoamericana.

[8] G. David and S. Semmes, *Analysis of and on uniformly rectifiable sets*, A.M.S. Series of Mathematical surveys and monographs, vol. 38, 1993.

[9] A. M. Davie, *Analytic capacity and approximation problems*, Transac. Amer. Math. Soc 171 (1972), 409–444.

[10] A. M. Davie and B. Oksendal, *Differentiability properties for finely harmonic functions*, Acta Math. 149 (1982), 127–152.

[11] J. Garnett, *Positive length but zero analytic capacity*, Proceedings A.M.S. 21 (1970), 696–699.

[12] ———, *Analytic capacity and measure*, Lecture Notes in Math. 297, Springer-Verlag, Berlin, 1972.

[13] L. D. Ivanov, *Variations of sets and functions*, Nauka, 1975 (Russian).

[14] ———, *On sets of analytic capacity zero*, in *Linear and Complex Analysis, Problem Book 3*, Part II, V. P. Havin and N. K. Nikloski, eds., Lecture Notes in Math. 1574, Springer-Verlag, Berlin, 1994.

[15] P. Jones, *Rectifiable sets and the traveling salesman problem*, Inventiones Mathematica 102, 1 (1990), 1–16.

[16] P. Jones and T. Murai, *Positive analytic capacity but zero Buffon needle probability*, Pacific J. Math. 133 (1988), 99–114.

[17] J.-C. Léger, *Menger curvature and rectifiability*, to appear, Ann. of Math.

[18] P. Mattila, *Smooth maps, null-sets for integral geometric measure and analytic capacity*, Ann. of Math. 123 (1986), 303–309.

[19] ———, *Cauchy singular integrals and rectifiability of measures in the plane*, Advances in Math. 115, 1 (1995), 1–34.

[20] ———, *Geometry of Sets and Measures in Euclidean Space*, Cambridge University Press, Cambridge, 1995.

[21] ———, *On the analytic capacity and curvature of some Cantor sets with non-σ-finite length*, preprint.

[22] P. Mattila, M. Melnikov, and J. Verdera, *The Cauchy integral, analytic capacity and uniform rectifiability*, Ann. of Math. 144 (1996), 127–136.

[23] M. Melnikov, *Analytic capacity: discrete approach and curvature of measure*, Math. Sbornik 186:6 (1995), 827–846.

[24] M. Melnikov and J. Verdera, *A geometric proof of the L^2-boundedness of the Cauchy integral on Lipschitz graphs*, to appear, International Research Notices.

[25] T. Murai, *Comparison between analytic capacity and the Buffon needle probability*, Trans. Amer. Math. Soc. 304 (1987), 501–514.

[26] ———, *A real variable method for the Cauchy transform and analytic capacity*, Lecture Notes in Math. 1307, Springer-Verlag, New York, 1988.

[27] ———, *The power 3/2 appearing in the estimate of analytic capacity*, Pacific Journal of Math. 143 (1990), 313–340.

[28] F. Nazarov, S. Treil, and S. Volberg, *Cauchy integral and the Calderón-Zygmund operator on non-homogeneous spaces*, preprint, available at fedj@math.msu.edu.

[29] ———, *Pulling ourselves up by the hair*, preprint, available at fedj@math.msu.edu.

[30] H. Pajot, *Conditions quantitatives de rectifiabilité*, Bulletin de la société mathématique de France, vol. 125 (1997), 15–53.

[31] ———, *Sous-ensembles de courbes Ahlfors-régulières et nombres de Jones*, Publications Mathématiques, vol. 40 (1996), 497–526.

[32] N. X. Uy, *Removable sets of analytic functions satisfying a Lipschitz condition*, Arkiv. Math. 17 (1979), 19–27.

[33] J. Verdera, *Removability, capacity and approximation*, Universitat Autònoma de Barcelona, 1994.

Chapter 12

Symplectic Subunit Balls and Algebraic Functions

Charles Fefferman

From the work of Stein and others (see, e.g., [7]), one knows that subelliptic partial differential equations are controlled by the geometry of certain non-Euclidean balls. For the last several years, A. Parmeggiani [8] and I have tried to understand an analogous family of balls associated to pseudodifferential operators. By studying the geometry of these balls, we have been led to questions about the smoothness and growth of algebraic functions; we had to show that certain algebraic functions behave like polynomials. R. Narasimhan and I have been studying these questions on algebraic functions, and we now understand them. However, the study of non-Euclidean balls for pseudodifferential operators remains at a primitive level. In this expository paper, I would like to begin by recalling the standard notion of subunit balls for differential operators, and then explain its proposed extension to pseudodifferential operators. Next, I will state the relevant questions on algebraic functions and say a little about their solution. Finally, I will show how the questions on algebraic functions arise in studying subunit balls. It is a pleasure to dedicate this paper to Alberto P. Calderón.

To recall the standard notion of non-Euclidean balls, let me restrict attention to the simple case of a second-order self-adjoint differential operator with real, smooth coefficients. Say $L = -\sum_{j,k} \frac{\partial}{\partial x_j} a_{jk}(x) \frac{\partial}{\partial x_k}$, where $(a_{jk}(x))$ is positive semidefinite for each $x \in \mathbb{R}^n$, but not strictly positive definite. A tangent vector $X = \sum_j g_j \frac{\partial}{\partial x_j}$ at a point $x_0 \in \mathbb{R}^n$ is called *subunit* for L if we have $(g_j g_k) \le (a_{jk}(x_0))$ as matrices. A *subunit path* $\gamma \colon [0, T] \to \mathbb{R}^n$ is a Lipschitz path whose velocity vector $\frac{d\gamma(t)}{dt}$ is subunit for almost every t. The *subunit ball* $B_L(x_0, \rho)$ consists of all points in $\mathbb{R}^n$ that can be joined to x_0 by a subunit path in time $T \leq \rho$. If the operator L is subelliptic, then the geometry of the $B_L(x, \rho)$ may be understood completely and used to control the regularity properties and fundamental solution of L. (See [6], [7].)

A basic special case is the Hörmander operator $L = -\sum_{j=1}^{N} X_j^2$ (which is self-adjoint modulo a negligible error). Here, $X_1, \ldots, X_N$ are smooth, real vector fields whose repeated commutators span the tangent space at every point of $\mathbb{R}^n$. In effect, a subunit tangent vector is simply a linear combination of the X_j with bounded coefficients; and the ball $B_L(x_0, \rho)$ consists of all points that can be reached from x_0 in time ρ by a broken path that flows first along one, then along another of the vector fields $\pm X_j$ ($1 \leq j \leq N$).

The basic geometric fact about the ball $B_L(x_0, \rho)$ is that it may be made comparable to a rectangular box (a Cartesian product of intervals) by performing a smooth coordinate change with controlled bounds. (See [6] for the precise statement.) As an easy consequence, one sees that the volume of $B_L(x_0, 2\rho)$ is at most a constant multiple of that of $B_L(x_0, \rho)$, and that $B_L(x, 2\rho)$ may be covered by a bounded number of balls $B_L(x_\nu, \rho)$. Thus, the $B_L(x, \rho)$ behave like Euclidean balls $\{y \in \mathbb{R}^n : |x - y| < \rho\}$ for many purposes.

A. Parmeggiani and I hope to find, understand, and use analogous subunit balls for pseudodifferential operators. These balls should live in the cotangent space and behave naturally under canonical transformations, just as standard subunit balls behave naturally under coordinate changes.

Let us start with a second-order nonnegative symbol $p(x, \xi)$ ($x, \xi \in \mathbb{R}^n$). Thus, p satisfies the estimates $|\partial_x^\alpha \partial_\xi^\beta p(x, \xi)| \leq C_{\alpha\beta}(1 + |\xi|)^{2-|\beta|}$. We say that q is a *subunit symbol* for p if it satisfies

$$|\partial_x^\alpha \partial_\xi^\beta q(x, \xi)| \leq C_{\alpha\beta}(1 + |\xi|)^{1-|\beta|} \quad \text{for} \quad |\alpha| + |\beta| \leq 2 \tag{12.1}$$

and

$$q^2 \leq p. \tag{12.2}$$

In (12.1), note that we restrict to $|\alpha| + |\beta| \leq 2$. Next, suppose $X = \sum_{j=1}^{n} g_j \frac{\partial}{\partial x_j} + \sum_{j=1}^{n} h_j \frac{\partial}{\partial \xi_j}$ is a tangent vector to $\mathbb{R}^{2n} = T^\star \mathbb{R}^n$ at the point $(x^0, \xi^0) \in \mathbb{R}^{2n}$. We say that X is a *subunit vector* for p if it agrees at (x^0, ξ^0) with the Hamiltonian vector field of a subunit symbol q. That is, $g_j = \frac{\partial q}{\partial \xi_j}(x^0, \xi^0)$ and $h_j = -\frac{\partial q}{\partial x_j}(x^0, \xi^0)$, with q satisfying (12.1) and (12.2). For example, if p is a sum $p = \sum_{j=1}^{n} q_j^2$ of squares of real first-order symbols, then each q_j is a subunit symbol, and its Hamiltonian vector field $\sum_{j=1}^{n} \left(\frac{\partial q}{\partial \xi_j} \frac{\partial}{\partial x_j} - \frac{\partial q}{\partial x_j} \frac{\partial}{\partial \xi_j} \right)$ is a subunit vector field.

As in the differential case, the subunit ball $B_p((x^0, \xi^0), 1)$ may be defined as the set of all points that can be reached in time 1 from (x^0, ξ^0) by a Lipschitz path whose velocity vector is subunit almost everywhere. The ball $B_p((x^0, \xi^0), \delta)$ of radius δ may then be defined as the unit ball $B_{\delta^2 p}((x^0, \xi^0), 1)$ arising from the symbol $\delta^2 p$.

Parmeggiani and I hope to understand the geometry of $B_p((x^0, \xi^0), \rho)$, and to use them to prove theorems about pseudodifferential operators. However, the geometry of these symplectic subunit balls seems much harder to understand than that of standard subunit balls. Examples suggest the following conjecture. Given $(x^0, \xi^0) \in \mathbb{R}^{2n}$, the ball $B_p((x^0, \xi^0), \delta)$ becomes comparable to a rectangular box after we apply a suitable canonical transformation, *unless* δ is comparable to one of finitely many bad values $\delta_1 < \delta_2 < \cdots < \delta_K$. The number K of bad δ_k's is bounded independently of (x^0, ξ^0), but there are definitely some bad δ's. For a bad δ, the ball of radius 2δ is much larger than the ball of radius δ, and the shape of the ball may be wild. These bad δ's have no analogues for the standard subunit balls arising from differential operators. We know from examples that they may occur. Our upper bound on the number of bad δ's is only a conjecture. Parmeggiani and I have ideas for a proof but do not have a complete argument even for pseudodifferential operators on $\mathbb{R}^2$.

Let me now explain the problems on algebraic functions that arise when one studies symplectic subunit balls. We need to know that certain algebraic functions satisfy the standard growth and smoothness properties of polynomials. Recall that if $F(x)$ is a polynomial of degree D on the real line, then we have the familiar estimates

$$\text{(Bernstein-Markov inequality)} \quad \max_I |F'| \leq \frac{C_\star}{|I|} \max_I |F| \tag{12.3}$$

$$\text{(Doubling Condition)} \quad \max_{I^\star} |F| \leq C_\star \max_I |F| \tag{12.4}$$

$$\text{(Equivalence of Norms)} \quad \max_I |F| \leq \frac{C_\star}{|I|} \int_I |F| dx . \tag{12.5}$$

Here, I is any interval, $I^\star$ denotes the double of I, and $C_\star$ depends only on D. Usually one studies the exact dependence of the constants in (12.3), (12.4), (12.5) on the degree D, but this issue will not matter for symplectic subunit balls. Rather, we need to extend these inequalities from polynomials to the following class of algebraic functions.

Let $S = [-1, 1] \times [-1, 1]$ be the unit square in $\mathbb{R}^2$, and let $\Gamma \subset S$ be an algebraic arc defined by $\Gamma = \{(x, y) \in S : Q(x, y) = 0\}$, where Q is a polynomial of degree D on $\mathbb{R}^2$. To make sure Γ is as nice as possible, we make the following assumptions.

For each $x \in [-1,1]$, there is exactly one $y \in [-1,1]$ with $Q(x,y) = 0$, $\qquad(12.6)$

$$\frac{\partial Q}{\partial y} \geq c > 0 \quad \text{for all} \quad (x,y) \in S, \qquad(12.7)$$

$$|Q| \leq C \quad \text{for all} \quad (x,y) \in S . \qquad(12.8)$$

Let $y = \psi(x)$ be the solution of $Q(x,y) = 0$ given by (12.6). If $P(x,y)$ is any polynomial of degree D on $\mathbb{R}^2$, then we set $F(x) = P(x, \psi(x))$. Thus, in effect, F is the restriction of a polynomial to the arc Γ.

The simplest estimates needed for symplectic subunit balls are given by the following innocent-sounding results.

Theorem 12.1 (Bernstein Theorem). *Let Γ and F be as above. Then F satisfies estimates (12.3), (12.4), (12.5) for all intervals I with $I^* \subset [-1,1]$. The constant C_* in (12.3), (12.4), (12.5) may be taken to depend only on the degree D and on the constants C, c in (12.7), (12.8).*

See Fefferman and Narasimhan [3].

When I first thought about this result, it gave me a lot of trouble. My first reaction was to try to extend polynomials from Γ to S with bounds. More precisely, I made the following:

Conjecture: Given Γ and P as above, there exists another polynomial $\widetilde{P}(x,y)$, with the following properties

The degree of $\widetilde{P}$ is bounded a priori in terms of the degree D, $\qquad(12.9)$

$$\widetilde{P} \text{ agrees with } P \text{ on } \Gamma, \qquad(12.10)$$

$\max_S |\widetilde{P}| \leq C_* \max_\Gamma |P|$, with C_* depending only on

the degree D and on the constants C, c in $(12.7), (12.8)$. $\qquad(12.11)$

This conjecture easily reduces the above Bernstein theorem to the familiar properties of polynomials on the unit square. However, as Narasimhan quickly pointed out to me, the conjecture is false. Here is a simple counterexample. For $\epsilon > 0$, let $\Gamma_\epsilon = \{(x,y) \in S : y(1+x^2) = \epsilon\}$, and let $P_\epsilon(x,y) = y/\epsilon$. Since $\Gamma_\epsilon = \{(x,y) : P_\epsilon(x,y) = \frac{1}{1+x^2}\}$ on Γ_ϵ, we have $\max_{\Gamma_\epsilon} |P_\epsilon| = 1$. Note that Γ_ϵ tends to the interval $[-1,1] \times \{0\} \subset S$. If we could find a polynomial $\widetilde{P}_\epsilon$ as in the conjecture, then we could find a sequence $\{\epsilon_j\} \to 0$ for which the $\widetilde{P}_{\epsilon_j}$ converge to a limiting polynomial $\widetilde{P}(x,y)$ on S. Recalling that $\widetilde{P}_{\epsilon_j}(x,y) = P_{\epsilon_j}(x,y) = \frac{1}{1+x^2}$ on Γ_{ϵ_j} and passing to the limit as $j \to \infty$, we find that $\widetilde{P}(x,0) = \frac{1}{1+x^2}$ for all $x \in [-1,1]$. This contradicts the fact that $\widetilde{P}(x,y)$ is a polynomial, so the conjecture must be false.

Fortunately, Narasimhan's counterexample suggests the correct modification of the false conjecture. In the example, although $P_\epsilon(x,y) = y/\epsilon$ may be as large as $1/\epsilon$ on S, we saw that $P_\epsilon(x,y) = \frac{1}{1+x^2}$ on Γ_ϵ. Evidently, $\max_{(x,y)\in S} \frac{1}{1+x^2}$ is bounded, uniformly in ϵ. Thus, P_ϵ extends, not to a polynomial $\widetilde{P}_\epsilon$, but to a rational function with a harmless denominator. Narasimhan and I showed in [3] that this happens in general. Our result is as follows.

Theorem 12.2 (Extension Theorem). *Let Q be a polynomial of degree D on $\mathbb{R}^2$. Suppose Q satisfies (12.6), (12.7), (12.8), and let $\Gamma = \{(x,y) \in S: Q(x,y) = 0\}$. Finally, let P be another polynomial of degree D on $\mathbb{R}^2$. Then there exists a rational function $\widetilde{P}/\widetilde{G}$ on $\mathbb{R}^2$, with the following properties:*

$$P = \widetilde{P}/\widetilde{G} \text{ on } \Gamma, \tag{12.12}$$

$$\widetilde{P} \text{ on } \widetilde{G} \text{ have degree at most } D_\star, \tag{12.13}$$

$$c_\star < \widetilde{G} < C_\star \text{ on } S, \tag{12.14}$$

$$\max_S |\widetilde{P}| \leq C_\star \max_\Gamma |P|. \tag{12.15}$$

Here, $D_\star$ depends only D, while $c_\star$, $C_\star$ are positive constants depending only on D, c, C in (12.6), (12.7), (12.8).

The Extension Theorem substitutes well for my false conjecture. In particular, it easily implies the above Bernstein Theorem. The proof of the extension theorem is not so easy. To study symplectic subunit balls, Parmeggiani needed a generalization of the Bernstein Theorem, in which Γ is a smooth patch in a real algebraic variety, rather than an arc. (See [8].) With considerable difficulty, Narasimhan and I were able to generalize our Bernstein and extension theorems to this case. (See [4].) The proof of the generalized extension theorem is formidable. It uses $\bar{\partial}$ technology, the Koszul complex, the theory of semialgebraic functions, and an induction on the dimension of an exceptional set of algebraic varieties. We believe that these tools will be needed in any foreseeable proof of the extension theorem of [4].

However, after doing all that work, Narasimhan and I discovered that there is a much simpler route to the Bernstein theorems, bypassing the extension theorem. In fact, all the Bernstein theorems in [3], [4] are contained in the following general result of functional analysis. (See [5].)

Theorem 12.3 (Abstract Bernstein Theorem). *Let* $i: X \to Y$ *be an injection of Banach spaces, and let* $f_1, f_2, \ldots, f_N$ *be real-analytic maps from an open set* $U \subset \mathbb{R}^n$ *into* X. *For* $\lambda \in U$, *write* V_λ *for the vector subspace* span $\{f_1(\lambda), f_2(\lambda), \ldots, f_N(\lambda)\} \subset X$. *Then, given* $K \subset U$, *we can find a constant* C_K *s.t.* $\|x\|_X \leq C_K \|ix\|_Y$ *whenever* $x \in V_\lambda$ *and* $\lambda \in K$.

To see a typical application of the abstract Bernstein theorem, take $X = C^1[-1, 1]$, $Y = C^0[-1, 1]$, $U = \mathbb{R}^{N+2}$, and for $\lambda = (\lambda_1, \lambda_2, \ldots, \lambda_N, x_0, r) \in U$, define $f_j(\lambda) \in C^1[-1, 1]$ by $(f_j(\lambda))(x) = \exp(\lambda_j \cdot [rx + x_0])$.

The abstract Bernstein theorem then says that $\max_I |F'| \leq \frac{C_*}{|I|} \max_I |F|$ whenever $F(x) = \sum_{j=1}^{N} A_j \exp(\lambda_j x)$ and $I = [a, b]$, with C_* depending only on bounds for $N, |\lambda_j|, a, b$. The abstract Bernstein theorem easily implies the Bernstein theorems in [3], [4], but not the extension theorems given there.

The proof of the abstract Bernstein theorem is a double induction on the dimension of the parameter space U and the number of real-analytic maps $f_1, \ldots, f_N$. The argument is rather simple, but it uses a powerful algebraic tool from Bierstone-Milman [1], essentially equivalent to Hironaka's theorem on resolution of singularities in the special case of hypersurfaces.

Before leaving algebraic functions, I would like to point out that Roytvarf and Yomdin [9] recently settled the question of how the constants C_* in (12.3), (12.4), (12.5) depend on the degree of the polynomial P. Although this is not needed in studying symplectic subunit balls, it is a very natural question in view of the classical estimates for polynomials.

Let me conclude this paper by explaining the connection between algebraic functions and symplectic subunit balls.

For simplicity, I will restrict attention to pseudodifferential operators on $\mathbb{R}^1$, even though this leaves out significant issues, e.g., there are no bad δ's in one dimension. We want to understand how the symbol $p(x, \xi) \geq 0$ looks near a point $(x^0, \xi^0) \in \mathbb{R}^2$. By Taylor-expanding p to high order about (x^0, ξ^0), we may assume without loss of generality that p is a polynomial of some high degree D. A suitable Calderón-Zygmund decomposition (see [2]) breaks up $\mathbb{R}^2$ into rectangles $S_\nu = I_\nu \times J_\nu$, on which p is either elliptic or "nondegenerate." The elliptic S_ν are easy to understand. On a nondegenerate S_ν, after making a canonical transformation, we may assume that $\frac{\partial^2}{\partial \xi^2} p(x, \xi)$ is bounded below by a positive constant. The implicit function theorem then shows that p may be written on S_ν in the form $p(x, \xi) = e(x, \xi) \cdot (\xi - \psi(x))^2 + V(x)$, where e is an elliptic 0th order symbol (bounded above and below by positive constants). Here, $\xi = \psi(x)$ is the solution of $\frac{\partial p}{\partial \xi}(x, \xi) = 0$, and $V(x) = p(x, \psi(x))$. Since p is a polynomial of degree D, $\Gamma = \{\xi = \psi(x)\} \subset S_\nu$ is a smooth arc on an algebraic curve, and $V(x) = p(x, \psi(x))$ is the restriction of a polynomial to Γ. Our Bernstein theorems show that V and ψ have the growth and smoothness properties of polynomials. For instance, the maximum of $|V(x)|$ on I_ν (the x-component of S_ν) is comparable to the maximum of $|V(x)|$ on the double of I_ν. This is a crucial ingredient in showing that the symplectic subunit ball of radius 2δ about (x^0, ξ^0) is comparable to that of radius δ. See [8]. Note that our Bernstein inequalities must be uniform in Γ to be of any use, since the arc Γ depends on the symbol p. Thus, Bernstein inequalities for algebraic functions enter into the study of symplectic subunit balls.

References

[1] E. Bierstone and P. Milman, *Semianalytic and subanalytic sets*, I.H.E.S. Publications 67 (1988), 5–42.

[2] C. Fefferman, *The uncertainty principle*, Bull. A.M.S. 9–2 (1983), 126–206.

[3] C. Fefferman and R. Narasimhan, *Bernstein's inequality on algebraic curves*, Ann. Inst. Fourier 43–5 (1993), 1319–1348.

[4] ———, *On the polynomial-like behavior of certain algebraic functions*, Ann. Inst. Fourier 44–4 (1994), 1091–1179.

[5] ———, *Bernstein's inequality and the resolution of spaces of analytic functions*, Duke Math. J. 81–1 (1995), 77–98.

[6] C. Fefferman and A. Sanchez-Calle, *Fundamental solutions for second-order subelliptic operators*, Ann. Math. 124 (1986), 247–272.

[7] A. Nagel, E. M. Stein, and S. Wainger, *Balls and metrics defined by vector fields I: Basic properties*, Acta Math. 155 (1985), 103–147.

[8] A. Parmeggiani, *Subunit balls for symbols of pseudodifferential operators*, Advances in Math., to appear.

[9] N. Roytvarf and Y. Yomdin, *Bernstein Classes*, to appear.

Chapter 13

Multiparameter Calderón-Zygmund Theory

Robert A. Fefferman

I would like to take this opportunity to express my deep gratitude to my teacher, colleague, and friend, Alberto P. Calderón. Professor Calderón's brilliance, warmth, and generosity have greatly enriched my life at the University, and, on the occasion of his seventy-fifth birthday, it gives me the greatest pleasure to acknowledge this.

The purpose of this article is to give something of an overview of those extensions of classical Calderón-Zygmund theory dealing with classes of operators which are invariant under dilation groups larger than the classical ones. It will be as nontechnical as possible in the hope that analysts not already familiar with the material will be able to gain a good understanding of the main issues and features of this theory here without being burdened by the details. The interested reader can consult the references to find precise proofs of the results we present.

The main example of nonclassical dilations, and the only ones whose theory is understood well at the present time, are the product dilations. These are defined as follows: For $x = (x_1, x_2, \ldots, x_n) \in \mathbb{R}^n$, we set

$$\rho_{\delta_1 \delta_2 \cdots \delta_n}(x_1, x_2, \ldots, x_n) = (\delta_1 x_1, \delta_2 x_2, \ldots, \delta_n x_n),$$

$\delta_i \geq 0$, $i = 0, 1, 2, \ldots, n$. Some basic examples of operators naturally associated to these dilations are as follows:

Example 13.1. The relevant maximal function is the strong maximal function, M_S, given by

$$M_S f(x) = \sup_{x \in R} \frac{1}{|R|} \int_{\mathbb{R}} |f(y)| \, dy,$$

where the supremum is taken over the family of all rectangles with sides parallel to the axes. According to the well-known Jessen-Marcinkiewicz-Zygmund Theorem [1], M_S is bounded for the space $L(\log^+ L)^{n-1}(Q)$ to weak $L^1(Q)$, where Q denotes the unit cube in $\mathbb{R}^n$.

Example 13.2. The multiple Hilbert transform,

$$H_{\text{prod}}(f) = f * \frac{1}{x_1 x_2 \cdots x_n},$$

is the simplest example of a product singular integral and commutes with the product dilations. The fact that the convolution kernel for this operator has a higher-dimensional singularity set makes its analysis delicate.

Example 13.3. If we regard $\mathbb{R}^{n+m}$ as a product $\mathbb{R}^n \times \mathbb{R}^m$, with the dilations $\rho_{\delta_1,\delta_2}(x,y) = (\delta_1 x_1, \delta_2 x_2)$, $\delta_1, \delta_2 > 0$ ($x \in \mathbb{R}^n$, $y \in \mathbb{R}^m$) then we have a class of Marcinkiewicz-type multipliers, $\mathcal{M}$ invariant under these dilations. The class $\mathcal{M}$ is defined as follows: Suppose

$$A = \left\{ (\xi, \eta) \in \mathbb{R}^n \times \mathbb{R}^m : \frac{1}{2} < |\xi| \leq 1, \frac{1}{2} < |\eta| \leq 1 \right\}.$$

Then a multiplier $m(\xi, \eta) \in \mathcal{M}$ provided m is infinitely differentiable away from $\{(\xi, \eta) : \xi = 0 \text{ or } \eta = 0\}$ and

$$\left| \partial_\xi^\alpha \partial_\eta^\beta [m(\delta_1 \xi, \delta_2 \eta)] \right| \leq C$$

for all $(\xi, \eta) \in A$, $|\alpha| \leq [\frac{n}{2}] + 1$, $|\beta| = [\frac{m}{2}] + 1$, uniformly for all $\delta_1, \delta_2 > 0$. This definition simply states that $m(\xi, \eta) \in \mathcal{M}$ provided m behaves like a product $m_1(\xi) m_2(\eta)$ of classical (Hörmander) multipliers.

Example 13.4. The product Littlewood-Paley square function. Let $\eta_1 \in C_c^\infty(\mathbb{R}^n)$, $\eta_2 \in C_c^\infty(\mathbb{R}^m)$ and assume that both η_1 and η_2 have integral zero. Set

$$\Psi_{\delta_1,\delta_2}(x,y) = \delta_1^{-n} \delta_2^{-m} \eta_1\left(\frac{x}{\delta_1}\right) \eta_2\left(\frac{y}{\delta_2}\right) \quad \text{for } \delta_1, \delta_2 > 0.$$

Then

$$S_{\text{prod}}^2(f)(x,y) = \int_{\substack{|u-x|<t_1 \\ |v-y|<t_2 \\ t_1,t_2>0}} \int |f * \Psi_{t_1,t_2}(u,v)|^2 \, du \, dv \frac{dt_1 \, dt_2}{t_1^{n+1} t_2^{m+1}}.$$

Then S_{prod} is a vector-valued product singular integral and plays very much the same role for the product theory that the classical Littlewood-Paley function plays in the classical Calderón-Zygmund theory. For example, as we shall see, S_{prod} will define Hardy spaces in the setting of product dilations, and also we shall have Stein's pointwise majorization of $S_{\text{prod}}(Tf)$ by an appropriate version of $g_\lambda^*(f)$ whenever T is a multiplier operator whose associated multiplier belongs to $\mathcal{M}$.

Example 13.5. Rubio de Francia's Square Function. If R is a rectangle with sides parallel to the axes define the partial sum operator S_R by

$$S_R(f)\widehat{} = \chi_R \hat{f}.$$

Then let $\mathcal{S}(f) = [\sum_k (S_{R_k} f)^2]^{1/2}$, where $\{R_k\}$ is a collection of pairwise disjoint rectangles whose union is all of $\mathbb{R}^2$. Then $\mathcal{S}$ is (the product version of) Rubio de Francia's square function.

All of these operators can be analyzed rather completely at this point. Let us attempt here to identify some of the highlights of the theory.

In the first place, the theory of Hardy spaces was extended by Gundy and E. M. Stein [3]. Taking the product dilation on $\mathbb{R}^2$ as the simplest example, it was shown in [3] that for $\phi \geq 0$, $\phi \in C_c^\infty(\mathbb{R}^2)$. Then setting

$$f_{\text{prod}}^*(x, y) = \sup_{\substack{|u-x| < t_1 \\ |v-y| < t_2 \\ t_1, t_2 > 0}} |f * \phi_{t_1, t_2}(u, v)|,$$

we have:

Theorem 13.6 (Gundy-Stein). *The product space $f_{\text{prod}}^* \in L^p(\mathbb{R}^2)$ if and only if $S_{prod}(f) \in L^p(\mathbb{R}^2)$ for all $p > 0$ and the L^p norms of these maximal and square functions are comparable.*

We may then unambiguously define $H^p(\mathbb{R}^1 \times \mathbb{R}^1)$, product Hardy spaces as $\{f | f_{\text{prod}}^* \in L^p(\mathbb{R}^2)\} = \{f | S_{\text{prod}}(f) \in L^p\}$ and set $||f||_{H^p(\mathbb{R}^1 \times \mathbb{R}^1)}$ to be either $||f_{\text{prod}}^*||_{L^p(\mathbb{R}^2)}$ or $||S_{\text{prod}}(f)||_{L^p(\mathbb{R}^2)}$.

Now, as is well known, there is a basic obstacle to the theory at this point. Thus, as was pointed out by Carleson [4], the dual spaces of $H^1(\mathbb{R}^1 \times \mathbb{R}^1)$ cannot be identified with the space of functions whose mean oscillation (defined appropriately) over rectangles is bounded. Also, one does not have the expected atomic decomposition of functions in $H^p(\mathbb{R}^1 \times \mathbb{R}^1)$. Instead one has a characterization of the dual of $H^1(\mathbb{R}^1 \times \mathbb{R}^1)$ in terms of Carleson conditions with respect to open sets in $\mathbb{R}^2$ rather than rectangles and an atomic decomposition of $H^p(\mathbb{R}^1 \times \mathbb{R}^1)$ into atoms supported on open sets. This theory, together with the theorems on interpolation between $H^p(\mathbb{R}^1 \times \mathbb{R}^1)$ and $L^2(\mathbb{R}^2)$ provide the correct framework in order to allow the almost complete circumvention of the failure of the classical theory to extend to product spaces (see S. Y. Chang and R. Fefferman [5], [6]). The key geometric fact which combines with, say, the atomic decomposition of Hardy space in order to make this happen is due to J. L. Journé [2], [7]. This fact concerns the geometry of how maximal dyadic subrectangles sit inside an open set in $\mathbb{R}^n$, $n > 1$. Taking the case $n = 2$, suppose $\Omega \subset \mathbb{R}^2$ is open and of finite measure. Let $\mathcal{M}(\Omega)$ denote the family of all maximal dyadic subrectangles of Ω. Then $\sum_{R \in \mathcal{M}(\Omega)} |R|$ may well diverge, and this divergence is a serious obstacle. However, in [7] it is shown that

$$\sum_{R \in \mathcal{M}(\Omega)} |R| \gamma(R)^{-\delta} \leq C_\delta |\Omega| \text{ for any } \delta > 0,$$

where $\gamma(R)$ is a factor which reflects how much R can be stretched and still remain inside the expansion of Ω, $\widetilde{\Omega} = \{M_S(\chi_\Omega) > 1/2\}$.

When this geometric result is combined in the right way with the atomic decomposition, the outcome is quite surprising. To describe it, we make the following definition:

Definition 13.7. A function $a(x, y)$ supported in a rectangle $R = I \times J \subset \mathbb{R}^2$ is called an H^p *rectangle atom* provided

$$\int_I a(x, y) x^\alpha \, dx = 0 \text{ for all } \alpha = 0, 1, 2, \ldots, N_p,$$

where N_p is sufficiently large (for example, $N_p > 1/p - 2$ suffices),

$$\int_J a(x, y) y^\alpha \, dy = 0 \text{ for all } \alpha = 0, 1, 2, \ldots, N_p,$$

and

$$||a||_{L^2(\mathbb{R})} \leq |R|^{1/2 - 1/p}.$$

This means that a behaves like $a_1(x)a_2(y)$, where the a_i are classical $H^p(\mathbb{R}^1)$ atoms, although, of course, $a(x, y)$ need not actually be of the form $a_1(x)a_2(y)$. According to Carleson's counterexample, the H^p rectangle atoms do not span $H^p(\mathbb{R}^1 \times \mathbb{R}^1)$ as was expected prior to his work. However, we have

Theorem 13.8. [8]. *Fix $0 < p \leq 1$. Let T be a linear operator which is bounded in $L^2(\mathbb{R}^2)$ and which satisfies*

$$\int_{\widetilde{R}_\gamma} |T(a)|^p \, dx \, dy \leq C\gamma^{-\delta} \quad \text{for all } \gamma \geq 2 \text{ and for some } \delta > 0$$

(here $\widetilde{R}_\gamma$ denotes the γ-fold concentric enlargement of R) whenever a is a rectangle H^p atom supported in R. Then T is bounded from $H^p(\mathbb{R}^1 \times \mathbb{R}^1)$ to $L^p(\mathbb{R}^2)$.

Thus, while the space $H^p(\mathbb{R}^1 \times \mathbb{R}^1)$ has a complicated structure, and its atoms are based on opens sets in $\mathbb{R}^2$ as far as operator theory is concerned, this difficulty can be avoided. To check whether an operator acts boundedly on product H^p, one can simply check its action on (nonspanning) rectangle atoms. So, in an appropriate sense, the operator theory is better than that of the underlying functions spaces. This theme plays a central role as well when we consider weighted estimates in product spaces. Let us turn now to our overview of this issue.

To begin with, a certain amount of the theory is straightforward: We define $w \in A^p(\mathbb{R}^1 \times \mathbb{R}^1)$ by the condition that

$$\left(\frac{1}{|R|} \int_R w\right) \left(\frac{1}{|R|} \int_R w^{-1/p-1}\right)^{p-1} \leq C$$

for all rectangles R with sides parallel to the axes.

It is very easy to see that this product A^p condition is completely equivalent to the requirement that $w(\cdot, y)$ belong to $A^p(\mathbb{R}^1)$ with an A^p norm which is bounded as a function of y, and also the symmetric requirement applied to $w(x, \cdot)$. With this in mind it is simple to use the Jessen-Marcinkiewicz-Zygmund iterative proof to show that M_S is bounded on $L^p(w \, dx \, dy)$, $1 < p < \infty$, if and only if $w \in A^p(\mathbb{R}^1 \times \mathbb{R}^1)$. However, the weighted estimates for singular integrals are a different matter entirely. The reason for this is that in most of the examples of product operators given above iteration completely fails, and we must look to prove the desired estimates by understanding how the strong maximal operator controls the product singular integral. Unfortunately, we do not have good λ inequalities at our disposal, so it is far from obvious how to proceed.

It turns out that the solution involved an extension of the classical estimate

$$(Tf)^{\#}(x) \leq C[M(f^2)(x)]^{1/2},$$

where T is a classical Calderón-Zygmund singular integral, M is the Hardy-Littlewood maximal operator, and $\#$ is the (Charles) Fefferman-Stein sharp function. In order to extend this to product spaces, we have a similar obstacle to that facing us from product H^p or BMO theory. That is, we would be tempted to define a product version of the sharp function $f^{\#}(x, y)$ as $\sup_{(x,y) \in R} \text{osc}_R(f)$, where $\text{osc}_R(f)$ would denote some suitable notion of the (product) mean oscillation. Unfortunately, it can be shown [9] that it is possible to have $f^{\#} \in L^p(\mathbb{R}^2)$ but $f \notin L^p(\mathbb{R})$ for any $p > 2$. Once again, the operator theory is more satisfactory than the function theory, in the following sense:

Suppose T is a bounded linear operator on $L^2(\mathbb{R}^n)$. We say that a positive operator S is a *sharp operator* for T provided S has the property that not only is $Sf(x, y) \geq \sup_{(x,y) \in R} \text{osc}_R(Tf)$, but has the additional property that whenever R is a rectangle which contains (x, y) and f is a function supported outside of $\tilde{R}_\gamma$, $\gamma \geq 2$, then

$$Sf(x, y) \geq \gamma^\delta \text{osc}_R(Tf).$$

For example, for product singular integrals like the product commutators, we have

$$T^{\#} = M_S(f^2)^{1/2}.$$

The point is that, whereas for an individual function the condition $f^{\#} \in L^p$ is useless, we have the following:

Theorem 13.9 (R. Fefferman [8]). *If $T^{\#}$ is bounded in $L^p(\mathbb{R}^2)$, then T is bounded on $L^p(\mathbb{R}^2)$, for all $2 < p < \infty$.*

The proof of this consists of a duality estimate:

$$\int_{\mathbb{R}^2} S^2_{\mathrm{prod}}(Tf)\phi\,dx\,dy, \leq C \int_{\mathbb{R}^2} [I + T^{\#}](f)^2 M_S^{(N)}(\phi)\,dx\,dy,$$

where $M_S^{(N)}$ stands for the strong maximal operator composed with itself N times.

This estimate, which in the special case when $T = I$ is due to Wilson [10] is sufficiently strong to imply that whenever $T^{\#} = M_S(f^2)^{1/2}$ we have T bounded on $L^p(w\,dx\,dy)$ whenever $w \in A^{p/2}(\mathbb{R}^1 \times \mathbb{R}^1)$. Because the assumption $T^{\#} = M_S(f^2)^{1/2}$ includes singular integrals only bounded on $L^p(\mathbb{R}^2)$ when $p > 2$, the class $A^{p/2}$ is best possible. However, when we consider singular integrals bounded on the full range of $L^p(\mathbb{R}^2)$, $1 < p < \infty$, such as product commutators, then it is possible to extend this result to the sharp one, where T is bounded on $L^p(w\,dx\,dy)$ when $w \in A^p(\mathbb{R}^1 \times \mathbb{R}^1)$ (see R. Fefferman [11]). Here, we shall be content to remark that this is done by using the $A^{p/2}$ result above to obtain weighted estimates for the action of T on $H^1(\mathbb{R}^1 \times \mathbb{R}^1)$ atoms and then applying Rubio de Francia's Extrapolation Theorem.

It is interesting to note that, while one might interpret the discussion of the product theory above to mean that spaces of functions defined in terms of mean oscillations over rectangles or satisfying BMO conditions uniformly in each variable separately are of no use, this is actually quite wrong. In fact, several interesting results in this direction have been obtained recently by Cotlar and Sadosky [12].

For example, although our product BMO space can be used to show that certain singular integrals are bounded on L^p spaces, we have lost the connection with the natural class of product A^p weights. In [12], Cotlar and Sadosky find the correct space (they name it *bmo*) to establish this connection. A function ϕ belongs to *bmo* provided for each rectangle R with sides parallel to the axes,

$$\frac{1}{|R|}\int_R\int |\phi(x,y) - \phi_R|\,dx\,dy \leq C,$$

where ϕ_R denotes the mean value of ϕ over R. This is shown to be equivalent to the condition that $\phi(\cdot, y)$ belongs to $BMO(\mathbb{R}^1)$ with norm bounded as a function of y, and $\phi(x, \cdot) \in BMO(\mathbb{R}^1)$ with norm bounded in x. The point is then that bmo has the same relation to the classes $A^p(\mathbb{R}^1 \times \mathbb{R}^1)$ that classical BMO has to $A^p(\mathbb{R}^1)$.

Cotlar and Sadosky also introduce another space which extends to product dilations, the classical space $BMO(\mathbb{R}^1)$. This space, called restricted BMO, is defined as follows: Suppose P_x and P_y denote the analytic projection operators in the x and y variables, respectively. Then a function ϕ on $\mathbb{R}^2$ is in restricted BMO provided there exist L^∞ functions ϕ_0, ϕ_1 and ϕ_2 so that

$$(I - P_x)\phi = (I - P_x)\phi_1, \quad (I - P_y)\phi = (I - P_y)\phi_2, \text{ and } P_x P_y \phi = P_x P_y \phi_0.$$

Then this space of functions in restricted BMO is shown in [12] to be the natural space to consider when characterizing the product versions of Hankel operators. The same cannot be said of the Chang-Fefferman $BMO(\mathbb{R}^1 \times \mathbb{R}^1)$. So it is very clear that the latter space does not tell the whole story of extending BMO to product spaces.

The eventual goal of the program is to extend harmonic analysis past the realm of product spaces, considering other dilation groups, and the operators associated to them. This theory is just at its start, and it seems very difficult indeed at this point. Nevertheless, we would like to indicate here a few results in this direction. They are delicate, and, in some cases relate directly to issues which are new, even for singular integrals on $\mathbb{R}^1$.

The setting will be the next simplest after product space dilations, and those are as follows: In $\mathbb{R}^3$ consider the family of dilations $\{\rho_{\delta_1, \delta_2}\}_{\delta_1, \delta_2 > 0}$ given by

$$\rho_{\delta_1, \delta_2}(x, y, z) = (\delta_1 x, \delta_2 y, \delta_1 \delta_2 z).$$

Then the maximal operator invariant under these dilations was first considered by A. Zygmund:

$$M_s f(x, y, z) = \sup_{(x,y,z) \in R_s} \frac{1}{|R|} \int_R |f|,$$

where R_s denotes the family of rectangles with sides parallel to the axes, with side lengths of the form $\delta_1, \delta_2, \delta_1 \delta_2$ for some $\delta_1, \delta_2 > 0$. The singular integrals associated with the dilations were introduced by Ricci and Stein [13] in connection with problems involving singular integrals along surfaces. In order to motivate their definition of these singular integrals, let us consider a classical singular integral kernel $K(x)$ of $\mathbb{R}^1$. By examining this kernel on intervals of the dyadic decomposition of $\mathbb{R}^1$, it is easy to see that there exist functions $\{\phi^k\}_{k \in \mathbb{Z}}$ so that the ϕ^k are all supported in $[-1, 1]$, they all have mean value 0, and they satisfy a uniform smoothness condition (such as $|(\phi^k)'(x)| \leq C$ for all x) while

$$K(x) = \sum_{k \in \mathbb{Z}} 2^{-k} \phi^k \left(\frac{x}{2^k} \right).$$

In analogy with this, Ricci and Stein introduced convolution kernels of the form

$$K(x, y, z) = \sum_{k,j \in \mathbb{Z}} 2^{-2(k+j)} \phi^{k,j} \left(\frac{x}{2^k}, \frac{y}{2^k}, \frac{z}{2^k 2^j} \right),$$

where the functions $\phi^{k,j}$ are supported in the unit cube of $\mathbb{R}^3$, have a certain amount of uniform smoothness and each satisfy the cancellation condition

$$\int_{\mathbb{R}^2} \phi^{k,j}(x, y, z)\, dx\, dy = 0 \text{ for all } z \in \mathbb{R}^1,$$

$$\int_{\mathbb{R}^2} \phi^{k,j}(x, y, z)\, dx\, dz = 0 \text{ for all } y \in \mathbb{R}^1,$$

$$\int_{\mathbb{R}^2} \phi^{k,j}(x, y, z)\, dy\, dz = 0 \text{ for all } x \in \mathbb{R}^1.$$

It is shown in [13] that the requirement of having mean value zero in each pair of variables fixing any value of the remaining variable is the correct cancellation condition to place on the $\phi^{k,j}$ because without this cancellation it is shown that one does not have (when all $\phi^{k,j}$ are the same) boundedness of $T_j f = f * K$ on $L^2(\mathbb{R}^2)$ and with it, one has T_j bounded on $L^p(\mathbb{R}^3)$, for all $1 < p < \infty$.

The idea required to prove the boundedness of T_j on $L^p(\mathbb{R}^3)$ is a clever decomposition of each $\phi^{k,j}$ into a sum of functions that have "product space cancellation," i.e., mean value zero in each variable separately. The functions will, however, only be supported in various rectangles of arbitrary dimension, and so what one realizes in the end is that T_j can be viewed as a product operator. Here, we shall discuss problems where the product theory cannot be applied in this fashion. The first such example is A. Cordoba's solution [14] of Zygmund's conjecture for the maximal function M_j:

$$m\{(x, y, z) \in Q : M_j(f)(x, y, z) > \alpha\} \leq \frac{C}{\alpha} \|f\|_{L \log^+ L(Q)}, \tag{13.1}$$

where Q denotes the unit cube in $\mathbb{R}^3$. Cordoba's theorem means that M_j is a maximal operator with respect to a two-parameter family of rectangles (i.e., a two-parameter family of dilations acting on the unit cube) and so M_j behaves as though it were the strong maximal operator in $\mathbb{R}^2$, the model two-parameter operator. It would clearly be of no use to appeal to the majorization of $M_{\mathfrak{z}}$ be the three-dimensional strong maximal operator which, of course, fails to satisfy (13.1).

Instead, what is used is the method (started in Cordoba and R. Fefferman [15] and perfected in Stromberg [16] and Cordoba and R. Fefferman [17]) of controlling more complicated maximal operators by simpler ones by means of a covering lemma.

The next example of such a problem occurs in obtaining weighted estimates for $M_{\mathfrak{z}}$ and $T_{\mathfrak{z}}$. The reason that the weighted problems cannot be solved by a direct appeal to the product theory becomes clear after a cursory glance at the problem.

To begin with, there is an obvious guess at what is the correct class of weights to consider. This is the class of $A^p(\mathfrak{z})$ defined by

$$\left(\frac{1}{|R|}\int_R w\right)\left(\frac{1}{|R|}\int_R w^{-1/p-1}\right)^{p-1} \leq C \text{ for all } R \in R_{\mathfrak{z}},$$

where $R_{\mathfrak{z}}$ denotes Zygmund's class of rectangles, i.e., those whose z side length equals the product of their x and y side length. It will turn out that this $A^p(\mathfrak{z})$ condition is necessary and sufficient for a weight w so that $M_{\mathfrak{z}}$ and $T_{\mathfrak{z}}$ are bounded on $L^p(w)$, but one cannot get this directly from product theory because there are many examples of $w \in A^p(\mathfrak{z})$ which are not product space weights, i.e., which fail to satisfy the A^p condition on all rectangles with sides parallel to the axes. Thus if $M_s^{(3)}$ denotes the strong maximal operator on $\mathbb{R}^3$, $M_s^{(3)}$ will be unbounded (in general) on $L^p(w)$. So knowing that $M_{\mathfrak{z}}f \leq M_s^{(3)}f$ or that $T_{\mathfrak{z}}$ can be viewed as a product singular integral is of no use here. Let us describe the solution of this problem from R. Fefferman and J. Pipher [18]:

We begin with the weighted estimates for $M_{\mathfrak{z}}$. It turns out that while $M_{\mathfrak{z}} \leq M_s^{(3)}$ does us no good, the key observation is that $M_{\mathfrak{z}}^w \leq M_s^w$, solves the problem, where the superscript w means that w measure is substituted for Lebesgue measure everywhere in the definition of the operator. (For example,

$$M_s^w f(x,y,z) = \sup_{(x,y,z)\in R} \frac{1}{w(R)}\int_R |f|w,$$

and the sup is taken over all R with sides parallel to the axes.) The reason for this is that M_s^w is bounded on $L^p(w)$ when $w \in A^p(\mathfrak{z})$ even though M_s is unbounded there. The reason for this boundedness of M_s^w is not obvious, but lies in the theory of the covering lemma techniques in [15]. In order to prove M_s^w is bounded on $L^p(w)$, we think of proving an appropriate covering lemma as in [15], but with w measure replacing Lebesgue measure. Now, in the covering lemma, the main idea is to order the given rectangles in decreasing order of their z side length. Then the rectangles are sliced by a plane parallel to the (x, y) plane, and the two-dimensional slices are then analyzed. So it is not surprising that, in the w-measure version of the covering lemma, one only needs w to be well behaved in the x and y variables only (not in z).

However, by simply observing that intervals oriented in the x direction are collapsed versions of rectangles in Zygmund's class, it follows from the definition of $A^p(\mathfrak{z})$ that weights $w \in A^p(\mathfrak{z})$ are classical A^p weights in the x variable for each y and z and similarly they are uniformly in A^p of the y variable for each fixed x and z. Therefore each $w \in A^p(\mathfrak{z})$ is, in fact, well behaved in the required two (out of three) directions, proving the desired weighted estimates for $M_{\mathfrak{z}}$.

Now, there are no good λ estimates to show how $M_{\mathfrak{z}}$ controls $T_{\mathfrak{z}}$, so there is no immediate way to derive weighted estimates of $T_{\mathfrak{z}}$ from those of $M_{\mathfrak{z}}$. We therefore proceed as follows: Our idea, as always in Calderón-Zygmund theory, is to see why L^2 theory is special and somehow simpler than L^p theory, $p \neq 2$. Once we do this, we shall apply Rubio de Francia's extrapolation theorem in order to establish that $T_{\mathfrak{z}}$ is bounded on $L^p(w)$ when $w \in A^p(\mathfrak{z})$. We should make it clear that Rubio's extrapolation is not a completely abstract machine which works in the setting of an arbitrary dilation group. So it turns out that in order to apply this theorem, we need to know exactly one fact in advance: the weighted norm theory for the appropriate maximal operator.

This fits our intuition on Calderón-Zygmund theory in the sense that the maximal operator's behavior is what is necessary to know in order to understand properly the singular integral. Rather than the decomposition into good and bad parts of a function using the maximal operator to pass from L^2 to L^p, we use the maximal operator (in our case $M_{\mathfrak{z}}$) to be able to apply Rubio's theorem.

Now we return to the question of what is so special about the L^2 theory. For the case of product dilations, the weighted norm inequalities for convolution operators analogous to $T_{\mathfrak{z}}$ can be done on $L^p(w)$, $1 < p < \infty$ by iteration. In the case of other dilations such as Zygmund's, iteration only works when $p = 2$. This is done as follows:

We introduce the appropriate square function $S_{\mathfrak{z}}f$.

$$S_{\mathfrak{z}}f = \left(\int_{\Gamma_{\mathfrak{z}}(0)} \cdots \int |f * \Psi_{s,t}(x+u, y+v, z+w)|^2 \, \frac{du\, dv\, dw\, ds\, dt}{s^3 t^3} \right)^{1/2},$$

where the Zygmund cone

$$\Gamma_{\mathfrak{z}}(0) = \{(u, v, w, z, t) : |u| < s, |v| < t, \text{ and } |w| < st \text{ while } s, t > 0\},$$

and where $\Psi \in C_c^\infty(\mathbb{R}^3)$ has mean value zero in each pair of variables, fixing any value of the other, and finally, where

$$\Psi_{s,t}(x, y, z) = s^{-2} t^{-2} \Psi\left(\frac{x}{s}, \frac{y}{t}, \frac{z}{st}\right).$$

It turns out that $S_3(T_3 f)$ is very well controlled by $S_3(f)$ so that what we really require is the weighted Littlewood-Paley theorem,

$$\frac{||S_3(f)||_{L^p(w)}}{||f||_{L^p(w)}}$$

is bounded above and below by constants depending only on p, $1 < p < \infty$ and on $||w||_{A^p(3)}$.

The Littlewood-Paley theorem is obtainable by iteration for L^2 only. This is accomplished by an observation regarding functions which have the appropriate cancellation, for the definition of T_3, namely having vanishing mean taken in each pair of variables separately. It turns out that any $\Psi(x, y, z)$ having this cancellation can be written as a sum

$$\Psi(x, y, z) = \Psi_1(x, y, z) + \Psi_2(x, y, z),$$

where Ψ_1 and Ψ_2 have stronger cancellation than does Ψ: Ψ_1 has mean zero in the x variable for all values of y and z and has mean zero in the (y, z) variables (taken together) for each fixed x; Ψ_2 will have mean zero in the y variable and in the (x, z) variables. We can then write $T_3 = T_3^1 + T_3^2$ where the bump functions defining the convolution kernel for T_3^1 have the extra cancellation of Ψ_1 above and similarly the kernel for T_3^2 will be formed out of dilates of functions with the extra cancellation of Ψ_2 above. How does this help us? This makes it possible to control T_3^1 using a Littlewood-Paley function S_3 defined by means of a Ψ of the form $\Psi(x, y, z) = \eta(x)\theta(y, z)$ where η and θ have mean 0 in the x and (y, z) variables, respectively. This "separation of variables" makes it possible to control

$$\frac{||S_3(f)||_{L^p(w)}}{||f||_{L^p(w)}}$$

when $p = 2$ directly by integration, and this completes the proof of the weighted estimates for T_3 (for details, see R. Fefferman and J. Pipher [18]).

There is another deep problem concerning the singular integral T_3 that should be mentioned. Here we seek (Lebesgue measure) L^p estimates which express the fact that T_3 is a singular integral associated with a two-parameter family of dilations. Thus in the spirit of Cordoba's estimate for M_3 we might suspect that the operator norm of T_3 on $L^p(\mathbb{R}^3)$ is $O(p^2)$ as $p \to \infty$. In [18], we are not able to get this, but we are able to improve upon the bound $O(p^3)$ which follows from the Ricci-Stein approach using an appeal to product theory in three parameters. In [18] we obtain

$$||T_3||_{L^p(\mathbb{R}^3)} = O(p^{5/2}) \text{ as } p \to \infty,$$

and here we would just like to sketch one of the main ingredients in the proof of this estimate, because it involves some new observations about classical operators, such as the Hilbert transform, and because it has applications in other settings as well.

In attempting to estimate $||T_\delta||_{L^p(\mathbb{R}^3)}$, one is faced with the following question: For a classical operator, such as the Hilbert transform, does there exist an L^2 estimate which implies the sharp estimate of the operator norm of H on L^p as $p \to \infty$? (Of course we require the sharp estimate $||H||_{L^p(\mathbb{R}^1)} = O(p)$ as $p \to \infty$.)

The same question can be asked for the classical square function, where

$$||Sf||_{L^p(\mathbb{R}^1)} \le Cp^{1/2}||f||_{L^p(\mathbb{R}^1)} \text{ for large } p.$$

This is best seen via a duality inequality in Chang, Wilson, and Wolff [19]

$$\int_{\mathbb{R}^1} (Sf)^2(x)\phi(x)\,dx \le C \int_{\mathbb{R}^1} f^2(x)M(\phi)(x)\,dx,$$

and one could hope to answer the question for the Hilbert transform by proving

$$\int (Hf)^2(x)\phi(x)\,dx \le C \int f^2(x)M(M(\phi))(x)\,dx.$$

Unfortunately this is false, but we do have, for example,

$$\int (Hf)^2(x)\phi(x)\,dx \le C \int f^2(x)M(M(M(\phi)))(x)\,dx$$

(see Wilson [10]). This, however, does not yield sharp information about $||H||_{L^p(\mathbb{R}^1)}$. The way around this lies in the idea of what might be called "sharp weighted inequalities."

We know very well that H is bounded on $L^2(w)$ if $w \in A^2(\mathbb{R}^1)$, and

$$||Hf||_{L^2(w)} \le C_w||f||_{L^2(w)},$$

where C_w depends only on $||w||_{A^2(\mathbb{R}^1)}$. The point is to ask precisely for the sharp dependence of C_w upon $||w||_{A^2}$ or at least some norm of w in an appropriate weight space. In [18] we show that

$$\left(\int_{\mathbb{R}^1} f^2(x)w(x)\,dx \right)^{1/2} \le C||w||_{A^1(\mathbb{R}^1)}^{1/2} \left(\int_{\mathbb{R}^1} (Sf)^2(x)w(x)\,dx \right)^{1/2}$$

and

$$\left(\int_{\mathbb{R}^1} (Hf)^2(x)w(x)\,dx \right)^{1/2} \le C||w||_{A^1(\mathbb{R}^1)} \left(\int_{\mathbb{R}^1} f^2(x)w(x)\,dx \right)^{1/2}.$$

Considering, for example, the second of these estimates, we claim that it easily implies the sharp estimate for $||H||_{L^p(\mathbb{R}^1)}$:

Let $p > 2$, and $||\phi||_{(p/2)'} = 1$. Then, following Rubio de Francia, define

$$v = \phi + \frac{M(\phi)}{2||M||_{L^{(p/2)'}}} + \frac{M \circ M(\phi)}{(2||M||_{L^{(p/2)'}})^2} + \cdots$$

Then $||v||_{L^{(p/2)'}} \le 2$ and $||v||_{A^1(\mathbb{R}^1)} \le 2||M||_{L^{(p/2)'}} = O(p)$. Taking into account the sharp weighted estimate above gives

$$
\begin{aligned}
\int_{\mathbb{R}^1} (Hf)^2(x)\phi(x)\,dx &\le \int_{\mathbb{R}^1} (Hf)^2(x)v(x)\,dx \\
&\le C^2||v||^2_{A^1(\mathbb{R}^1)} \int_{\mathbb{R}^1} f^2(x)v(x)\,dx \\
&\le C^2 p^2 \left(\int |f(x)|^p\,dx \right)^{2/p},
\end{aligned}
$$

which proves that $||H||_{L^p(\mathbb{R}^1)} = O(p)$.

It is interesting that the first of the sharp weighted estimates above dealing with the control of f by Sf with constant $O(||w||^{1/2}_{A^1})$ is shown in [18] to be completely equivalent to the estimates by Chang, Wilson, and Wolff which show that if $Sf \in L^\infty(T)$, then f^2 is exponentially integrable. And this allows us to give a very simple treatment of the problem in the thesis of Jill Pipher: If S_{prod} denotes the product square function on T^2 then in [20] Pipher shows that the Chang-Wilson-Wolff result extends to T^2: If $S_{\text{prod}}(f) \in L^\infty(T^2)$, then $|f|$ is exponentially (not square exponentially) integrable.

The reason that this is nontrivial is that one seeks to reduce this, by an iteration argument, to the one-dimensional case. This can be done by a standard procedure, but with the complication that one needs the one-dimensional result for Hilbert space valued functions $\vec{f}(x)$, and if one looks at the Chang-Wilson-Wolff estimate, this means we need to handle $\int e^{|\vec{f}(x)|}$ where $|\cdot|$ denotes the Hilbert space norm. This can be done, but it is delicate and is not a direct consequence of the scalar valued case. However, by observing that the Chang-Wilson-Wolff theorem in completely equivalent to the (quadratic) sharp weighted inequalities, the extension to the Hilbert space valued functions is absolutely immediate, and so we have an extremely simple proof of the extension of this theorem to product spaces.

References

[1] B. Jessen, J. Marcinkiewicz, and A. Zygmund, *Notes on the differentiability of multiple integrals*, Fund. Math. 24 (1935).

[2] J. L. Journe, *Calderón-Zygmund operators on product spaces*, Revista Math. Iberamericana 1 (1985).

[3] R. Gundy and E. M. Stein, H^p *theory for the polydisk*, Proc. Nat. Acad. Sci. 76 (1979).

[4] L. Carleson, *A counterexample for measures bounded on H^p for the bidisk*, Mittag-Leffler Report, no. 7 (1974).

[5] S. Y. Chang and R. Fefferman, *A continuous version of the duality of H and BMO on the bi-disk*, Annals of Math 112, no. 2 (1980).

[6] ———, *The Calderón-Zygmund decomposition for the poly-disk*, Amer. Journal of Math 104, no. 3 (1982).

[7] J. L. Journe, *A covering lemma for product spaces*, Proc. AMS 96, no. 4 (1986).

[8] R. Fefferman, *Harmonic analysis on product spaces*, Annals of Math. 126, no. 1 (1987).

[9] ———, *On Functions of bounded mean oscillation on the poly-disk*, Annals of Math. 110, no. 2 (1979).

[10] M. Wilson, *Weighted norm inequalities for the continuous square function*, Trans. AMS 314 (1989).

[11] R. Fefferman, A^p *weights and singular integrals*, Amer. Journal of Math. 110 (1988).

[12] M. Cotlar and C. Sadosky, *Two Distinguished Subspaces of Product BMO, and the Nehari-AAK Theory for Hankel Operators on the Torus*, Integral Eq. and Operator Theory 26, no. 3, 1996.

[13] F. Ricci and E. M. Stein, *Multiparameter singular integrals and maximal functions*, Annals Inst. Fourier (Grenoble) 42 (1992).

[14] A. Cordoba *Maximal functions, covering lemmas, and Fourier multipliers*, Proc. Symposia in Pure Math. 35 (1979).

[15] A. Cordoba and R. Fefferman, *A geometric proof of the strong maximal theorem*, Annals of Math 102 (1975).

[16] J. O. Stromberg, *Weak estimates on maximal functions with rectangles in certain directions*, Arkiv. für Math. 15 (1977).

[17] A. Cordoba and R. Fefferman, *On differentiation of integrals,* Proc. National Acad. of Sci. 74 (1977).

[18] R. Fefferman and J. Pipher, *Multiparameter operators and sharp weighted inequalities,* Amer. Journal of Math. 119, no. 2 (1997).

[19] S. Y. Chang, M. Wilson, and T. Wolff, *Some weighted norm inequalities concerning the Schrödinger operators,* Comment. Math. Helv. 60 (1985).

[20] J. Pipher, *Bounded double square functions,* Ann. Inst. Fourier (Grenoble) 36, no. 2 (1986).

Chapter 14

Nodal Sets of Sums of Eigenfunctions

David Jerison and Gilles Lebeau

14.1 Introduction

In the late 1980s Harold Donnelly and Charles Fefferman [7] showed how to estimate the rate of vanishing of eigenfunctions and the size of their zero sets. Their results say roughly that eigenfunctions resemble polynomials of degree comparable to the square root of the eigenvalue. But the set of polynomials up to a fixed degree is closed under addition. So it is natural to ask whether sums of eigenfunctions have the same polynomial-like properties as single eigenfunctions.

This paper extends the estimates of Donnelly and Fefferman from single eigenfunctions to sums. As in the original proofs, the key tools in our approach are Carleman inequalities. Carleman inequalities or closely related techniques are at the heart of most approaches to uniqueness of solutions to partial differential equations. It is entirely fitting to discuss the Carleman method at this conference since the fundamental paper of Alberto Calderón [5] on uniqueness for partial differential equations is a spectacular extension of the Carleman method from equations in two variables to systems in several variables.

The point of view that we adopt will lead us naturally to some recent work and unsolved problems in the subject of Carleman inequalities. Namely, we will discuss problems related to uniqueness for nonlocal (pseudodifferential) operators with minimal smoothness. These minimal smoothness questions

Research by David Jerison partially supported by NSF grant DMS-9401355.

are directly inspired by the Calderón program for singular integral operators with minimal smoothness. There are other new results that we will not discuss. In particular, we alert the reader to important works of Thomas Wolff surveyed in [28, 29].

The approach here simplifies slightly the original proof of Donnelly and Fefferman by eliminating the need for a technically demanding version of Carleman inequalities. But we wish to emphasize that this paper does not prove any new theorems. Rather it recovers results of Fang-Hua Lin [22]. Lin was led to similar questions through an attempt to extend results of Donnelly and Fefferman to solutions of the heat equation. The details of the present method are different only because Lin does not rely on Carleman inequalities. There are also more recent and more general estimates due to Igor Kukavica [18, 19, 20].

14.2   Statements of Results

Let M be a compact C^∞ manifold. Consider an eigenfunction φ satisfying

$$\Delta\varphi = -\omega^2\varphi.$$

If the boundary of M is empty, then there are no boundary conditions, but if the boundary is nonempty, then we impose either Dirichlet boundary conditions ($\varphi = 0$ on ∂M) or Neumann boundary conditions ($\partial\varphi/\partial\nu = 0$ on ∂M). The results of Donnelly and Fefferman can be stated as follows.

Theorem 14.1 (Doubling). *There are constants C_1 and C_2 depending only on M such that*

$$\max_{B(2r)} |\varphi| \le e^{C_1\omega+C_2} \max_{B(r)} |\varphi|,$$

where $B(r)$ and $B(2r)$ represent concentric balls of M. In particular, φ vanishes at most at a rate $r^{C_1\omega+C_2}$.

Theorem 14.2. *If M is real-analytic, then there are constants C_1 and C_2 depending only on M such that the $(n-1)$-dimensional Hausdorff measure $\mathcal{H}_{n-1}$ of the zero set of φ satisfies*

$$\mathcal{H}_{n-1}(\{x \in M : \varphi(x) = 0\}) \le C_1\omega + C_2.$$

These theorems describe the sense in which φ resembles a polynomial of degree at most $C_1\omega + C_2$. As already observed in [6], one cannot hope for a lower bound for the Hausdorff measure of the nodal set because in the case of sums of eigenfunctions the nodal sets can be the empty set.

Let φ_k be an eigenfunction with eigenvalue ω_k^2, that is, $\Delta\varphi_k = -\omega_k^2\varphi$, with Dirichlet or Neumann boundary conditions in the case that the boundary is nonempty. We will prove

Theorem 14.3. *Theorems 14.1 and 14.2 are valid for any sum of the form*

$$\sum_{\omega_k \leq \omega} a_k \varphi_k.$$

How should one characterize a sum of eigenfunctions? Donnelly [6] uses the fact that φ satisfies

$$(\Delta + \omega_1^2)(\Delta + \omega_2^2) \cdots (\Delta + \omega_m^2)\varphi = 0.$$

But then he is forced to deal with a higher-order operator. He obtains the same type of estimates as above for sums of eigenfunctions, but his constants depend not only on ω, but also on the number of terms in the sum. More recently Kukavica [18, 19, 20] has proved estimates for Δ^k that are uniform in k and he has deduced Theorem 14.3 .

We will follow an approach similar to Lin [22]. Namely, consider a function F defined on $\mathbb{R} \times M$ by

$$(\partial_t^2 + \Delta)F \text{ in } \mathbb{R} \times M; \quad F(0, x) = 0; \quad \partial_t F(0, x) = \varphi(x).$$

The solution to this problem is

$$F(t, x) = \sum a_k \frac{\sinh(\omega_k t)}{\omega_k} \varphi_k(x).$$

Denote $X_T = [-T, T] \times M$. Then the quantitative property that characterizes the "degree" of our sum of eigenfunctions is (for, say, $T = 1, 2$)

$$\|F\|_{H^1(X_T)} \approx e^{T\omega} \|\varphi\|_{L^2(M)}. \tag{14.1}$$

(Here $H^1(X_T)$ denotes the Sobolev space of functions with first derivatives in $L^2(X_T)$. This discussion is not very sensitive to the particular choice of function spaces.)

The case of the constant function (with zero eigenvalue) has to be discussed separately:

$$\lim_{\omega \to 0} \frac{\sinh(\omega t)}{\omega} = t.$$

Therefore, in the case $\omega_k = 0$, the expression $\omega_k^{-1} \sinh(\omega_k t)$ is replaced by the function t. The constant function is always an eigenfunction in the Neumann problem and in the case of empty boundary. This gives the extra dividend that Theorem 14.3 applies not only to the zero set, but also any level set $\{\varphi = \alpha\}$. In fact, we can even add the constant function in the Dirichlet case because the comparison inequality (14.1) is still valid. (This remark will be proved in an appendix.) Thus in all cases the theorem applies to level sets as well as the zero set.

14.3   Carleman Inequalities

Let $L = \partial_t^2 + \Delta$. The first Carleman-type inequality that we need is

$$\int |\rho^{-\beta} Lg|^2 \geq c \int \left[|\rho^{-\beta} \nabla g|^2 + |\rho^{-\beta} g|^2 \right] \text{ for all } g \in C_0^\infty(B \backslash \{0\}). \quad (14.2)$$

Here B is the ball of fixed small radius around the origin 0 and ρ is a carefully chosen function comparable to the distance to the origin. The main point is that the inequality is valid uniformly as $\beta \to \infty$. This inequality was proved by Aronszajn [3] in 1957 in the C^∞ case. When the coefficients of the operator L are of class $C^{0,1}$, it was proved by Aronszajn, Krzywicki, and Szarski in 1963 [4]. The case of nonsmooth coefficients is used in the proof in the case in which the boundary is nonempty. (See [8] and the remarks later on.)

The proofs in [4] are not easy to read. The best reference for a proof of (14.2) is Proposition 2.10 of [7] (smooth case) and [8] ($C^{0,1}$ case). The statement omits the gradient term on the right-hand side. But inequality (2.9) of [8] and the subsequent inequality $K \geq I$ leading to the proof of Proposition 2.10 in that paper show that the gradient term can be included. As mentioned before, there is the slight simplification that we need only the case $\lambda = 0$.

It is not hard to deduce from (14.2) that the following inequality holds.

Lemma 14.4. *For any nonempty open set V, $V \subset X_1$, there exists $\alpha > 0$ and C such that*

$$\|F\|_{H^1(X_1)} \leq C \|F\|_{H^1(V)}^\alpha \|F\|_{H^1(X_2)}^{1-\alpha}.$$

Rather than carry out the proof of Lemma 14.4, we refer to the analogous proof of Lemma 14.5 below. An immediate consequence of Lemma 14.4 is the *unique continuation property* for the operator L, namely, if $F = 0$ on V, then F is identically zero on all of X_1.

For the remainder of the paper, we use the shorthand M for the subset $\{0\} \times M$ in X_T. The next lemma makes the connection between F and its values on M.

Lemma 14.5. *Let B be a ball in X_2 centered on M and disjoint from ∂M. Let $B^+ = \{(t, x) \in B : t > 0\}$. Suppose that $V \subset B^+$ and that $F(0, x) = 0$. There exists $\alpha > 0$ and C depending only on V, B and M such that*

$$\|F\|_{H^1(V)} \leq C \|\partial_t F\|_{L^2(B \cap M)}^\alpha \|F\|_{H^1(B^+)}^{1-\alpha}.$$

This lemma implies another unique continuation result; namely, if F and $\partial_t F$ are zero on $B \cap M$, then F is identically zero in B. It can be deduced from a boundary Carleman-type inequality

$$\int \int_{B^+} |(\partial_t^2 + \Delta)g|^2 e^{2\beta\psi} \, dt \, dx \quad + \quad \beta \int_{B\cap M} (\partial_t g)^2 e^{2\beta\psi} \, dx$$
$$\geq \quad c \int \int_{B^+} (\beta |\nabla_{t,x} g|^2 + \beta^3 g^2) e^{2\beta\psi} \, dt \, dx$$

for all $g \in C_0^\infty(B)$ such that $g = 0$ on M. The hypothesis on ψ that is needed will be discussed later. This Carleman inequality was proved by Lebeau and Robbiano [21] and used there for an application to the theory of optimal control for solutions of the heat equation.[1] Closely related inequalities are proved in [16]. We are indebted to Louis Nirenberg for pointing out what may be the earliest instance of inequalities of the type in Lemma 14.5 (in different function spaces and for equations with real analytic coefficients) in the work of Fritz John [14,15]. John's main theme is that the extra hypothesis of boundedness of the solution over the larger set X_2 restores something resembling well-posedness of what otherwise is an ill-posed Cauchy problem in the sense of Hadamard.

14.4   Main Idea of the Carleman Method

The idea of the Carleman method is to prove inequalities with weights $e^{\beta\psi}$ with ψ largest on the set from which the uniqueness propagates. Thus we want ψ to take its largest values on $B \cap M$.[2] On the other hand, there is a kind of convexity constraint on ψ, to be specified later, that will be needed in order for inequality (14.3) to be true.

Suppose, without loss of generality, that $(0,0)$ is the center of B. We let

$$\psi(t,x) = -t + t^2/2 - |x|^2/4. \tag{14.3}$$

This function was chosen so that (14.3) is valid, as we shall see below. Pick three numbers $0 > \psi_1 > \psi_2 > \psi_3$. Define $S_i = \{\psi > \psi_i\}$ and $S_i^+ = S_i \cap \{t > 0\}$. Thus $S_1 \subset S_2 \subset S_3$. Let $\chi \in C_0^\infty(S_3)$ satisfy $\chi = 1$ on S_2^+ and define $g = \chi F$. Assume for simplicity that Δ is the ordinary Laplace operator (for

---

[1]They show that given any nonempty open set U in (t,x)-space with $0 < t < T$ and any initial condition $u(0,x) = f(x)$ one can choose an inhomogeneous term Q supported in U so that the solution to $(\partial_t + \Delta)u = Q$ satisfies $u(T,x) = 0$. In other words, any disturbance (initial condition) can be restored to zero using a "control" (inhomogeneous term) supported in an arbitrarily small region of space-time.

[2]Similarly, in (14.2) the key feature for applications of the function $\rho^{-\beta}$ is that it is largest ($+\infty$ in fact) at the central point. Thus the Aronszajn inequality actually proves what is called strong unique continuation; namely, a function satisfying an elliptic equation that vanishes to infinite order at a point must vanish identically.

a general Laplace-Beltrami one has to use normal coordinates and make a somewhat more detailed computation):

$$(\partial_t^2 + \Delta)g = 2\nabla_{t,x}\chi \cdot \nabla_{t,x}F + ((\partial_t^2 + \Delta)\chi)F = 0 \text{ on } S_2^+.$$

Substituting into (14.3), we have

$$e^{2\beta\psi_2}||F||_{H^1(S_3^+\backslash S_2^+)}^2 + ||\partial_t F||_{L^2(S_3\cap M)}^2 \geq ce^{2\beta\psi_1}||F||_{H^1(S_1^+)}.$$

The term on the left is increased if one replaces $H^1(S_3^+\backslash S_2^+)$ with $H^1(S_3^+)$. (It may seem as if this gives away too much. But we are retaining the very small factor $e^{2\beta\psi_2}$.) Now by ordinary arithmetic, selecting the best choice of β, one can find $\alpha > 0$ depending on $\psi_1 > \psi_2$ for which the conclusion of Lemma 14.5 holds.

Now we come to the proof of (14.3). Define

$$L_\beta f = e^{\beta\psi}(\partial_t^2 + \Delta)(e^{-\beta\psi}f).$$

This conjugated operator is the same as the one treated by Gunther Uhlmann in chapter 19 and by coincidence we have used the same notation L_β.[3] We change variables by replacing g by the function $f = e^{-\beta\psi}g$. Then

$$\int\int[(\partial_t^2 + \Delta)g]^2 e^{2\beta\psi} = \int\int(L_\beta f)^2.$$

Decompose L_β into its self-adjoint and skew-adjoining parts.

$$L_\beta = P + Q, \quad P^* = P, \quad Q^* = -Q.$$

Then

$$\begin{aligned}
||L_\beta f||^2 &= ((P+Q)f, (P+Q)f) \\
&= ||Pf||^2 + ||Qf||^2 + (Pf, Qf) + (Qf, Pf)
\end{aligned}$$

and

$$\begin{aligned}
(Pf, Qf) &= (f, PQf) + \text{ boundary terms} \\
&= (f, [P,Q]f) + (f, QPf) + \text{ boundary terms} \\
&= (f, [P,Q]f) - (Qf, Pf) + \text{ boundary terms}.
\end{aligned}$$

---

[3]Paul Malliavin informed us in his lecture that in 1954 Alberto Calderón had already essentially proved his result on the inverse problem discussed by Gunther Uhlmann, although his paper on that subject appeared much later. Thus very early in his career Calderón was preoccupied with the analysis of operators conjugated by real as well as imaginary exponentials, that is, Fourier analysis in the complex domain. The fact that he was considering Carleman inequalities and this inverse problem at about the same time makes the connections between the two problems seem far from coincidental.

In all,

$$||L_\beta f||^2 = ||Pf||^2 + ||Qf||^2 + (f, [P,Q]f) + \text{ boundary terms.}$$

Therefore, in order to find a lower bound for $||L_\beta f||^2$ we will want some condition on ψ that guarantees that

$$[P,Q] >> 0 \text{ modulo } P \text{ and } Q.$$

This is a convexity condition on ψ. The condition can be phrased in terms of the symbol of the operators P and Q. It is important to require uniform control with respect to the parameter β, which is similar to an extra dual variable.

Let us carry out the computations explicitly. Recall that ψ was given in (14.3) and Δ is the ordinary Laplace operator. Then

$$P = \partial_t^2 + \Delta + \beta^2 |\nabla_{t,x}\psi|^2 + \beta(\partial_t^2 + \Delta)\psi$$

$$Q = -2\beta\nabla_{t,x}\psi \cdot \nabla_{t,x} = \beta(\partial_t^2 + \Delta)\psi.$$

Note that ψ was chosen so that $\nabla_{t,x}\psi(0,0) = (-1,0)$ and the Hessian of ψ at $(0,0)$ is the diagonal matrix with first entry 1 and $-1/2$ as the remaining nonzero diagonal elements. Therefore

$$[P,Q] = -2\beta\partial_t^2 + \beta\Delta + 4\beta^3 + \text{ lower order terms.}$$

The first and last terms on the right-hand side are positive as operators, but the middle term $\beta\Delta$ is not. But we only need positivity modulo P and Q. This "bad" term can be absorbed by adding and subtracting a suitable multiple of P to obtain

$$[P,Q] = 2\beta P - 4\beta\partial_t^2 - \beta\Delta + 2\beta^3 + \text{ lower order terms.}$$

The term $4\beta^3$ was very helpful and the multiple of P that was subtracted needed to be small enough to cancel only part of the $4\beta^3$ term. Furthermore, because $f = 0$ on $M = \{t = 0\}$,

$$\text{boundary term} = -2\beta \int_{t=0} (\partial_t f)^2 \, dx.$$

Combining the formulas above one obtains

$$||L_\beta f||^2 + 2\beta \int_{t=0} (\partial_t f)^2 \, dx$$
$$\geq ||Pf||^2 + ||Qf||^2 + \beta^3 ||f||^2 + (f, -4\beta\partial_t^2 f) + (f, -\beta\Delta f) + 2\beta(f, Pf)$$
$$\geq \beta ||\nabla_{t,x} f||^2 + \tfrac{1}{2}\beta^3 ||f||^2$$
$$\geq \tfrac{1}{4} \left[\beta ||(\nabla_{t,x} g)e^{\beta\psi}||^2 + \tfrac{1}{2}\beta^3 ||ge^{\beta\psi}||^2 \right].$$

This proves the Carleman-type inequality (14.3). Arguments of the kind just given, involving commutators and integration by parts, can be found in Hörmander's text [12, p. 182] where they are attributed to F. Treves [26]. They are closely related to the arguments of Calderón [5]. The same device was used by J. J. Kohn to make subelliptic estimates of operators arising in analysis in several complex variables [17].

14.5   Applications to Sums of Eigenfunctions

Lemma 14.4 says

$$||F||_{H^1(X_1)} \leq C||F||_{H^1(V)}^\alpha ||F||_{H^1(X_2)}^{1-\alpha},$$

and Lemma 14.5 says

$$||F||_{H^1(V)} \leq C||\partial_t F||_{L^2(B \cap M)}^\alpha ||F||_{H^1(B^+)}^{1-\alpha}.$$

Therefore,

$$||F||_{H^1(X_1)} \leq C||\partial_t F||_{L^2(B \cap M)}^{\alpha^2} ||F||_{H^1(X_2)}^{1-\alpha^2}.$$

Recall that $\varphi = \partial_t F$. Recall further that, for $T = 1, 2$,

$$||F||_{H^1(X_T)} \approx e^{T\omega} ||\varphi||_{L^2(M)}.$$

Combining these inequalities, one obtains the following theorem.

Theorem 14.6. *Let φ be a sum of eigenfunctions as in Theorem 14.3. For any ball B there exist constants C_1 and C_2 depending only on M and the radius of B such that*

$$||\varphi||_{L^2(M)} \leq e^{C_1\omega + C_2} ||\varphi||_{L^2(B \cap M)}.$$

Theorem 14.6 is the crucial global estimate. The remainder of the proof of Theorem 14.3 follows [7, 8].

Theorem 14.7 ([7], Proposition 6.7). *Let B_r denote the ball of radius r in $\mathbb{C}^n$. Let $\mathbb{R}^n$ be the real n-dimensional subspace of purely real vectors of $\mathbb{C}^n$. Let H be a holomorphic function on B_2 satisfying*

$$||H||_{L^\infty(B_2)} \leq e^\mu ||H||_{L^\infty(B_1 \cap \mathbb{R}^n)}.$$

Then there are dimensional constants C_1 and C_2 such that

$$\mathcal{H}_{n-1}(\{H = 0\} \cap B_{1/2} \cap \mathbb{R}^n) \leq C_1\mu + C_2.$$

Theorems 14.6 and 14.7 and real-analytic hypoellipticity for the operator $\partial_t^2 + \Delta$ on $\mathbb{R} \times M$ ([24, 27]) give the following corollary.

Corollary 14.8. *If M and ∂M are real analytic, and φ is as in Theorem 14.3, then*

$$\mathcal{H}_{n-1}\{\varphi = 0\} \le C_1\omega + C_2.$$

Next, we wish to deduce the doubling property for φ. If one replaces a ball with a slightly smaller ball of comparable radius, elliptic regularity implies that the Sobolev norm of F controls the L^∞ norm of F and vice versa. In particular, Theorem 14.6 and elliptic regularity imply that F satisfies doubling at *unit* scale. Next, Corollary 3.5 of [7] (applied to L with $\lambda = 0$) says that if F satisfies doubling at unit scale then it satisfies doubling at all smaller scales.

Finally, we need to pass from doubling for F to doubling for φ. Lemma 14.5 says that

$$||F||_{H^1(V)} \le C||\partial_t F||^\alpha_{L^2(B\cap M)}||F||^{1-\alpha}_{H^1(B+)}.$$

One can rescale this estimate to sets of size $r < 1$. At the same time we wish to use more appropriate function spaces for this scaling, which is possible because elliptic regularity estimates are valid independent of scale. To state a rescaled version of Lemma 14.5, consider a ball B_r or radius r centered on M and a ball V_r contained in $B_r^+ = B_r \cap \{t > 0\}$. Then

$$||F||_{L^\infty(V_r)} \le C||r\partial_t F||^\alpha_{L^\infty(B_r\cap M)}||F||^{1-\alpha}_{L^\infty(B_r^+)}.$$

The doubling condition for F proved in the preceding paragraph implies

$$||F||_{L^\infty(B_{10r})} \le e^{C_1\omega+C_2}||F||_{L^\infty(V_r)}.$$

Combined with the scale-invariant version of Lemma 14.5, this yields

$$||r\varphi||_{L^\infty(B_{2r}\cap M)} \le C||F||_{L^\infty(B_{10r})} \le e^{C_1\omega+C_2}||r\varphi||_{L^\infty(B_r\cap M)},$$

which is the desired doubling condition for φ.

We need to make a few more comments to complete the proof in the case in which the boundary of M is nonempty. Lemma 14.5 was proved only for balls that are disjoint from the boundary of M. Nevertheless, we shall see that the doubling property for φ is still valid up to the boundary. This is because, as described in [8], one can glue an isomorphic copy of M onto M at the boundary and consider even functions on the "doubled" manifold in the case of Neumann boundary conditions. In the case of Dirichlet boundary conditions, one takes the odd extension of the sum of eigenfunctions plus the constant (even) extension of the constant. Then the $C^{0,1}$ coefficient case estimates of [4] apply, and one can prove the doubling property for F for balls that do not touch the boundary. In this way, the estimate up to the boundary of the type in Lemma 14.5 is never needed. Instead it suffices to have such an estimate for balls whose diameters are comparable to their distances to the boundary.

The estimate of Theorem 14.6 is best possible in the following sense.

Proposition 14.9. *For any* $\Omega \subset M$ *such that* $\overline{\Omega} \neq M$, *and any* $\omega \geq 1$, *there exists a sum of eigenfunctions* φ *of eigenvalue at most* ω *such that*

$$||\varphi||_{L^2(M)} > e^{c\omega}||\varphi||_{L^2(\Omega)}.$$

Proof. Consider the heat kernel

$$p(t, x, y) = \sum_k e^{-t\omega_k^2}\varphi_k(x)\varphi_k(y),$$

where φ_k is a complete system of normalized eigenfunctions. Then choose $y \in \overset{\circ}{M} \setminus \overline{\Omega}$ such that $\varphi_1(y) \neq 0$. It is not hard to show that

$$\sup_{z \in \overline{\Omega}} p(t, x, y) \leq A e^{-a/t} \text{ for } 0 < t \leq 1.$$

Let $t = 1/\omega$ and define

$$\varphi(x) = \sum_{\omega_k \leq \omega} \left[e^{-t\omega_k^2}\varphi_k(y) \right] \varphi_k(x).$$

Then

$$||p(t, \cdot, y) - \varphi(\cdot)^2||_{L^2(M)} \leq \sum_{\omega_k > 1/t} e^{-2t\omega_k^2}(1 + \omega_k)^n \leq C e^{-c/t}.$$

Hence

$$||\varphi||_{L^2(\Omega)} \leq C e^{-c/t}.$$

On the other hand,

$$||\varphi||_{L^2(M)} \geq c|\varphi_1(y)| > 0,$$

independent of t. This proves the proposition. While Theorem 14.6 is best possible for sums of eigenfunctions, it is not optimal for single eigenfunctions. For example, the eigenfunctions $\sin(\pi k x)$ on the unit interval are equidistributed.   □

Finally, let us note that the estimates hold for infinite sums of eigenfunctions with exponentially decreasing coefficients.

Theorem 14.10. *Let $\epsilon > 0$. Then there exist constants C_1 and C_2 depending only on M and ϵ such that if*

$$\sum_{k=1}^{\infty} a_k^2 = 1, \quad \varphi(x) = \sum_{k=1}^{\infty} a_k \varphi_k(x),$$

and

$$\mu = \log \left(\sum a_k^2 e^{\epsilon \omega_k} \right) < \infty$$

then Theorems 14.3 and 14.6 are valid with ω replaced with μ.

The proof is similar. One need only replace the sets X_1 and X_2 with $X_{\epsilon/4}$ and $X_{\epsilon/2}$. (Lin gives a more precise dependence on a quantity related to μ in [22].)

14.6 Unique Continuation

Let u be a solution to the equation $Lu = 0$ on a connected subset Ω of a manifold M. Recall that L is said to satisfy the unique continuation property if $u = 0$ on a nonempty open set of Ω implies that $u = 0$ on all of Ω. A central ingredient of the proof of estimates of level sets of eigenfunctions was Lemma 14.5, a boundary version of uniqueness. Thus we are led to uniqueness questions for operators on a boundary. We will discuss both qualitative and quantitative forms of uniqueness for boundary operators in the case of Lipschitz graphs.

Let f be a Lipschitz function on $\mathbb{R}^n$, that is, $\|\nabla f\|_{L^\infty}$. Let M be the graph of f in $\mathbb{R}^{n+1}$,

$$M = \{(x, f(x)) : x \in \mathbb{R}^n\}.$$

We will discuss three operators on M. First, let U satisfy $\Delta U = 0$ in $x_{n+1} > f(x)$ and $U = g$ on M. The Dirichlet to Neumann operator is defined by

$$\Lambda g = \partial U / \partial \nu.$$

Second, define the single-layer potential

$$Sg(X) = \int_M |X - Y|^{1-n} g(Y) \, d\sigma(Y).$$

Third, consider the Laplace-Beltrami operator L on M.

As was first pointed out to us by Carlos Kenig, the operator Λ does satisfy the unique continuation property. This can be proved as follows. Suppose that $\Lambda g = 0$ and $g = 0$ on an open subset V of M. Extend U to the other side of V by defining it to be zero. The two boundary conditions imply that the equation $\Delta U = 0$ is valid in a neighborhood in $\mathbb{R}^{n+1}$ of any point of V. Therefore, by unique continuation for the ordinary Laplace operator, U is identically zero.

The qualitative uniqueness property for Λ led Kenig to conjecture quantitative forms of unique continuation such as Carleman inequalities. It is useful to formulate several quantitative properties that are closely related, or perhaps equivalent. For example, one variant is the substitute for well-posedness proved in Lemma 14.7. Another is the conjecture that if $g = 0$ on an open set, then $|\Lambda g|$ satisfies a doubling condition. More precisely, the doubling constant on a unit ball controls the doubling constant on all smaller concentric balls. Furthermore, the density $|\Lambda g|$ should be an A_∞ weight. (For a discussion of the weight condition for solutions of elliptic equations, see the work of Garofalo and Lin [11].) In particular, if $|\Lambda g|$ were to satisfy the weight property, then one could prove the following version of uniqueness: If $g = 0$ on an open set V and $\Lambda g = 0$ on a set of positive measure in V, then U is identically zero. Results of this type have been proved by Adolfsson, Escauriaza, Kenig, and Wang [1, 2, 16] for convex domains and for $C^{1,\alpha}$ domains for any $\alpha > 0$. But as we have learned from the work of Calderón, the natural scale-invariant question is the one for Lipschitz graphs.

Let us turn to the second operator, the single-layer potential. This is an integral operator with an explicit kernel that is smoothing of first order. (As is well known, the operator is equal to Λ^{-1} in the case that M is a hyperplane and has the same principal part in general.) The analogous question is whether solutions to $Sg = 0$ satisfy a doubling condition or if g is an A_∞ weight on any bounded subset of M. As far as we know, this question has not been raised before. Some evidence in favor of such bounds can be found in [13], where certain Carleman-type estimates for fractional powers of the Laplace operator are proved.

Finally, we consider the Laplace-Beltrami operator on M. If M were $C^{1,1}$, then the coefficients would be $C^{0,1}$ and the theorem of [4] would imply unique continuation for this operator. But since we are only assuming that M is $C^{0,1}$, the coefficients are merely bounded and measurable, an entire derivative short of the hypothesis of [4]. The hypothesis that the coefficients are $C^{0,1}$ is known to be best possible. Extending earlier examples of Plis [25], Miller [23] gave an example in three variables of an elliptic second-order operator with C^α coefficients for any $\alpha < 1$ having nonzero compactly supported solutions. Thus there is evidence against any unique continuation result for this operator. On the other hand, the examples of Plis and Miller are not graphs and the assumption that M is a Lipschitz graph is known to suffice for other estimates in harmonic analysis. (See [9, 10].) So we can still ask the question whether unique continuation is valid for the Laplace-Beltrami operator for M and whether solutions satisfy a doubling or A_∞ property.

14.7 Appendix

Here is a proof of (14.1). In the empty boundary case and the Neumann condition case, we use orthogonality of the eigenfunctions to compute the norms.

$$
\begin{aligned}
||F||^2_{H^1(X_T)} &= \int_{-T}^{T} \int_M |\partial_t F|^2 + |\nabla_x F|^2 + |F|^2 \, dx \, dt \\
&= 2 \int_0^T \sum |a_k|^2 \left[\cosh^2 \omega_k t + \sinh^2 \omega_k t + \frac{\sinh^2 \omega_k t}{\omega_k^2} \right] dt.
\end{aligned}
$$

It follows that

$$
2T \sum |a_k|^2 \le ||F||^2_{H^1(X_T)} \le 2T \sum |a_k|^2 (1 + T^2) e^{2\omega_k T}.
$$

In particular, for $0 < T \le 10$,

$$
T \sum |a_k|^2 \le ||F||^2_{H^1(X_T)} \le C e^{2T\omega} \sum |a_k|^2.
$$

(In the case of Theorem 14.10, ω is replaced by μ and T is a suitable multiple of ϵ.)

In the case of the Dirichlet problem, we wish to prove the variant of (14.1) suitable for Theorem 14.10, namely, comparability of norms up to a factor of the form $e^{c\mu}$. In order that Theorems 14.2 and 14.3 apply to level sets and not just zero sets of sums of eigenfunctions, we wish to consider a linear combination of a sum of eigenfunctions and the constant function. The only additional difficulty beyond the case of the Neumann problem is that the constant is not orthogonal to the eigenfunctions. Define η to be the normalized constant function on M,

$$
\eta(x) = 1/\sqrt{\operatorname{vol} M}.
$$

Let φ_k denote normalized Dirichlet eigenfunctions. Suppose that

$$
\sum_{k=1}^{\infty} a_k^2 = 1 \text{ and } \sum_{k=1}^{\infty} a_k^2 e^{\epsilon \omega_k} = e^{\mu} < \infty.
$$

Let

$$
\varphi = \sum_{k=1}^{\infty} a_k \varphi_k(x).
$$

We will prove that

$$
1 + a_0^2 + ||\varphi||^2_{L^2(M)} + ||a_0 \eta||^2_{L^2(M)} \le C e^{\mu} ||\varphi + a_0 \eta||^2_{L^2(M)}. \tag{14.4}
$$

This gives the upper bound

$$\|F + a_0 t\eta\|_{H^1(X_{\epsilon/2})} \leq e^\mu \|\varphi + a_0\eta\|_{L^2(M)}.$$

The lower bound is similar, since a similar argument shows that any sum of eigenfunctions with exponentially decreasing coefficients on sets $t = t_0$ has the same very weak orthogonality to the constant function.

We claim that there is a dimensional constant m and a constant C depending on M such that

$$|\varphi_k(x)| \leq C\omega_k^m \operatorname{dist}(x, \partial M) \tag{14.5}$$

for all x such that $\operatorname{dist}(x, \partial M) < 1/\omega_k$. To prove (14.5), choose local coordinates so that the boundary is a hyperplane and fix a cube with one face on ∂M of side length $1/\omega_k$. Dilate by the factor ω_k to rescale the cube to a unit cube. Denote by f the rescaled eigenfunction φ_k. Then f satisfies an elliptic equation of the form $Lf = cf$, with a constant c comparable to 1 in a unit cube with zero boundary conditions on one side. Moreover, since φ_k has L^2 norm 1, the rescaled function has norm at most comparable to $\omega_k^{n/2}$. Now by elliptic interior and boundary regularity, we see that on half the cube f is bounded by $\omega_k^{n/2}$ times the distance to the boundary. This translates into a bound on φ_k of the form above with $m = n/2 + 1$.

Next, using (14.5), we have

$$\begin{aligned}
|\varphi(x)| &\leq \sum_{k=1}^\infty |a_k| |\varphi_k(x)| \leq C \sum_{k=1}^\infty |a_k| \omega_k^m \operatorname{dist}(x, \partial M) \\
&\leq C \left(\sum_{k=1}^\infty a_k^2 e^{\epsilon\omega_k} \right)^{\frac{1}{2}} \left(\sum_{k=1}^\infty \omega_k^{2m} e^{-\epsilon\omega_k} \right)^{\frac{1}{2}} \operatorname{dist}(x, \partial M) \\
&\leq C e^{\mu/2} \operatorname{dist}(x, \partial M).
\end{aligned}$$

In order to prove (14.4), note first that it follows from the triangle inequality when a_0 is not comparable to 1. If, on the other hand, a_0 is comparable to 1, then the estimate above shows that there is a neighborhood of ∂M of volume comparable to $e^{-\mu/2}$ on which $|\varphi| < |a_0\eta|/2$. Therefore,

$$\|\varphi + a_0\eta\|_{L^2(M)} \geq ce^{-\mu/4},$$

which proves (14.4).

References

[1] V. Adolfsson and L. Escauriaza, $C^{1,\alpha}$ *domains and unique continuation at the boundary*, Comm. Pure Appl. Math., vol. 50 (1997), 935–969.

[2] V. Adolfsson, L. Escauriaza, and C. E. Kenig, *Convex domains and unique continuation at the boundary*, Revista Matemática Iberoamericana, vol. 11 (1995), 513–525.

[3] N. Aronszajn, *A unique continuation theorem for solutions of elliptic partial differential equations or inequalities of second order*, J. Math. Pures et Appliquées 36 (1957), 235–249.

[4] N. Aronszajn, A. Krzywicki, and J. Szarski, *A unique continuation theorem for exterior differential forms on Riemannian manifolds*, Arkiv für Matematik 4 (1963), 417–453.

[5] A. P. Calderón, *Uniqueness in the Cauchy problem for partial differential equations*, Am. J. Math. 80 (1958), 16–36.

[6] H. Donnelly, *Nodal sets for sums of eigenfunctions on Riemannian manifolds*, Proc A.M.S. 121 (1994), 967–973.

[7] H. Donnelly and C. Fefferman, *Nodal sets of eigenfunctions on Riemannian manifolds*, Invent. Math. 93 (1988), 161–183.

[8] ――――, *Nodal sets of eigenfunctions: Riemannian manifolds with boundary*, Analysis, et cetera (Moser volume) (P. H. Rabinowitz, E. Zehnder, ed.) Academic Press, Boston, 1990, pp. 251–262.

[9] E. B. Fabes, D. Jerison, and C. E. Kenig, *Necessary and sufficient conditions for absolute continuity of elliptic-harmonic measure*, Ann. Math. 119 (1984), 121–141.

[10] R. A. Fefferman, C. E. Kenig, and J. Pipher, *The theory of weights and the Dirichlet problem for elliptic equations*, Ann. Math. 134 (1991), 65–124.

[11] N. Garofalo and F.-H. Lin, *Monotonicity properties of variational integrals, A_p weights and unique continuation*, Indiana Univ. Math. J. 35 (1986), 245–267.

[12] L. Hörmander, *Linear Partial Differential Operators*, Springer-Verlag, Berlin, 1963.

[13] D. Jerison and C. E. Kenig, *Unique continuation and absence of positive eigenvalues for Schrödinger operators*, Ann. Math. 121 (1985), 463–488.

[14] F. John, *A note on "improper" problems in partial differential equations*, Comm. Pure Appl. Math. 8 (1955), 591–594.

[15] ——, *Continuous dependence on data for solutions of partial differential equations with a prescribed bound,* Comm. Pure Appl. Math. 13 (1960), 443–492.

[16] C. E. Kenig and W. Wang, *A note on boundary unique continuation for harmonic functions in nonsmooth domains,* Potential Analysis, vol. 8 (1998), 143–147.

[17] J. J. Kohn, *Boundaries of complex manifolds,* Proceedings Conf. on Complex Manifolds, Minneapolis, 1964, pp. 81–94.

[18] I. Kukavica, *Hausdorff measure of level sets for solutions of parabolic equations,* Internat. Math. Research Notes 13 (1995), 671–682.

[19] ——, *Nodal volumes for eigenfunctions of analytic regular elliptic problems,* J. d'Analyse Math., vol. 67 (1995), 269–280.

[20] ——, *Quantitative uniqueness for second-order elliptic operators,* Duke Math. J., vol. 91 (1998), 225–240.

[21] G. Lebeau and L. Robbiano, *Contrôle exact de l'équation de la chaleur,* Comm. in P.D.E. 20 (1995), 335–356.

[22] F.-H. Lin, *Nodal sets of solutions of elliptic and parabolic equations,* Comm. Pure Appl. Math. 44 (1991), 287–308.

[23] K. Miller, *Non-unique continuation for certain ODE's in Hilbert Space and for uniformly parabolic and elliptic equations in self-adjoint divergence form,* Symposium on Non-Well-Posed Problems and Logarithmic Convexity. Springer Lecture Notes 316 (R. J. Knops, ed.), Springer-Verlag, Berlin, 1973, pp. 85–101.

[24] C. B. Morrey and L. Nirenberg, *On the analyticity of the solutions of linear elliptic systems of partial differential equations,* Comm. on Pure Appl. Math. 10 (1957), 271–290.

[25] A. Plis, *On non-uniqueness in the Cauchy problem for an elliptic second order differential equation,* Bull. Acad Sci Polon., Ser. Sci Math. Astro. Phys. 11 (1963), 95–100.

[26] F. Treves, *Relations de domination entre opérateurs différentiels,* Acta Math. 101 (1959), 1–139.

[27] ——, *Introduction to Pseudodifferential Operators and Fourier Integral Operators, Vol. 1,* Plenum Press, New York, 1980, Theorem 5.4.

[28] T. H. Wolff, *Counterexamples with harmonic gradients in* $\mathbb{R}^3$, Essays on Fourier Analysis in Honor of Elias M. Stein (C. Fefferman et al., eds.), Princeton University Press, Princeton, NJ, 1995, pp. 321–384.

[29] ———, *Recent work on sharp estimates in second-order elliptic unique continuation problems*, J. Geometric Analysis 3 (1993), 621–650.

Chapter 15

Large-time Behavior and Self-similar Solutions of Some Semilinear Diffusion Equations

Yves Meyer

In his Ph.D. dissertation, M. Cannone constructed self-similar solutions to the Navier-Stokes equations. Then F. Planchon proved that these self-similar solutions provide the asymptotic behavior at large scales of some global solutions ([3], [18]).

Here we want to show that these results are not specific to the 3-D Navier-Stokes equation but are also valid for the nonlinear heat equation

$$\frac{\partial u}{\partial t} = \Delta u + \gamma u^3$$

whatever be the sign of γ and for the nonlinear Schrödinger equation.

This survey paper is organized as follows. We first review some well-known material concerning the nonlinear heat equation $\frac{\partial u}{\partial t} = \Delta u + u^3$ (blow up in finite time and self-similar solutions) in §§15.1 to 15.6. We show that these classical results are better understood when one is using a strategy due to T. Kato [9]. This strategy amounts to finding a solution as a vector-valued function $u(x, t)$ of the time variable $t \geq 0$. More precisely, u will belong to $\mathcal{C}([0, \infty); E)$, and the novelty of our approach consists in replacing the Lebesgue space $E = L^3(\mathbb{R}^3)$ which was used by T. Kato by a suitable Besov space E. Using such a Besov space offers many advantages. It permits us to explain the role played by the oscillations of the initial value $u_0(x)$ in the lifetime of the corresponding solution. This would not be possible if the L^3-norm were used. Moreover, the Besov spaces we will use contain functions which are homogeneous of degree -1. This is not the case for $L^3(\mathbb{R}^3)$. Such homogeneous u_0 initial values will generate self-similar solutions.

In §§15.7 and 15.8 we show that the self-similar solutions which are constructed in the preceding sections are driving the large-scale behavior of most of the global solutions to our nonlinear heat equation. In §15.9 we treat the nonlinear heat equation with the opposite sign.

Sections 15.10 and 15.12 are devoted to Navier-Stokes equations which obey the same scaling laws as our model nonlinear heat equations. The same type of methods applies.

Finally in §15.13 we review some quite recent results obtained by T. Cazenave and F. Weissler. These authors constructed self-similar solutions for the nonlinear Schrödinger equation. The Besov spaces which we used in the previous examples are now replaced by a new Banach space which looks quite exciting.

15.1   A First Model Case: the Nonlinear Heat Equation

Our first model case will be the following nonlinear heat equation

$$
\begin{cases}
\frac{\partial u}{\partial t} = \Delta u + u^3 \text{ on } \mathbb{R}^3 \times (0, \infty), \\
u(x, 0) = u_0(x)
\end{cases}
\tag{15.1}
$$

where $u = u(x, t)$ is a real-valued function of $x \in \mathbb{R}^3$ and $t \geq 0$. Below we will be much more specific about the functional spaces in which the solutions will be constructed. For the time being we are considering classical solutions to (15.1) with enough regularity and with appropriate size estimates. For example, for the time being, all L^p-norms in the x variable are supposed to exist.

Multiplying (15.1) by u and integrating over $\mathbb{R}^3$ yields the equation $\frac{1}{2}\frac{d}{dt}\|u\|_2^2 = -\|\nabla u\|_2^2 + \|u\|_4^4$ which means that the evolution results from a competition between $\|u\|_4^4$ and $\|\nabla u\|_2^2$. This remark makes the following theorem plausible.

Theorem 15.1 (J. Ball [1], H. A. Levine [13], and L. Payne [15]). *If* $u_0 \in C_0^\infty(\mathbb{R}^3)$, $u_0 \neq 0$ *and*

$$
\|\nabla u_0\|_2 \leq \frac{1}{\sqrt{2}}\|u_0\|_4^2,
\tag{15.2}
$$

then the corresponding solution to (15.1) blows up in finite time.

It means that there exists a $T_0 < \infty$ such that

$$\limsup_{t \uparrow T_0} \|u(\cdot, t)\|_2 = +\infty.$$

Let us make a few comments about this theorem. If $u(x,t)$ is a solution to (15.1), so are $u_\lambda(x,t) = \lambda u(\lambda x, \lambda^2 t)$ for any $\lambda > 0$. The L^4-norm is not invariant under this rescaling and a condition of the type $\|u_0\|_4 < \eta$ for some small η cannot imply the existence of a global solution. Indeed, such a condition would always be satisfied by one among the rescaled solutions u_λ, $\lambda > 0$. However, the L^3-norm is invariant, and we will see below that $\|u_0\|_3 < \eta$, η small enough, implies that the corresponding solution exists globally in time.

A second remark is the following. If $\varphi \in C_0^\infty(\mathbb{R}^3)$ then for $\varepsilon > 0$ small enough, $\varphi(\varepsilon x) = u_0(x)$ will satisfy (15.2).

In contrast for any $\varphi \in C_0^\infty(\mathbb{R}^3)$, there exists a positive $\omega_0 = \omega_0(\varphi)$ such that $|\omega| > \omega_0$, and $u_0(x) = e^{i\omega \cdot x} \varphi(x)$ imply that the corresponding solution $u(x,t)$ to (15.1) is global in time.

Our next step is to find sufficient conditions on the initial value $u_0(x)$ that imply that the corresponding solution $u(x,t)$ will be global in time. Such sufficient conditions will be of the type $\|u_0\|_B < \eta$, where B is some convenient Besov space which we will now describe.

15.2 Some Besov Spaces

We consider a function φ belonging to the Schwartz class $\mathcal{S}(\mathbb{R}^3)$ with the following two properties: $\widehat{\varphi}(\xi) = 1$ on $|\xi| \leq 1$ and $\widehat{\varphi}(\xi) = 0$ on $|\xi| \geq 4/3$. Here and in what follows, $\widehat{f}$ will denote the Fourier transform of f defined by $\widehat{f}(\xi) = \int e^{-ix\cdot\xi} f(x)\, dx$. Next $\varphi_j(x) = 2^{3j}\varphi(2^j x)$, $S_j(f) = f * \varphi_j$, $\Delta_j = S_{j+1} - S_j$, and $\psi(x) = 2^3\varphi(2x) - \varphi(x)$ in such a way that $\Delta_j(f) = f * \psi_j$.

If $\alpha > 0$, $1 \leq q \leq \infty$, the homogeneous Besov space $\dot{B}_q^{-\alpha,\infty}$ is defined by the condition

$$\|S_j(f)\|_q \leq C 2^{j\alpha}, \ j \in \mathbb{Z}. \tag{15.3}$$

The usual definition (which applies as well when α is any real number) is $\|\Delta_j(f)\|_q \leq C 2^{j\alpha}$ $(j \in \mathbb{Z})$, but this standard definition is equivalent to (15.3) if $\alpha > 0$. The drawback of the standard definition is the fact that any polynomial P satisfies $\Delta_j(P) = 0$, which forces us to add some requirement to get rid of such polynomials. This is not the case for the "clean definition" given by (15.3).

When $\alpha > 0$, $\dot{B}_q^{-\alpha,\infty}$ is a Banach space consisting of tempered distributions. This Banach space is not separable but is the dual space of $\dot{B}_p^{\alpha,1}$, which is defined by

$$\sum_{-\infty}^{\infty} 2^{j\alpha} ||\Delta_j(f)||_1 < \infty \qquad (1/q + 1/p = 1).$$

If $S(t) = \exp(t\Delta)$ is the heat semigroup, then $\dot{B}_q^{-\alpha,\infty}$ is characterized by

$$\sup_{t>0} t^{\alpha/2} ||S(t)f||_q = C < \infty. \qquad (15.4)$$

Definition 15.2. *We will denote by B_q the homogeneous Besov space $\dot{B}_q^{-\alpha,\infty}$ when $\alpha = 1 - \frac{3}{q}$, $3 < q \leq \infty$.*

It is easily checked that

$$L^3(\mathbb{R}^3) \subset L^{3,\infty}(\mathbb{R}^3) \subset B_q \subset B_{q'} \subset \dot{B}_\infty^{-1,\infty}, \qquad (15.5)$$

where $3 \leq q \leq q' \leq +\infty$.

The Banach space $\dot{B}_\infty^{-1,\infty}$ consists of $f = \Delta g$, where g belongs to the Zygmund class defined by

$$|g(x+y) + g(x-y) - 2g(x)| \leq C|y| \qquad (15.6)$$

for $x \in \mathbb{R}^n$, $y \in \mathbb{R}^n$.

In the next section, we will show that $||u_0||_{B_q} < \eta(q)$ for some $q \in (3,9)$ suffices for the existence of a global solution. Here $\eta(q)$ is a continuous function of q, which tends to 0 as q tends to 9. It means that we are far from obtaining an optimal sufficient condition for global solutions. For the sake of simplicity we will concentrate on the special case $q = 6$.

15.3 A Sufficient Condition for Global Solutions

The following theorem is implicit in some joint work between M. Cannone, F. Planchon, and the author of these notes:

Theorem 15.3. *There exists a positive constant $\eta > 0$ such that for any $u_0 \in C_0^\infty(\mathbb{R}^3)$ and*

$$||u_0||_{B_6} < \eta, \qquad (15.7)$$

there exists a global solution $u(x,t)$ to (15.1) such that $u(x,0) = u_0(x)$.

Let us stress that Theorem 15.3 cannot be valid without (15.7). Such a claim would contradict Theorem 15.1. A second observation is given by the following Gagliardo-Nirenberg-type inequality.

Lemma 15.4. *There exists a constant C such that for any f in Sobolev space $H^1(\mathbb{R}^3)$*

$$\|f\|_4^2 \leq C\|\nabla f\|_2\|f\|_{B_\infty}. \tag{15.8}$$

A fortiori (15.7) implies $\|u_0\|_4^2 \leq C\eta\|\nabla u_0\|_2$, which is a strong form of the negation of (15.2).

We should also observe that $u_0(x)$ and $\lambda u_0(\lambda x)$, $\lambda > 0$, have the same B_6 norm. Therefore Theorem 15.3 is in full agreement with the rescaling of (15.1).

Theorem 15.3 explains some observations we made after stating Theorem 15.1. If $\varphi \in C_0^\infty(\mathbb{R}^3)$ and $\varepsilon > 0$ tends to 0, then the norm in B_6 of $\varphi(\varepsilon x)$ is $\frac{C(\varphi)}{\varepsilon}$ and tends to infinity. Therefore (15.7) is violated, and we know that the solution to (15.1) blows up whenever ε is small enough. In contrast, if g is fixed in the Schwartz class $\mathcal{S}(\mathbb{R}^3)$ we consider $u_0(x) = e^{i\omega \cdot x}g(x)$. Then

$$\|u_0\|_{B_6} = \frac{\|g\|_6}{|\omega|^{1/2}} + O(|\omega|^{1/2}) \qquad \text{as} \qquad |\omega| \to +\infty \tag{15.9}$$

and (15.7) is satisfied for $|\omega|$ large enough.

15.4   Self-Similar Solutions to the Nonlinear Heat Equation

The following theorem was found by A. Haraux and F. Weissler.

Theorem 15.5. *There exists a (nontrivial) function $w(x)$ in the Schwartz class $\mathcal{S}(\mathbb{R}^3)$ such that*

$$u(x,t) = \frac{1}{\sqrt{t}}w\left(\frac{x}{\sqrt{t}}\right), \quad t > 0, \quad x \in \mathbb{R}^3. \tag{15.10}$$

is a global solution to (15.1).

This self-similar solution is invariant under the rescaling $\lambda u(\lambda x, \lambda^2 t) = u(x,t)$.

The corresponding initial value $u_0(x) = 0$ identically. It means that in Theorem 15.3, $u \in \mathcal{C}([0,\infty), L^3(\mathbb{R}^3))$ cannot be replaced by the weaker condition

$$\sup_{t>0} \|u(\cdot,t)\|_3 < \infty \tag{15.11}$$

without losing uniqueness. But the stronger condition does not imply uniqueness either [19].

In the seminal paper [9] by A. Haraux and F. Weissler, $w(x)$ was a radial function. A more systematic treatment can be found in a remarkable paper by M. Escobedo and O. Kavian [7].

15.5   Multilinear Operators Arising in the Proof of Theorem 15.3

The following treatment of (15.1) is due to T. Kato, who converted (15.1) into an integral equation. Instead of writing $\frac{\partial u}{\partial t} = \Delta u + u^3$ with $u(x, 0) = u_0(x)$, he wrote

$$u(t) = S(t)u_0 + \int_0^t S(t - s)u^3(s)\, ds \qquad (15.12)$$

where $S(t) = e^{t\Delta}$ is the heat semigroup and $u(x, t)$ is viewed as a vector-valued function of the time variable t. Kato's program consisted in looking for solutions $u(t)$ belonging to $C([0, \infty); E)$, where E is a suitable Banach space of functions of x. The norm of $u(t)$ in $C([0, \infty); E)$ is $\sup_{t \geq 0} ||u(\cdot, t)||_E$. The collection of Banach spaces which are being used splits into two classes. If E belongs to the first class, E will be a separable Banach space and the condition $u \in C([0, \infty); E)$ means that u, as a function of t, is continuous when E is equipped with the topology defined by the norm. When E belongs to the second class, E will always be the dual F^* of a separable Banach space F, and the continuity requirement will concern the $\sigma(E, F)$ topology.

Examples of this situation are given by $E = L^{3,\infty}$ (the weak L^3-space), which is the dual of $F = L^{3/2,1}$ (the corresponding Lorentz space), or by $E = \dot{B}_q^{-\alpha,\infty}$ (the homogeneous Besov space), which is the dual of $F = \dot{B}_q^{\alpha,1}$ (when $1/p + 1/q = 1$).

T. Kato's program consists in solving (15.12) through Picard's fixed-point theorem. The norm of u in the Banach space $X = C([0, \infty); E)$ is defined as

$$||u||_X = \sup_{t > 0} ||u(\cdot, t)||_E. \qquad (15.13)$$

Then the main issue is to prove the estimate

$$||\Gamma(u_1, u_2, u_3)||_X \leq C||u_1||_X ||u_2||_X ||u_3||_X \qquad (15.14)$$

for the trilinear operator Γ defined by

$$\Gamma(u_1, u_2, u_3) = \int_0^t S(t - s)u_1(s)u_2(s)u_3(s)\, ds. \qquad (15.15)$$

We wrote $u(s) = u(\cdot, s)$ and so on for keeping the notation as simple as possible.

A solution $u \in X$ of (15.12) is called a "mild solution" of (15.1). Whenever (15.14) holds, Picard's fixed-point scheme can be applied and yields a mild solution as long as the norm of u_0 in E is small enough. More precisely we should have $||u_0||_E < \frac{2}{3\sqrt{3}} C^{-1/2}$ where C is defined by (15.14).

T. Kato and F. Weissler applied this program to $E = L^3(\mathbb{R}^3)$, but the fundamental trilinear estimate is definitely incorrect. T. Kato was able to offer a substitute for the missing estimate. Indeed, he considered a subspace $T \subset X$ consisting of function $u(x, t)$ which are continuous on $[0, \infty)$ with values in $L^3(\mathbb{R}^3)$ and such that $t^{\frac{1}{4}} u(\cdot, t)$ is continuous on $[0, \infty)$ with values in $L^6(\mathbb{R}^3)$. Moreover, one imposes that $\lim_{t \to 0} t^{1/4}||u(\cdot, t)||_6 = 0$ as well as $\lim_{t \to +\infty} t^{1/4}||u(\cdot, t)||_6 = 0$.

Finally one writes

$$||u||_* = \sup\{t^{1/4}||u(\cdot, t)||_6 : t > 0\} \tag{15.16}$$

and

$$||u||_\tau = ||u||_X + ||u||_*. \tag{15.17}$$

Then the following estimates are easily verified

$$||\Gamma(u_1, u_2, u_3)||_\tau \leq C||u_1||_*||u_2||_*||u_3||_* \tag{15.18}$$

and

$$||S(t)u_0||_\tau \leq ||u_0||_3. \tag{15.19}$$

Therefore Picard's scheme can be used to solve (15.12) inside the Banach space T whenever $||u_0||_3$ is small enough. The drawback of this approach is the lack of uniqueness inside the "natural space" X. Indeed we get uniqueness inside the smaller space T.

In [19], E. Terraneo proved the lack of uniqueness inside the "natural space" X.

Our Theorem 15.3 is a slight improvement on Kato's ideas. We consider the Banach space Y of functions $u(x, t)$ which are continuous on $(0, \infty)$ with values in $L^6(\mathbb{R}^3)$ such that $||u||_*$ is finite and $\lim_{t \to 0} t^{1/4}||u(\cdot, t)||_6 = \lim_{t \to +\infty} t^{1/4}||u(\cdot, t)||_6 = 0$. If u_0 belongs to $L^3(\mathbb{R}^3)$, then $S(t)u_0$ belongs to Y and we have $||S(t)u_0||_* = ||u_0||_{B^6} < \eta$.

Then Picard's scheme can be applied inside Y and yields a mild solution to (15.12). Then it is easy to check that

$$\left\| \int_0^t S(t - s)u^3(s)\, ds \right\|_6 \to 0 \text{ as } t \to 0, \tag{15.20}$$

which will imply that the solution which is constructed belongs to X. If u_0 belongs to $C_0^\infty(\mathbb{R}^3)$, we want to prove that the "mild solution" arising from Kato's program is a classical solution to (15.1). The reader is referred to [7], [9] where the issue is discussed.

Let us remark that Theorem 15.3 is not optimal and better sufficient conditions can be found in [11]. However, Theorem 15.3 paves the way to the general methodology that will be used later in these notes.

15.6   More Self-Similar Solutions to the Nonlinear Heat Equation

We already observed that if $u(x,t)$ is a solution to (15.1), so are $u_\lambda(x,t) = \lambda u(\lambda x, \lambda^2 t)$ for $\lambda > 0$. A solution $u(x,t)$ is self-similar if $u_\lambda = u$. In other terms

$$u(x,t) = \frac{1}{\sqrt{t}} V\left(\frac{x}{\sqrt{t}}\right). \tag{15.21}$$

In contrast with the approach indicated by A. Haraux and F. Weissler, we want to construct our self-similar solutions by solving an initial-value problem. This approach will exclude the Haraux-Weissler's self-similar solution for which the initial value is 0. For this trivial initial value, our approach will uniquely yield the trivial solution to (15.12). If in the distributional sense, $\lim_{t\downarrow 0} \frac{1}{\sqrt{t}} V\left(\frac{x}{\sqrt{t}}\right) = u_0(x)$ exists, then this distribution $u_0(x)$ will be homogeneous of degree -1. This remark implies that the functional space E which will be used in our methodology should contain homogeneous functions of degree -1. This excludes the L^p spaces but includes the Besov over spaces $\dot{B}_q^{-\alpha,\infty}$, when $\alpha = 1 - 3/q$ or the Lorentz space setting.

Theorem 15.6. *There exist two positive constants $\beta > 0$ and $\gamma > 0$ with the following properties: if $u_0 \in L^{3,\infty}(\mathbb{R}^3)$ satisfies $||u_0||_{(3,\infty)} < \beta$ and is homogeneous of degree -1, there exists a unique solution $u(x,t)$ of (15.12) such that*

$$u(x,t) = S(t)[u_0](x) + \frac{1}{\sqrt{t}} W\left(\frac{x}{\sqrt{t}}\right), \tag{15.22}$$

where

$$W(x) \in L^3(\mathbb{R}^3) \cap L^6(\mathbb{R}^3) \text{ and } ||W||_6 < \beta. \tag{15.23}$$

Indeed, (15.22) implies that $u(\cdot,t)$ is a continuous function of t with values in $L^{3,\infty}$ equipped with its $\sigma(L^{3,\infty}, L^{3/2,1})$ topology.

Since $u_0(x)$ is homogeneous of degree -1,

$$S(t)[u_0](x) = \frac{1}{\sqrt{t}} V_0 \left(\frac{x}{\sqrt{t}} \right) \tag{15.24}$$

and $u(x,t)$ is a self-similar solution to (15.12).

If $u_0 = 0$, Theorem 15.5 yields $u(x,t) = 0$. It means that the Haraux-Weissler self-similar solution $\frac{1}{\sqrt{t}} w \left(\frac{x}{\sqrt{t}} \right)$ satisfies $||w||_6 > \beta$. This was observed by M. Escobedo and O. Kavian: all the self-similar solutions to the nonlinear heat equation they constructed have "large" norms.

To illustrate Theorem 15.5, let us consider the trivial example where $u_0(x) = \eta |x|^{-1}$, $\eta > 0$ being small. This example was treated in the Haraux-Weissler seminal paper, and they found that the corresponding solution to (15.12) was given by

$$u(x,t) = \frac{1}{\sqrt{t}} u \left(\frac{|x|}{\sqrt{t}} \right) \tag{15.25}$$

where $0 < u(r) \leq C_0$ on $[0, \infty)$ and $\lim_{r \to +\infty} r u(r) = L > 0$.

It means that $u(x,t) \sim \frac{L}{|x|}$ at infinity (t being frozen). Therefore $u(x,t)$ belongs to $L^{3,\infty}$ as expected from Theorem 15.5 and $u(x,t) \in L^6(\mathbb{R}^3)$.

The Lorentz space $L^{3,\infty}(\mathbb{R}^3)$ was one example of functional space which is adapted to finding self-similar solutions to the nonlinear heat equation. In the next section the Besov space $\overset{.}{B}_q^{-\alpha,\infty}$, $\alpha = 1 - 3/q$, $3 < q < 9$, will be used for the same goal.

Theorem 15.5 was obtained in a joint work with O. Barraza [2].

15.7 Another Approach to Self-Similar Solutions

Instead of using the Lorentz space $L^{3,\infty}(\mathbb{R}^3)$, we will be using the Besov space $\overset{.}{B}_q^{-\alpha,\infty} = B_q$ where $3 < q < 9$. We first construct general solutions inside a functional setting which is adapted to the existence of nontrivial self-similar solutions. This approach will be used for Navier-Stokes equations.

Theorem 15.7. *Let us assume that* $3 < q < 9$. *Then there exist two positive constants* $\eta(q)$ *and* $\beta(q)$ *with the following properties: if* $||u_0||_{B_q} < \eta(q)$, *then there exists a solution* $u(x,t)$ *to (15.12) which satisfies the following properties*

$$u(x,t) = S(t)[u_0](x) + w(x,t), \tag{15.26}$$

$$\sup_{t>0} ||w(\cdot, t)||_3 < \infty, \tag{15.27}$$

$$\sup_{t>0} t^{\alpha/2} ||w(\cdot, t)||_q < \beta(q). \tag{15.28}$$

Moreover such a solution to (15.12) is uniquely defined by (15.26), (15.27), and (15.28).

In order to find self-similar solutions, it suffices to assume that $u_0(x)$ is homogeneous of degree -1. Then the scaling invariance of the nonlinear heat equation implies that whenever $u_0(x)$ is replaced by $\lambda u_0(\lambda x)$, then $u(x, t)$ is replaced by $\lambda u(\lambda x, \lambda^2 t)$.

Finally, the uniqueness of the solution implies $\lambda u(\lambda x, \lambda^2 t) = u(x, t)$, and our $u(x, t)$ is a self-similar solution. This approach is due to M. Cannone.

Theorem 15.6 both generalized Theorem 15.3 and Theorem 15.5. Let us explain why. Indeed, Theorem 15.3 is a special case of Theorem 15.6 ($q = 6$) and for obtaining Theorem 15.5 it suffices to observe that the Lorentz space $L^{3,\infty}(\mathbb{R}^3)$ is contained in each Besov space B_q. Indeed $L^{3,\infty}(\mathbb{R}^3)$ is contained inside the smallest one, i.e., $\dot{B}_3^{0,\infty}$.

The reason why Theorem 15.5 was stated before is the following. Any analyst will tell you that (in three dimensions) $|x|^{-1}$ belongs to $L^{3,\infty}(\mathbb{R}^3)$ (i.e., any weak L^3), and only a few would consider this function as belonging to our fancy Besov spaces.

15.8   Convergence to a Self-Similar Solution

In this section, the exponent q will always belong to the open interval $(3, 9)$.

If $u(x, t)$ is a global solution to (15.12), so is

$$u_\lambda(x, t) = \lambda u(\lambda x, \lambda^2 t) \text{ for } \lambda > 0.$$

In order to study the large-scale behavior of $u(x, t)$, it becomes natural to raise the following problem. Does there exist a limit function $v(x, t)$ such that

$$\lim_{\lambda \uparrow +\infty} ||u_\lambda(\cdot, t) - v(\cdot, t)||_{L^q(\mathbb{R}^3)} = 0 \tag{15.29}$$

uniformly in $t \in [t_0, t_1]$ whenever $t_1 \geq t_0 > 0$.

We then have $v(x, t) = \frac{1}{\sqrt{t}} V\left(\frac{x}{\sqrt{t}}\right)$ and

$$\lim_{t \uparrow +\infty} ||\sqrt{t} u(\sqrt{t} x, t) - V(x)||_q = 0. \tag{15.30}$$

Another way of raising the same problem is to ask for an asymptotic expansion

$$u(x,t) = \frac{1}{\sqrt{t}} V\left(\frac{x}{\sqrt{t}}\right) + \frac{1}{\sqrt{t}} R\left(\frac{x}{\sqrt{t}}, t\right), \qquad (15.31)$$

where $V \in L^q(\mathbb{R}^3)$ and

$$\lim_{t \to +\infty} ||R(\cdot, t)||_q = 0 \qquad (t \to +\infty). \qquad (15.32)$$

The following theorem was proved in a joint work with F. Planchon ([17]). We want to know whether (15.31) holds for a global solution to the nonlinear heat equation. An obvious necessary condition is $||\sqrt{t}u(\sqrt{t}x, t)||_q \leq C$ for $t > T$. We will modify our problem by imposing the stronger condition

$$||\sqrt{t}u(\sqrt{t}x, t)||_q \leq C, \qquad 0 < t < \infty, \qquad (15.33)$$

which is no longer necessary to the problem we have in mind. Then our problem can be efficiently solved whenever $3 < q < 9$.

Theorem 15.8. *Let* $u(x,t)$, $x \in \mathbb{R}^3$, $t > 0$, *be a global solution to (15.33) and*

$$u(x,t) \to u_0(x) \quad as \ t \ tends \ to \ 0. \qquad (15.34)$$

The limit in (15.34) is taken in the distributional sense. Then $u_0(x) \in \dot{B}_q^{-\alpha,\infty}$, $\alpha = 1 - 3/q$, $3 < q < 9$. *Moreover, if (15.31) and (15.32) hold, then* $\frac{1}{\sqrt{t}} V\left(\frac{x}{\sqrt{t}}\right)$ *is a self-similar solution to the nonlinear heat equation. Conversely there exists a positive constant* $\beta(q)$ *such that whenever* $||u_0||_{\dot{B}_q^{-\alpha,\infty}} < \beta(q)$, *the following properties (a) and (b) are equivalent ones:*

a. *(15.31) and (15.32) hold.*

b.

$$\lambda u_0(\lambda x) \to v_0(x) \quad as \ \lambda \to +\infty \ and \qquad (15.35)$$

$$t^{\alpha/2}||S(t)[u_0 - v_0]||_q \qquad (t \to +\infty) \qquad (15.36)$$

It is easily checked that the second requirement (15.36) implies the first one. Both of them concern the large-scale behavior of the initial value $u_0(x)$ and tell that the infrared limits of $u_0(x)$ and $v_0(x)$ are the same.

For stressing the difference between (15.35) and (15.36) let us consider a wavelets expansion of $f(x) = v_0(x) - u_0(x)$. One uses the wavelets which are constructed in [12] and forgets the index of the mother wavelet ψ (this index ranges from 1 to 7). Then

$$f(x) = \sum_j \sum_{k \in \mathbb{Z}^3} \alpha(j, k) 2^j \psi(2^j x - k)$$

(the normalization of the wavelets is not the standard one). The following observations will clarify the conditions (15.35) and (15.36). First $f \in \overset{\cdot}{B_q^{-\alpha,\infty}}$ means $\sigma_j \leq C < \infty$, $-\infty < j < \infty$, where

$$\sigma_j = \sum_k |\alpha(j, k)|^q. \tag{15.37}$$

Then $\lambda f(\lambda x) \to 0$ $(\lambda \to +\infty)$ is equivalent to

$$\sup_k |\alpha(j, k)| \to 0 \qquad (j \to -\infty) \tag{15.38}$$

while $t^{\alpha/2} \|S(t)f\|_q \to 0$ $(t \to +\infty)$ is equivalent to the stronger condition $\sigma_j \to 0$ $(j \to +\infty)$.

Now it is obvious that this condition implies (15.38), and it is also obvious that the converse is not true.

15.9   The Second Model Case

Our second model case is the nonlinear heat equation with the oppposite sign:

$$\frac{\partial u}{\partial t} = \Delta u - u^3, \tag{15.39}$$

where $u = u(x, t)$, $x \in \mathbb{R}^3$, $t > 0$.

In that situation, $\|u\|_2$ is a decreasing function of t which excludes self-similar solutions $u(x, t) = \frac{1}{\sqrt{t}} U\left(\frac{x}{\sqrt{t}}\right)$, where U belongs to $L^2(\mathbb{R}^3)$. For such a solution $\|u\|_2 = \|U\|_2 \sqrt{t}$. However, Theorem 15.5 is still valid. This is not surprising since Theorem 15.5 is obtained through a perturbation method in which the nonlinearity is treated as a small error term when compared to the linear term. Therefore the sign of this small error term does not play any role.

15.10 The Navier-Stokes Equations

Once more we follow Kato's approach with a small twist at the end. Navier-Stokes equations are

$$\begin{cases} \frac{\partial v}{\partial t} = \Delta v - (v \cdot \nabla)v - \nabla p \\ \\ \nabla \cdot v = 0 \\ \\ v(x,0) = v_0(x) \end{cases} \tag{15.40}$$

where $v = (v_1, v_2, v_3)$ is the velocity, p is the pressure, both the velocity and the pressure are defined in $\mathbb{R}^3 \times [0, \infty)$, $v \cdot \nabla$ is the derivative in the direction v ($v \cdot \nabla = v_1 \frac{\partial}{\partial x_1} + \cdots + v_3 \frac{\partial}{\partial x_3}$) and $\nabla \cdot v$ is the divergence of v. These equations are invariant under the rescaling

$$(v, p) \to (v_\lambda, p_\lambda),$$

where $\lambda = 0$, $v_\lambda(x,t) = \lambda v(\lambda x, \lambda^2 t)$, and $p_\lambda(x,t) = \lambda^2 p(\lambda x, \lambda^2 t)$. The rescaling of the velocity is the same as the one we used in the two model cases and our approach to the Navier-Stokes equations will be quite similar to what we did in the previous cases.

The drawback in what follows is the lack of understanding of the cancellations which occur in the nonlinear term $(v \cdot \nabla)v$ and result from $\nabla \cdot v = 0$. In other words, our approach to Navier-Stokes equations will use perturbation arguments where the bilinear term will be treated as a perturbation of the "main term" which is Δv. That is why we only will obtain local results where the smallness of initial condition will play a crucial role.

Following T. Kato, (15.40) will be rewritten as an integral equation. Let R_1, R_2, R_3 be the three Riesz transformations, and let P be the orthogonal projection on the divergence-free vector fields:

$$P \begin{pmatrix} v_1 \\ v_2 \\ v_3 \end{pmatrix} = \begin{pmatrix} v_1 - R_1(\sigma) \\ v_2 - R_2(\sigma) \\ v_3 - R_3(\sigma) \end{pmatrix}, \tag{15.41}$$

where $\sigma = R_1(v_1) + R_2(v_2) + R_3(v_3)$.

Then (15.40) is equivalent to

$$v(x,t) = S(t) - \int_0^t PS(t-s)\partial_j(v_j v)(s)\, ds \tag{15.42}$$

whenever $v(x,t)$ is sufficiently regular.

More generally, a solution to (15.42) is called a mild solution. These mild solutions will be considered as vector-valued functions of the time variable.

More precisely, $v(x,t) \in \mathcal{C}([0,\infty); E)$, where E is a suitable space of functions of $x \in \mathbb{R}^3$.

The norm of v is

$$\sup\{||v(\cdot,t)||_E : 0 < t < \infty\}. \tag{15.43}$$

The main issue concerns the bilinear operators

$$B_{j,k,l}(f,g) = \int_0^t R_j R_k S(t-s)\partial_t(fg)\,ds, \tag{15.44}$$

where j,k,l belong to $\{1,2,3\}$, $f = f(x,s)$, $g = g(x,s)$.

If f and g belong to $\mathcal{C}([0,\infty); E)$ then two difficulties might arise.

In many instances which will be treated, the pointwise product between two functions f and g in E does not have any meaning (even in a weak sense). This "divergence" needs to be fixed.

A second divergence is coming from the competition between the differentiation $\frac{\partial}{\partial x_l} = \partial_l$, which destroys smoothness, and $S(t-s)$, which improves smoothness. As s tends to t, the differentiation will win!

The Banach space which will be considered is our friend $\dot{B}_q^{-\alpha,\infty}$. With this choice the bilinear operators (15.44) are not continuous. This difficulty was already present in §15.5 and was solved by introducing the "artificial norm" $||u||_\tau = ||u||_X + ||u||_*$ in (15.17).

T. Kato applied the same program in the Navier-Stokes case and could prove the existence of positive constants β and γ such that, if div $v_0 = 0$ and $||v_0||_3 < \beta$, there exists a solution $v(x,t)$ to the Navier-Stokes equations with the following properties:

$$v(\cdot,t) \in \mathcal{C}([0,\infty); L^3(\mathbb{R}^3)), \tag{15.45}$$

$$t^{1/4}||v(\cdot,t)||_6 < \gamma. \tag{15.46}$$

Moreover, such a solution is unique.

We will improve on this result by relaxing the hypothesis $||v_0||_3 < \beta$ into a much weaker one. The following theorem was obtained in a collaboration with M. Cannone when $3 < q \leq 6$ and then improved by F. Planchon in his Ph.D. dissertation, where he got rid of the limitation $q \leq 6$.

Theorem 15.9. *There exists a continuous function $\eta(q)$, defined on $(3,\infty)$, with values in $(0,\infty)$ with the following property: if $v_0 \in L^3(\mathbb{R})$, div $v_0 = 0$ and if, for some $q \in (3,\infty)$ we have*

$$||v_0||_{\dot{B}_q^{-(1-3/q),\infty}} < \eta(q), \tag{15.47}$$

then there exists a unique solution $v(x,t)$ to the Navier-Stokes equations such that

$$v(\cdot,t) \in \mathcal{C}([0,\infty); L^3(\mathbb{R}^3)) \qquad (15.48)$$

$$\lim_{t\downarrow 0} \sqrt{t}\|u(\cdot,t)\|_\infty = 0. \qquad (15.49)$$

Since the Besov spaces $\overset{\cdot}{B_q}^{-(1-3/q),\infty}$ are increasing with q one might be tempted to believe that the largest q yields the best result in (15.47). However, $\eta(q) = 0$ as $q \to +\infty$, and this limitation is due to the proof.

15.11   Self-Similar Solutions to Navier-Stokes Equations

A self-similar solution is a solution $v(x,t)$ such that $v = v_\lambda$ for $0 < \lambda < \infty$. This implies $v(x,t) = \frac{1}{\sqrt{t}} V\left(\frac{x}{\sqrt{t}}\right)$ as before. Now V is vector-valued.

It is well known that for any solution to (15.40) for which it makes sense we have $\frac{d}{dt}\|v\|_2^2 \leq 0$. It means that $V(x)$ cannot belong to $L^2(\mathbb{R}^3)$.

We will construct self-similar solutions to Navier-Stokes equations by solving an initial value problem. Therefore we want $u_0(x)$ to be the weak limit of $\frac{1}{\sqrt{t}} V\left(\frac{x}{\sqrt{t}}\right)$ as t tends to 0.

Then $u_0(x)$ will be homogeneous of degree -1, and the functional space which will be used should contain such functions. Once more we are led to using our Besov spaces $B_q = \overset{\cdot}{B_q}^{-(1-3/q),\infty}$. The first construction of self-similar solutions to Navier-Stokes equations was obtained in a joint work with M. Cannone and F. Planchon ([4], [5]).

Then F. Planchon could relax the condition imposed upon q and proved the following theorem ([18]).

Theorem 15.10. *There exist two continuous functions $\eta(q) > 0$ and $\beta(q) > 0$, defined on $(3,\infty)$, with the following property: if $u_0(x)$ belongs to B_q for some $q \in (3,\infty)$ and satisfies $\nabla \cdot u_0 = 0$, $u_0(\lambda x) = \lambda^{-1}u_0(x)$, $\lambda > 0$ and $\|u_0\|_{B_q} < \eta(q)$, then there exists a unique self-similar solution $u(x,t) = \frac{1}{\sqrt{t}} U\left(\frac{x}{\sqrt{t}}\right)$ to the Navier-Stokes equations such that*

$$u(x,t) = S(t)[u_0](x) + \frac{1}{\sqrt{t}} W\left(\frac{x}{\sqrt{t}}\right), \qquad (15.50)$$

$$\|W\|_3 < \beta(q), \qquad (15.51)$$

and, more precisely, $W \in \overset{\cdot}{B}_{3/2}^{1,2}$.

15.12   Convergence to Self-Similar Solutions

This section is almost identical to what we obtained for the nonlinear heat equation. As already stressed, these similarities come from the following two points (a) the scaling invariance is the same in the two cases, (b) the nonlinearity is treated by a perturbation argument without taking in account the subtle cancellations, which played a crucial role in the Navier-Stokes equations.

We want to know the large-scale behavior of the Navier-Stokes equations, which is related to studying

$$\lim_{\lambda\uparrow+\infty} \lambda v(\lambda x, \lambda^2 t) = \lim_{\lambda\uparrow+\infty} v_\lambda(x,t). \tag{15.52}$$

More precisely, we want to find necessary and sufficient conditions for the existence of $\tilde{v}(x,t)$ such that

$$\lim_{\lambda\uparrow+\infty} ||v_\lambda(\cdot,t) - \tilde{v}(\cdot,t)||_q = 0 \tag{15.53}$$

uniformly on $t \in [t_0, t_1]$ for $t_1 > t_0 > 0$.

As we already observed, this limit field $\tilde{v}(x,t)$ is self-similar: $\tilde{v}(x,t) = \frac{1}{\sqrt{t}} V\left(\frac{x}{\sqrt{t}}\right)$ if it exists.

Then our problem amounts to deciding whether

$$||\sqrt{t} v(\sqrt{t} x, t) - V(x)||_q \to 0 \qquad (t \to +\infty). \tag{15.54}$$

If it is the case, we necessarily have

$$||\sqrt{t} v(\sqrt{t} x, t)||_q \le C \text{ for } t \ge T. \tag{15.55}$$

Instead of (15.55) we will assume the stronger condition

$$||\sqrt{t} v(\sqrt{t} x, t)||_q \le C \text{ for } 0 < t < \infty. \tag{15.56}$$

We then have ([18]).

Lemma 15.11. *If $v(x,t)$ is a global solution to the Navier-Stokes equations, if (15.56) holds for some $q \in (3,\infty)$ and if $\lim_{t\downarrow 0} v(x,t) = v_0(x)$ exists in the weak sense, then $v_0(x) \in \dot{B}_q^{-(1-3/q),\infty}$*

Our next step will be to decide, for a given initial value $v_0(x)$, whether the corresponding solution to the Navier-Stokes equations is global and converges to a self-similar solution. This issue is answered by the following theorem (obtained in collaboration with F. Planchon).

For simplifying the notation, we will denote by $D_t : L^q(\mathbb{R}^3) \to L^q(\mathbb{R}^3)$ the normalized dilation defined as $D_t f(x) = \sqrt{t} f(\sqrt{t} x)$ and D_t will also act on functions $v(x,t)$ defined on $\mathbb{R}^3 \times (0,\infty)$ by $[D_t v](x,t) = \sqrt{t} v(\sqrt{t} x, t)$.

Our issue is to know whether

$$\lim_{t\uparrow+\infty} [D_t v](x,t) = V(x) \text{ in } L^q(\mathbb{R}^3). \tag{15.57}$$

Theorem 15.12. *There exists a continuous positive function $\eta(q)$ defined on $(3,\infty)$ with the following property: for every $v_0 \in B_q^{-(1-3/q),\infty}$ such that*

$$\|v_0\|_{B_q} < \eta(q) \text{ and } \nabla \cdot v_0 = 0, \tag{15.58}$$

the following two properties are equivalent ones:

$$\|[D_t v](x,t) - V(x)\|_q \to 0, \text{ as } t \to +\infty, \tag{15.59}$$

$$\|[D_t S(t) v_0](x,t) - V_1(x)\|_q \to 0, \text{ as } t \to +\infty. \tag{15.60}$$

Theorem 15.10 means that the asymptotic behavior of the full solution to (15.12) is governed by the asymptotic behavior of the linear evolution. It is easy to check that (15.60) implies $\lim_{\lambda\uparrow+\infty} \lambda v_0(\lambda x) = V_0(x)$ in the weak-star topology $\sigma(B_q^{-(1-3/q),\infty}, B_p^{(1-3/q),1})$, where $1/p + 1/q = 1$. However, as was mentioned before, (15.60) is a stronger statement. The relationship between V_1 and V_0 is given by $V_1 = S(1)V_0$. Moreover, $V_0(x)$ is homogeneous of degree -1, $\nabla \cdot V_0 = 0$, and the corresponding self-similar solution to (15.12) is precisely $\frac{1}{\sqrt{t}} V\left(\frac{x}{\sqrt{t}}\right)$.

A final observation is $\lim_{\lambda\uparrow+\infty} \lambda v(\lambda x, \lambda^2 t) = \frac{1}{\sqrt{t}} V\left(\frac{x}{\sqrt{t}}\right)$.

If $v_0(x)$ belongs to $L^3(\mathbb{R}^3)$, then $V_0 = 0$ and the theorem yields the well-known decay estimates for solutions of the Navier-Stokes equations.

15.13 The Nonlinear Schrödinger Equation

We will consider the following evolution equation

$$i\frac{\partial u}{\partial t} + \Delta u = \gamma u |u|^2, \quad u(x,0) = u_0(x) \tag{15.61}$$

where $u = u(x,t)$ is a complex-valued function of $x \in \mathbb{R}^3$ and $t \geq 0$. The constant γ is real-valued.

T. Cazenave and F. Weissler [6] succeeded in constructing self-similar solutions to (15.61) by applying the general organization which was used before. However, the functional spaces which will play a crucial role are no longer Besov spaces.

Let us denote by $S(t) = \exp(it\Delta)$ the linear Schrödinger group. These operators are not smoothing but act unitarily on $L^2(\mathbb{R}^3)$. Then (15.61) can be rewritten

$$u(t) = S(t)u_0 - i\gamma \int_0^t S(t-s)u|u|^2(s)\,ds, \qquad (15.62)$$

and we are looking for solutions $u(t)$ to (15.61) which belong to $\mathcal{C}([0,\infty); E)$ where E is a suitable Banach space.

We then mimic our previous approach and define E by the following condition

$$\sup_{t>0} t^{1/8} \|S(t)[f]\|_4 = \|f\|_E < \infty. \qquad (15.63)$$

One easily checks that testing functions belong to E. An equivalent norm on E is defined by the condition

$$\|\mathcal{F}(e^{i\lambda|x|^2}f(x))\|_4 \leq C\lambda^{5/6}, \qquad (15.64)$$

where $\lambda \in (0,\infty)$ and

$$[\mathcal{F}f](x) = \int e^{-ix\cdot y}f(y)\,dy. \qquad (15.65)$$

It is easy to check that E does not contain any L^p space for $1 \leq p \leq \infty$.

But E contains some nonsmooth functions since Cazenave and Weissler [5] observed that $|x|^{-1} \in E$. What is most surprising is the following observation by F. Oru [15]: $f(x) \in E$ does not imply $\overline{f}(x) \in E$. Indeed, $e^{i|x|^2}|x|^{-1}$ belongs to E but $e^{-i|x|^2}|x|^{-1}$ does not. This is related to the fact that (15.62) is valid only for positive t. A final observation is the following: E is isometrically invariant under translations and modulations (i.e., multiplication by $e^{i\omega x}$). This implies that quite irregular functions may belong to E.

Cazenave and Weissler proved the following theorem [6]:

Theorem 15.13. *There exist two positive constants η and β with the following properties: let $u_0(x) \in S'(\mathbb{R}^3)$ satisfy the condition*

$$\sup_{t>0} t^{1/8} \|S(t)u_0\|_4 < \eta. \qquad (15.66)$$

Then there exists a unique solution $u \in \mathcal{C}([0,\infty); E)$ to (15.62) such that

$$\sup_{t>0} t^{1/8} \|u(.,t)\|_4 < \beta. \qquad (15.67)$$

We should warn the reader that E is not a separable Banach space, which implies that $u \in \mathcal{C}([0,\infty); E)$ means that $u(\cdot, t)$ is a continuous function of t when E is equipped with its weak-star topology (as it was the case for the Besov spaces). However, the solution which is being constructed will also be a continuous function of $t \in (0,\infty)$ with values in L^4 (where L^4 is given its strong topology defined by the L^4 norm).

After proving Theorem 15.9, Cazanave and Weissler applied Theorem 15.9 to the construction of self-similar solutions to (15.61). The rescaling is once more the same as in the nonlinear heat equation or as in the Navier-Stokes equations.

To obtain self-similar solutions to (15.61), it suffices to start with $u_0(x)$ with the following properties: $u_0(x) \in C^\infty(\mathbb{R}^3 \setminus \{0\})$ and $u_0(\lambda x) = \lambda^{-1} u_0(x)$, $\lambda > 0$. Then it is easily checked that such a u_0 belongs to the Banach space defined by (15.66). If the norm of u_0 in this Banach space exceeds η, it suffices to multiply u_0 by a small constant. We end with a large family of self-similar solutions to the nonlinear Schrödinger equation.

Addendum: With grief and sorrow we are returning to these notes which were written three years ago when A. Calderón was still among us. We would like to mention two important theorems which have been obtained recently. The first one is an improvement on Theorem 15.9. G. Furioli, P. G. Lemarié-Rieusset, and E. Terraneo obtained uniqueness without the artificial condition (15.49). The best reference is [14]. In contrast, E. Terraneo built two distinct solutions $u(x,t)$ and $v(x,t)$ to (15.1) such that $u(x,0) = v(x,0) = u_0(x)$ and $u, v \in C([0,\infty), L^3(\mathbb{R}^3))$. In her counterexample, u is the solution given by Theorem 15.3 while $\|v(\cdot, t)\|_6 = +\infty$ for $t > 0$ [19].

References

[1] J. M. Ball, *Remarks on blow-up and nonexistence theorems for non-linear parabolic evolution equations*, Quart J. Math. Oxford Ser. 28 (1977), 473–486.

[2] O. Barraza, *Self-similar solutions in weak L^p spaces of the Navier-Stokes equations*, Revista Matematica Iberoamericana, vol. 12, no. 2 (1996), 411–439.

[3] M. Cannone, *Ondelettes, paraproduits et Navier-Stokes*, Diderot Editeur, 20 rue N. D. de Nazareth, 75003 Paris (1995).

[4] M. Cannone, Y. Meyer, and F. Planchon, *Solutions autosimilaires des équations de Navier-Stokes*, Séminaire Équations aux Dérivées Partielles, Centre de Mathématiques École Polytechnique (1993–1994).

[5] M. Cannone and F. Planchon, *Self-similar solutions for Navier-Stokes equations in* $\mathbb{R}^3$, Comm. in P. D. E. 21 (1996), 179–193.

[6] T. Cazenave and F. Weissler, *Asymptotically self-similar global solutions of the non-linear Schrödinger and heat equations*, Math. Z. 228 (1998), no. 1, 83–120.

[7] M. Escobedo and O. Kavian, *Variational problems related to self-similar solutions to the heat equation*, Nonlinear analysis, theory, methods, and applications, 11 (1987), 1103–1133.

[8] M. Escobedo, O. Kavian, and H. Matano, *Large time behavior of solutions of a dissipative semilinear heat equation*, Comm. in P. D. E. 20 (1995), 1427–1452.

[9] A. Haraux and F. Weissler, *Non-uniqueness for a semilinear initial value problem*, Indian Univ. Math. J. 31 (1982), 166–189.

[10] T. Kato, *Strong* L^p *solutions of the Navier-Stokes equations in* $\mathbb{R}^m$ *with application to weak solutions*, Math. Zeit. 187 (1982), 167–189.

[11] O. Kavian, *Remarks on the large time behavior of a nonlinear diffusion equation*, Analyse non-linéaire, Annales IHP 4 (1987), 423–452.

[12] P. G. Lemarié and Y. Meyer, *Ondelettes et bases hilbertiennes*, Revista Matematica Iberoamericana, vol. 2 (1986), 1–18.

[13] H. A. Levine, *Some nonexistence and instability theorems for solutions of formally parabolic equations of the form* $P(d/dt)u = -Au + F(u)$, Arch. Rat. Mech. Anal. 51 (1973), 371–386.

[14] Y. Meyer, *Wavelets, paraproducts, and Navier-Stokes equations*, in Current Developments in Mathematics, International Press, Cambridge, MA, 1996, 105–212.

[15] F. Oru, *Rôle des oscillations dans quelques problèmes d'analyse non-lineàire*, Ph.D. (June 9, 1998), CMLA, ENS-CACHAN (France).

[16] L. Payne, personal communication to F. Weissler.

[17] F. Planchon, *Solutions globales et comportement asymptotique pour les équations de Navier-Stokes*, Thesis, École Polytechnique, Centre de Mathématiques (May 28, 1996).

[18] ———, *Asymptotic behavior of global solutions to the Navier-Stokes equations in* $\mathbb{R}^3$, Revista Matemática Iberoamericana, vol. 14, no. 1 (1998), 71–93.

[19] E. Terraneo, *On the non-uniqueness of weak solutions of the nonlinear heat equation with nonlinearity u^3*, Comptes Rendus de l'Académie des Sciences de Paris (1999), to appear.

Chapter 16

Estimates for Elliptic Equations in Unbounded Domains and Applications to Symmetry and Monotonicity

Louis Nirenberg

The maximum principle for second-order elliptic equations plays a fundamental role in their study. In recent years it has been used to derive properties of solutions of nonlinear equations, such as symmetry and monotonicity in some direction. For simplicity I will discuss only problems of the form

$$u > 0, \quad \Delta u + f(u) = 0 \quad \text{in } \Omega,$$
$$u = 0 \quad \text{on } \partial\Omega. \tag{16.1}$$

Here Ω is a domain (open connected set) in $\mathbb{R}^n$, and f is a Lipschitz continuous function. I begin with the simplest result, one in [14]:

Theorem 16.1. *Suppose u is a solution of (16.1) for Ω a ball: $\{|x| < R\}$. Then u is radially symmetric and $u_r < 0$ for $0 < r < R$.*

The proof uses the method of Moving Planes introduced by A. D. Alexandroff [2] and then used by J. Serrin [17].

It involves showing monotonicity and symmetry in some direction, say the x_1-direction. I describe it briefly: For $-R < \lambda < 0$, let $\Sigma_\lambda = \{x \in \Omega : x_1 < \lambda\}$; one shows that if $x = (x_1, x')$, $x' = (x_2, \ldots, x_n)$, is in Σ_λ then

Research partially supported by grant ARO-DAAL-03-92-6-0143 and by NSF grant DMS-9400912.

$$u(x) < u(x^\lambda) = u(2\lambda - x_1, x').$$ (16.2)

The point x^λ is the reflection of x in the plane $\{x_1 = \lambda\}$. This then yields the x_1-monotonicity of u for $x_1 < 0$, and also the symmetry about $\{x_1 = 0\}$. For if we let $\lambda \to 0$ in (16.2) we find that

$$u(x_1, x') \leq u(-x_1, x') \quad \text{if} \quad x_1 < 0 .$$ (16.3)

However, we can replace x_1 by $-x_1$ in the equation, use (16.3) again, and conclude that equality must hold in (16.3). Since the direction x_1 is arbitrary, radial symmetry of u then follows.

To prove (16.2) one uses the maximum principle. In Σ_λ the function $v(x) = u(x^\lambda)$ satisfies the same equation

$$\Delta v + f(v) = 0 .$$

Subtracting this from the equation for u one finds readily that

$$w_\lambda(x) := u(x) - v(x)$$

satisfies

$$\begin{aligned} \Delta w_\lambda + c_\lambda(x)\, w_\lambda &= 0 \quad \text{in } \Sigma_\lambda, \\ w_\lambda &\lesseqgtr 0 \quad \text{on } \partial\Sigma_\lambda . \end{aligned}$$ (16.4)

Here $|c_\lambda(x)| \leq k$, the Lipschitz constant of f. If the maximum principle were to hold we could conclude that $w_\lambda < 0$ in Σ_λ and (16.2) would be proved. However, one cannot be sure that the maximum principle holds, since about the coefficient c_λ we only know that it is in L^∞. To prove that $w_\lambda < 0$ in Σ_λ, one first considers $0 < \lambda + R$ small; the domain Σ_λ is then narrow in the x_1-direction, and this is a classical sufficient condition for the maximum principle. So in this case, $w_\lambda < 0$ in Σ_λ. Now one moves the plane $\{x_1 = \lambda\}$ to the right, and one has to show that for *every* λ in $(-R, 0)$, $w_\lambda < 0$ in Σ_λ. For this, the argument in [14], as in [17], uses the Hopf Lemma and a version of it at a corner. Berestycki and I [8] have given a different proof which avoids the use of the Hopf Lemma. It enables one to prove symmetry and monotonicity in some direction even for domains with no smoothness requirements on the boundary. Another method, the Sliding Method, was also used by us to prove monotonicity.

In connection with questions concerning regularity of free boundaries, Berestycki, Caffarelli, and I were led to extend the symmetry and monotonicity results to various unbounded domains. This chapter is devoted to some of our recent results. They required new inequalities for elliptic equations in unbounded domains, and I will present some here. I will also mention several problems which are still open. I should say that since [14] and [15], which use moving planes for $\Omega = \mathbb{R}^n$, many people have used the technique in unbounded as well as bounded domains to derive estimates, prove nonexistence of solutions, etc.

First, a few words about the maximum principle. Here we consider a general linear second-order elliptic operator

$$L = a_{ij}(x) \frac{\partial^2}{\partial x_i \, \partial x_j} + b_i(x) \frac{\partial}{\partial x_i} + c(x) \tag{16.5}$$

in a domain Ω in $\mathbb{R}^n$; i, j summed from 1 to n. We assume uniform ellipticity

$$c_0 |\xi|^2 \le a_{ij} \, \xi_i \xi_j \le \frac{1}{c_0} |\xi|^2 \quad \text{for some} \quad c_0 > 0 \quad \text{and all} \quad \xi \le \mathbb{R}^n . \tag{16.6}$$

The a_{ij} are assumed to be in $C(\Omega)$ while b_i, c are in L^∞, with

$$|b_i|, \, |c| \le b . \tag{16.7}$$

There are various forms of the maximum principle. Here we use the following

Definition 16.2. We say that the *maximum principle holds* for L in a bounded domain Ω in case any function $w \in C(\overline{\Omega}) \cap W^{2,n}_{\text{loc}}(\Omega)$ which satisfies

$$Lw \ge 0 \quad \text{in } \Omega,$$
$$w \le 0 \quad \text{on } \partial\Omega$$

necessarily satisfies

$$w \le 0 \quad \text{in } \Omega .$$

Of course, the Maximum Principle need not always hold: on $\Omega = [0, \pi]$ in $\mathbb{R}$, the function $w = \sin x$ satisfies $\ddot{w} + w = 0$, $w = 0$ on $\partial\Omega$, but w is positive, not negative. As mentioned before, a classical sufficient condition (see, for example, Protter Weinberger [16]) is that Ω is narrow in some, say the x_1, direction:

$$a < x_1 < a + \varepsilon \quad \text{in} \quad \Omega \quad \text{for } \varepsilon \text{ small.}$$

How small ε is to be depends only on the constants c_0, b in (16.6), (16.7). In treating domains with irregular boundaries, Berestycki and I made use of the fairly recent sufficient condition, that Ω has small measure:

$$\text{meas } \Omega = |\Omega| < \delta(n, c_0, b);$$

see Berestycki, Nirenberg, and Varadhan [9].

F. Bakelman first observed this, but with δ depending also on the diameter of Ω. In [9] we also proved a more general sufficient condition, that Ω is "narrow" in a more general sense; see also X. Cabre [11]. In case $a_{ij} \in C(\overline{\Omega})$ and $\partial\Omega$ is smooth, a long-known *necessary and sufficient* condition for the maximum principle to hold is that the principal eigenvalue λ_1 for $-L$, under Dirichlet boundary conditions, be positive. The principal eigenvalue is one for which there is a positive eigenfunction ϕ;

$$\phi > 0, \quad (L + \lambda_1)\phi = 0 \quad \text{in } \Omega,$$

$$\phi = 0 \quad \text{on } \partial\Omega .$$

Incidentally in [9] it was proved that for any bounded Ω – $\partial\Omega$ need not be smooth – there exists a principal eigenvalue λ_1 with positive eigenfunction. λ_1 is algebraically simple and $\lambda_1 < \operatorname{Re}\lambda$ for any other (possibly complex) eigenvalue λ. λ_1 is given by

$$\lambda_1 = \sup\left\{\lambda \in \mathbb{R} : \exists\,\phi > 0 \text{ in } W^{2,n}_{\text{loc}}(\Omega) \text{ with } (L + \lambda)\phi \leq 0\right\}$$

$$= -\inf_{\substack{\phi > 0 \\ \phi \in W^{2,n}_{\text{loc}}}} \sup_{x \in \Omega} \frac{L\phi(x)}{\phi(x)} .$$

If $a_{ij} \in C(\overline{\Omega})$ and $\partial\Omega$ is smooth, Agmon [1] proved the existence of infinitely many eigenvalues, and proved the completeness of the eigenfunctions (including generalized ones) – also for operators of any order.

Open Problem.   If $a_{ij} \in C(\Omega)$ and $\partial\Omega$ is not smooth, do there exist eigenvalues other than λ_1? (See [9] for a more precise formulation.)

A useful fact in connection with the maximum principle and principal eigenvalue is the following, which we leave as an exercise for the reader.

Lemma 16.3. *Suppose $u, v > 0$ in Ω, and satisfy*

$$Lu \leq 0 \leq Lv \quad \text{in } \Omega.$$

If $v \in C(\overline{\Omega})$ and $v = 0$ on $\partial\Omega$ then

$$u \equiv tv , \qquad t \text{ a positive constant,}$$

and u and v are principal eigenfunctions.

Returning to the problem (16.1), in [4] we considered the case that Ω is a half-space $\{x_n > 0\}$ in $\mathbb{R}^n$ and u is bounded:

$$0 < u \leq \sup u = M < \infty, \quad \Delta u + f(u) = 0 \quad \text{in } \{x_n > 0\},$$

$$u = 0 \quad \text{on } \{x_n = 0\} . \tag{16.8}$$

As always, f is Lipschitz continuous.

Theorem 16.4 ([4]). *If u is a solution of (16.8) then $f(M) \geq 0$. If $f(M) = 0$, then*

$$u = u(x_n) \quad and \quad u_{x_n} > 0 \quad for \quad x_n > 0 .$$

The proof used the Sliding Method. We also permitted f to depend on x_n. H. Tehrani [18] proved an analogous result if f also depends on $|\nabla u|$.

Conjecture 1. If there is a solution u of (16.8) then necessarily $f(M) = 0$, and so the conclusion of Theorem 16.4 holds.

We know that the conjecture is true in case $n = 2$ and f is in $C^1(\mathbb{R}^+)$, see [7].

We have not even been able to settle the conjecture for a very simple case: $f(u) = u - 1$. For this case we make the following

Conjecture 2. Problem 16.8 has no solution in case

$$f = u - 1 . \tag{16.9}$$

Note that there is a nonnegative solution,

$$v = 1 - \cos x_n ,$$

but is there a positive one? We can prove Conjecture 2 in case $n = 2$ or 3 — in fact, by a very simple argument.

E. N. Dancer [13] proved the following

Theorem 16.5 ([13]). *If u is a solution of (16.8) and if $f(0) \geq 0$ then $u_{x_n} > 0$ for $x_n > 0$.*

In [13], whether u is a function of x_n alone is not studied. His result suggested to us a weaker conjecture than Conjecture 1, namely

Conjecture 3. If u is a solution of (16.8) and u is increasing in x_n, then necessarily $f(M) = 0$.

We have proved this in the case $n = 2$ or 3 and $f \in C^1(\mathbb{R}^+)$. Our proof relies, surprisingly, on a result for a Schrödinger operator of the form

$$L = -\Delta + q \quad in \ \mathbb{R}^n, \qquad q \in L^\infty \ and, \ say, \ smooth. \tag{16.10}$$

Conjecture 4. Suppose v is a solution of

$$Lv = 0 \quad in \ \mathbb{R}^n, \quad |v| \ bounded$$

and v changes sign. Then L has some negative spectrum, i.e., $\exists \zeta \in C_0^\infty(\mathbb{R}^n)$ such that

$$\int_{\mathbb{R}^n} \zeta L \zeta = \int_{\mathbb{R}^n} |\nabla \zeta|^2 + q \zeta^2 < 0 .$$

In [7] we prove this conjecture in the case $n = 1$ or 2.

We describe next another monotonicity result in a more general domain Ω. Consider Ω in $\mathbb{R}^n$ bounded by a Lipschitz graph:

$$\Omega = \{x \in \mathbb{R}^n : x_n > \phi(x')\}.$$

Here $x' = (x_1, \ldots, x_{n-1})$, and ϕ is a Lipschitz continuous function defined on all of $\mathbb{R}^{n-1}$.

Theorem 16.6 ([6]). *Let $u \in C(\overline{\Omega}) \cap W^{2,n}_{\text{loc}}(\Omega)$ be a solution of*

$$0 < u \le \sup u = M < \infty, \quad \Delta u + f(u) = 0 \quad \text{in} \ \ \Omega, \tag{16.11}$$
$$u = 0 \quad \text{on} \ \ \partial\Omega .$$

Here f is a nonnegative Lipschitz function on $\mathbb{R}^+$, satisfying

$$f(s) > 0 \ \ \text{on} \ \ (0, \mu), \qquad f(s) \le 0 \ \ \text{for} \ \ s \ge \mu ,$$

for some $\mu > 0$, and for some positive δ_0 and $s_0 < s_1 < \mu$,

$$f(s) \ \ge \ \delta_0 \, s \ \ \text{on} \ \ [0, s_0],$$
$$f(s) \ \ \text{is nonincreasing on} \ \ (s_1, \mu).$$

Then

(a) $u_{x_n} > 0$ in $\ \Omega$,

(b) $u < \mu$ and $u(x) \to \mu$ uniformly as $\ \text{dist}\,(x, \partial\Omega) \to \infty$,

(c) u is the unique solution of (16.11).

The proof of the theorem uses the Sliding Method and is rather tricky.

We turn now to an unbounded domain with a quite different geometry, a cylinder,

$$\Omega = \mathbb{R} \times \omega ,$$

where ω is a smooth bounded domain in $\mathbb{R}^{n-1}$. More generally, we consider

$$\Omega = \mathbb{R}^{n-j} \times \omega$$

with ω a smooth bounded domain in $\mathbb{R}^j$. The problem we consider is

$$u > 0, \quad \Delta u + f(u) = 0 \quad \text{in} \ \ \Omega, \tag{16.12}$$
$$u = 0 \quad \text{on} \ \ \partial\Omega,$$

where f is a Lipschitz continuous function. Denote the coordinates in $\mathbb{R}^{n-j}$ by x and those in $\omega \subset \mathbb{R}^j$ by y. No conditions on the behavior of u as $|x| \to \infty$ are required.

Question: Let u be a solution of (16.12) for ω a ball: $\{|y| < R\}$. Is it true that for $|y| = r$,

$$u = u(x,r) \quad \text{and} \quad u_r < 0 \quad \text{for} \quad 0 < r < R \,? \tag{16.13}$$

Theorem 16.7. *The answer is yes in the following cases:*

> *(i)* $\quad j > 1,$
> *(ii)* $\quad j = 1 \quad$ and $\quad n = 2,$
> *(iii)* $\quad j = 1, \quad n > 2 \quad$ and $\quad f(0) \geq 0 \,.$

Open Problem: What happens if $j = 1$, $n > 2$ and $f(0) < 0$?

The proofs of the different cases in Theorem 16.7 are surprisingly tricky. For cases (i) and (iii) the proofs are in [5]; case (ii) is treated in [7]. In [5] we use both the Method of Moving Planes and the Sliding Method. We move planes $\{y_1 = \lambda\}$ to prove symmetry and monotonicity in y_1. Here I will not describe the proofs, just some ingredients.

First, it is important to have some form of the maximum principle in unbounded domains like Ω above, i.e., results of Phragmèn–Lindelöf type. Here is a form that we use in

$$\Omega = \mathbb{R}^{n-j} \times \omega$$

with ω a smooth bounded domain in $\mathbb{R}^j$ (coordinates y).

Lemma 16.8. *Suppose a function w in $C(\overline{\Omega}) \cap W^{2,n}_{\text{loc}}(\Omega)$ satisfies*

$$\begin{cases} \Delta w + c(x,y)w \;\geq\; 0 \quad in \quad \Omega, \qquad |c(x,y)| \leq \gamma, \\ \qquad\qquad w \;\leq\; 0 \quad on \quad \partial\Omega, \\ \qquad\qquad w \;\leq\; C\,e^{\mu|x|} \,. \end{cases} \tag{16.14}$$

Here γ, C, μ are positive constants. There exists a constant $\delta = \delta(n,j,\gamma,\mu)$ such that if

$$meas \; \omega = |\omega| < \delta$$

then

$$w \leq 0 \quad in \quad \Omega.$$

Recently J. Busca [10] extended this result by considering in (16.14), in place of the Laplace operator Δ, any uniformly elliptic operator with coefficients depending on x and y. His proof is rather different from ours.

In Lemma 16.8, w is assumed to grow at most exponentially at infinity while in Theorem 16.7, no assumption on u is made near infinity. A basic step in the proof of Theorem 16.7 is

Theorem 16.9. *Consider a solution of (16.14) in* $\Omega = \mathbb{R}^{n-j} \times \omega$. *Here* ω *may be any bounded domain in* $\mathbb{R}^j$ *with smooth boundary. Then* $\exists\, C,\, \mu > 0$ *such that*

$$u \leq C\, e^{\mu|x|} \, . \tag{16.15}$$

The positivity of u is essential, if u changes sign the conclusion need not hold.

The proof of Theorem 16.9 relies on an inequality which has proved to be very useful. It is an extension of the Krylov–Safonov, Harnack principle, which we now recall. In a domain Ω in $\mathbb{R}^n$, consider a uniformly elliptic operator, i.e., satisfying (16.6),

$$M = a_{ij}(x)\, \frac{\partial^2}{\partial x_i\, \partial x_j} \, .$$

Krylov–Safanov Harnack inequality: If

$$u > 0, \quad |Mu| \leq A(|\nabla u| + u) \quad \text{in } \Omega$$

then, for any $x_0 \in \Omega$ and any compact subset K in Ω, there exists $C_1 = C_1(\Omega, K, x_0, c_0, A)$, such that

$$u(x) \leq C_1\, u(x_0) \quad \forall\, x \in K \, .$$

The useful extension of this result is one which is valid up to the boundary, under suitable conditions. After proving it we discovered that it had essentially been proved by P. Baumann [3] .

Theorem 16.10. *Let* Q_0 *be a cube of side length 2 in* $\mathbb{R}^n$ *centered at* $e_n = (0, \ldots, 0, 1)$,

$$Q_0 = \{x \in \mathbb{R}^n : |x_\alpha| < 1 \quad \text{for} \quad \alpha = 1, \ldots, n-1, \quad 0 < x_n < 2\} \, .$$

Set

$$\Gamma = \partial Q_0 \cap \{x_n > 0\} \, .$$

Let u *be a positive function in* $W_{\text{loc}}^{2,p}(Q_0) \cap C(\overline{Q}_0)$ *for some* $p > n$, *vanishing on the side* $x_n = 0$ *of* ∂Q_0, *and satisfying*

$$|Mu| \leq A(|\nabla u| + u + \kappa) \quad \text{in } Q_0$$

for some constants A, $\kappa \geq 0$. Then there are constants B, q depending only on n, c_0 and A (not on any modulus of continuity of the a_{ij}) such that

$$u(x) \leq \frac{B}{[\text{dist}(x, \Gamma)]^q} \left(u(e_n) + \kappa \right) \quad in \quad Q_0 .$$

The proof in [5] uses the Krylov–Safonov Harnack inequality and some ideas from [12]. Theorem 16.9 follows rather easily from Theorem 16.10. Here is another useful consequence of Theorem 16.10.

Let $\Omega = \mathbb{R}^{n-j} \times \omega$, where ω is a smooth bounded domain in $\mathbb{R}^j$. As before, x denotes coordinates in $\mathbb{R}^{n-j}$, y, coordinates in $\mathbb{R}^j$. Let L be a uniformly elliptic operator in Ω, as in (16.5), in the (x, y) variables, with coefficients depending on (x, y). Assume that the coefficients of the second-order terms belong to $C^\mu(\overline{\Omega})$ for some positive $\mu < 1$.

Theorem 16.11. *Let u be a solution in Ω of*

$$u > 0, \quad Lu = 0 \quad in \quad \Omega,$$
$$u = 0 \quad on \quad \partial\Omega .$$

Then there exists a constant $\overline{C}$ independent of u, such that

$$|\nabla_x u(x, y)| \leq \overline{C} u(x, y) \quad in \quad \Omega.$$

We conclude with some remarks about "principal" eigenfunction, i.e., a positive eigenfunction for a linear second-order elliptic operator. As we have remarked, in a bounded domain, the principal eigenvalue is real and algebraically simple. What about an unbounded domain? Is there some analogue of Lemma 16.3? Presumably one might have to require some condition at infinity. For example, in the strip in $\mathbb{R}^2$

$$\Omega = \{x, y : x \in \mathbb{R}, \ 0 < y < \pi\} ,$$

the two functions

$$u_\pm = e^{\pm x} \sin y$$

satisfy

$$u > 0, \quad \Delta u = 0 \quad in \quad \Omega,$$
$$u = 0 \quad on \quad \partial\Omega .$$

The eigenvalue is zero, and we have two linearly independent solutions. On the other hand, $\phi = \sin y$ satisfies

$$\phi > 0, \quad (\Delta + 1)\phi = 0 \quad in \quad \Omega,$$
$$\phi = 0 \quad on \quad \partial\Omega . \tag{16.16}$$

The eigenvalue is 1, greater than zero, but it is easy to see that any solution of (16.16) is a constant multiple of $\sin y$.

We have a general result, an analogue of Lemma 16.3, in $\Omega = \mathbb{R}^{n-j} \times \omega$, where ω is a smooth bounded domain in $\mathbb{R}^j$. In ω, consider an elliptic operator satisfying (16.6) and (16.7):

$$M = \sum_{i,k=1}^{n} a_{ik}(y) \frac{\partial^2}{\partial y_i\,\partial y_k} + b_i(y) \frac{\partial}{\partial y_i} + c(y)$$

with

$$a_{ik} \in C(\overline{\omega}) \ .$$

Theorem 16.12. *Suppose $\phi \in C^1(\overline{\omega}) \cap W^{2,n-j}_{\mathrm{loc}}(\omega)$ satisfies*

$$\phi > 0, \quad M\phi \geq 0 \ \ in \ \omega,$$

$$\phi = 0 \ \ on \ \partial\omega.$$

Let $u \in W^{2,n}_{\mathrm{loc}}(\Omega)$ be a function in Ω satisfying

$$u > 0, \quad (\Delta_x + M)\,u \leq 0 \ \ in \ \Omega.$$

Then
$$u(x,y) \equiv t\phi(y) \ , \quad t \ constant, \ and \ M\phi = 0,$$

provided $n - j = 1$ *or* 2. *The conclusion need not hold if $n - j > 2$.*

If we strengthen the conditions in Theorem 16.12 we obtain a result valid in all dimensions:

Theorem 16.13. *Under the conditions of Theorem 16.12, the conclusion holds for any j, n, $j < n$, provided we require in addition that*

$$M\phi = 0 \ \ in \ \omega$$

and
$$u > 0, \quad (\Delta_x + M)\,u = 0 \ \ in \ \Omega,$$

$$u = 0 \ \ on \ \partial\Omega.$$

Finally, a result in a half cylinder

$$\Omega_+ = \mathbb{R}_+ \times \omega,$$

where ω is a smooth bounded domain in $\mathbb{R}^{n-1}$:

Theorem 16.14. *Let M and ϕ be as in Theorem 16.12. Suppose $u > 0$ in Ω_+ and continuous in $\overline{\mathbb{R}}_+ \times \omega$, with*

$$u(0, y) \geq \phi(y) \quad in \ \omega.$$

Suppose also that u satisfies

$$(\Delta_x + M) u \leq \gamma(x, y) u \quad in \ \Omega_+$$

with $\gamma(x, y) \to 0$ as $x \to +\infty$ uniformly in y. Then $u(x, y)$ decays more slowly than any exponential as $x \to +\infty$; more precisely, $\forall \alpha > 0$, $\exists C(\alpha)$ such that

$$u(x) \geq C(\alpha) e^{-\alpha x} \phi(y) .$$

Theorem 16.14 was used in our first proof of Theorem 16.7(ii). Since then, we have a simpler proof.

Recently, counterexamples to Conjecture 4 have been constructed, first by N. Ghoussoub and C. Gui for $n \geq 7$, in *On a conjecture of De Giorgi and some related problems*, Math. Ann. 31 (1998), 481–491, and then, for $n \geq 3$, by M. T. Barlow in *On the Liouville property for divergence form operators*, Canadian J. Math. 50 (1998), 487–496.

References

[1] S. Agmon, *On the eigenfunctions and, on the eigenvalues of general elliptic boundary value problems*, Comm. Pure Appl. Math. 15 (1962), 119–148.

[2] A. D. Alexandroff, *A characteristic property of the spheres*, Ann. Mat. Pura Appl. 58 (1962), 303–354.

[3] P. Baumann, *Positive solutions of elliptic equations in nondivergence form and their adjoints*, Ark. Mat. 22 (1984), 153–173.

[10] J. Busca, *Maximum principles in unbounded domains via Bellman equations*, preprint.

[4] H. Berestycki, L. A. Caffarelli, and L. Nirenberg, *Symmetry for elliptic equations in a half space*, Boundary Value Problems for Partial Differential Equations and Applications, ed. J. L. Lions et al., Masson, Paris (1993), 27–42.

[5] ————, *Inequalities for second-order elliptic equations with applications to unbounded domains*, I. Duke Math. J. 81 (1996), 467–494.

[6] ———, *Monotonicity for elliptic equations in unbounded Lipschitz domains*, Comm. Pure Appl. Math. 50 (1997), 1089–1111.

[7] ———, *Further qualitiative properties for elliptic equations in unbounded domains*, Annali della Scuola Normale Sup. di Pisa, ser. IV 25 (1997), 69–94.

[8] H. Berestycki and L. Nirenberg, *On the method of moving planes and the sliding method*, Bol. Soc. Brasil Mat. (N.S.) 22 (1991), 1–37.

[9] H. Berestycki, L. Nirenberg, and S. R. S. Varadhan, *The principal eigenvalue and the maximum principle for second-order elliptic operators in general domains*, Comm. Pure Appl. Math. 47 (1994), 47–92.

[10] J. Busca, *Maximum principles in unbounded domains via Bellman equations*, preprint.

[11] X. Cabre, *On the Alexandroff–Bakelman–Pucci estimate and the reversed Hölder inequality for solutions of elliptic and parabolic equations*, Comm. Pure Appl. Math. 48 (1995), 539–570.

[12] L. A. Caffarelli, E. Fabes, G. Mortola, and S. Salsa, *Boundary behavior of solutions of elliptic divergence equations*, Indiana J. Math. 30 (1981), 621–640.

[13] E. N. Dancer, *Some notes on the method of moving planes*, Bull. Austral. Math. Soc. 46 (1992), no. 3, 425–434.

[14] B. Gidas, W. M. Ni, and L. Nirenberg, *Symmetry and related properties via the maximum principle*, Comm. Math. Phys. 6 (1981), 883–901.

[15] ———, *Symmetry of positive solutions of nonlinear elliptic equations in* $\mathbb{R}^n$, Math. Anal. and Appls. Part A. Advances in Math. Suppl. Studies, vol. 7A, ed. L. Nachbin, Academic Press (1981), 369–402.

[16] M. H. Protter and H. Weinberger, *Maximum Principles in Differential Equations*, Prentice-Hall, Englewood Cliff, NJ, 1967.

[17] J. Serrin, *A symmetry theorem in potential theory*, Arch. Rat. Mech. Anal. 43 (1971), 304–318.

[18] H. Tehrani, *On some semilinear elliptic boundary value problems*, Ph.D. thesis, Courant Institute (1994).

Chapter 17

Asymptotic Expansions for Atiyah-Patodi-Singer Problems

Robert Seeley

17.1 Background

We begin with a very brief reminder of some famous results on asymptotics for the eigenvalues and eigenfunctions of the Laplace operator $\Delta\phi_j = \lambda_j\phi_j$. They go back at least to 1912, when Hermann Weyl [20] proved, for a bounded region in $\mathbb{R}^2$, that

$$\sum_{\lambda_j < \lambda} \phi_j(x)^2 \sim \frac{1}{4\pi}\lambda.$$

Later Carleman [7] proved it again by studying the asymptotic behavior of the resolvent kernel

$$(\Delta + \lambda)^{-1}(x, y)$$

near the diagonal, and then applying a Tauberian theorem. Although Weyl's asymptotics cannot in general be refined by adding more terms with lower powers of λ, the asymptotics of the resolvent *can* be, and so can other related expansions. The first result of this kind was due to Minakshisundaram and Pleijel in 1949 [15] for the heat kernel:

$$e^{-t\Delta}(x, x) \sim t^{-n/2}\sum_0^\infty c_{2j}(x)t^j, \ t \to 0^+.$$

Their aim was to obtain the metamorphic extension of the "zeta function" $\mathrm{tr}(\Delta^{-s})$ and determine the singularities of its kernel:

$$\Gamma(s)\Delta^{-s}(x, x) \sim \frac{-1}{s \cdot \mathrm{vol}(M)} + \sum_0^\infty \frac{c_{2j}(x)}{s + j - n/2}.$$

275

In these expansions, the coefficients $c_{2j}(x)$ are determined by the jets of the metric at x. Minakshisundaram and Pleijel obtained them by means of Hadamard's transport equations. On a compact manifold without boundary, each one can be integrated to give the expansion of the trace of the corresponding operator.

The resolvent, heat, and zeta expansions are all related. For example,

$$e^{-t\Delta} = \frac{1}{2\pi i} \int_\Gamma e^{-t\lambda}(\Delta - \lambda)^{-1} \, d\lambda,$$

where Γ is an appropriate contour in the complex plane, and

$$\Gamma(s)\Delta^{-s} = \int_0^\infty (e^{-t\Delta} - \Pi_0)t^{2-1} \, dt,$$

where Π_0 is projection on the nullspace of Δ.

17.2   The Index Question

Interest in these expansions was renewed by the work of Atiyah and Singer [3] on the index theorem. They considered the general situation of an elliptic pseudodifferential operator D acting between two vector bundles E and F over a compact manifold M,

$$D : C^\infty(E) \to C^\infty(F).$$

The index is

$$\text{ind}\,(D) = \text{null}\,(\Delta^+) - \text{null}\,(\Delta^-), \text{ with } \Delta^+ = D^*D \text{ and } \Delta^- = DD^*.$$

The nonzero elements of the spectra of Δ^+ and Δ^- coincide, and from this it follows directly that the index is given by the trace formula

$$\text{ind}\,(D) = \text{tr}\left(e^{-t\Delta^+} - e^{-t\Delta^-}\right) \text{ for all } t > 0. \tag{17.1}$$

Important geometric examples include the de Rham operator on forms,

$$D_{\text{deR}} = d + d^* : \lambda^{\text{even}}(M) \to \lambda^{\,\text{odd}}(M); \quad \text{ind}\,(D_{\text{deR}}) = \text{Euler}\,(M),$$

the $\overline{\partial}$ operator on a Riemann surface

$$\overline{\partial} : f \mapsto \frac{\partial f}{\partial \overline{z}} \, d\overline{z}; \quad \text{ind}\,(\overline{\partial}) = 1 - \text{genus}\,(M),$$

and the signature operator acting between certain spaces of forms λ^+ and λ^-,

$$D_{\text{sign}} : \lambda^+(M) \to \lambda^-(M); \quad \text{ind}\,(D_{\text{sign}}) = \text{signature}(M^{4k}).$$

The question was raised, is there in general a "Minakshisundaram and Pleijel expansion" for the Laplacians Δ^+ and Δ^-? It turns out that there is ([17], [19]). If D has order ω and M has dimension n, then, for a differential operator D, the heat kernels have an expansion

$$e^{-t\Delta}(x,x) \sim t^{-n/2\omega} \sum_0^\infty c_{2j}(x)t^{j/\omega}.$$

The coefficients $c_{2j}(x)$ depend locally on the jets of the symbol of D at x. For a pseudodifferential operator, the expansion includes other terms, first noted by Duistermaat and Guillemin [8]:

$$e^{-t\Delta}(x,x) \sim t^{-n/2\omega} \sum_0^\infty c_k(x)t^{k/2\omega} + \sum_0^\infty c_j'(x)t^{j/2\omega}\log t.$$

Here $c_k(x)$ is locally determined if $k \leq n$. Hence the *trace formula* (17.1) above gives the index of D as the integral of a naturally defined density, determined locally by the symbol of D:

$$\text{ind}\,(D) = \int_M \omega_D.$$

The density ω_D is obtained from the terms in t^0 in the expansions for $e^{-t\Delta^+}$ and $e^{-t\Delta^-}$, taking the fiber trace in the appropriate bundles:

$$\omega_D(x) = \text{tr}_E\left[c_n^+(x)\right] - \text{tr}_F\left[c_n^-(x)\right].$$

For the classic geometric examples, remarkable results of Patodi [16], Gilkey [9], and Getzler [4] show that $\omega_D(x)$ can be expressed in terms of characteristic polynomials for the metric on M. In the simple case of a 2-manifold, the index form for the de Rham operator is just the scalar curvature, and the index formula is the Gauss-Bonnet theorem.

17.3 Boundary Problems

For boundary problems there is a similar formula, but including a term integrated over the boundary. As before, we have bundles E and F over the manifold M, an "interior operator"

$$D : C^\infty(E) \to C^\infty(F)$$

and a "boundary operator"

$$B : C^\infty\left(E|_{\partial M}\right) \to C^\infty(G)$$

for some bundle G over ∂M. This gives us an operator D_B, which is D acting in the domain

$$\mathrm{dom}(D_B) = \{u : Du \in L^2, Bu|_{\partial M}\} = 0.$$

If D and B are differential and (D, B) is "well posed" then there is an expansion ([10], [18])

$$\mathrm{tr}\, e^{-t\Delta_B^{\pm}} \sim t^{-n/2} \sum_0^{\infty} t^j \int_M c_{2j}^{\pm}(x)\, dx + t^{-n/2} \sum_1^{\infty} t^{k/2} \int_{\partial M} b_k^{\pm}(y)\, dy.$$

So now the trace formula shows that there is a boundary density $\beta_{D,B}$ such that

$$\mathrm{ind}\,(D_B) = \int_M \omega_D(x)\, dx - \int_{\partial M} \beta_{D,B}(y)\, dy.$$

When $n = 2$ and D_B is the de Rham operator with appropriate boundary conditions, this gives the familiar formula

$$\mathrm{Euler}\,(M) = \frac{1}{2\pi} \int_M (\mathrm{scalar}\ K) + \frac{1}{2\pi} \int_{\partial M} (\mathrm{geodesic}\ \kappa).$$

All terms are locally determined by the jets of the symbols $\sigma(D)$ and $\sigma(B)$. Many have been computed by Branson and Gilkey [5]. However, for some operators (including the signature operator) there is no well-posed differential boundary operator B, and this presented a serious obstacle to carrying out the index program for boundary problems. The simplest example is on the unit disk:

$$\bar{\partial} = e^{i\theta}\left(\partial_r + \frac{1}{r}\partial_\theta\right).$$

The natural boundary operator for this is the projection on the positive Fourier coefficients

$$\sum_{-\infty}^{\infty} a_n e^{in\theta} \mapsto \sum_0^{\infty} a_n e^{in\theta}.$$

B is a pseudodifferential operator, precisely, the Calderón projection for $\bar{\partial}$ on the disk [6]. It is the projection on the space $\{i\partial_\theta \leq 0\}$ spanned by the eigenfunctions of $i\partial_\theta$ with eigenvalue ≤ 0. The boundary condition $Bu|_{\partial M} = 0$ eliminates the harmonic functions from the nullspace of $\bar{\partial}$.

Every elliptic differential operator has a similar Calderón projection, and so admits a well-posed boundary condition which is not differential, but pseudodifferential. This overcame the obstacle to the index program, and led to the class called *Atiyah-Patodi-Singer problems*. In these problems, the interior operator has the form

$$D = \gamma(\partial_x + A + xA_1(x)) \tag{17.2}$$

with ∂_x the normal derivative, $\gamma : E \to F$ a unitary morphism, and A a self-adjoint first-order differential operator on $E|_{\partial M}$. The boundary operator chosen by Atiyah, Patodi, and Singer is

$$B = \Pi_{\geq} = \text{ projection on } \{A \geq 0\},$$

essentially the Calderón projection for D. (The difference in sign between this and the $\bar{\partial}$ example above is due to the fact that r points toward the boundary, while x points away.) This gives an operator $D_{\geq}$ with domain

$$\text{dom}(D_{\geq}) = \left\{ u : Du \in L^2, \Pi_{\geq}(u|_{\partial M}) = 0 \right\}.$$

The index of $D_{\geq}$ is not local in $\sigma(D)$ and $\sigma(\Pi_{\geq})$. When $A_1(x) \equiv 0$ for small x, it is given by the *Atiyah-Patodi-Singer formula* [2]:

$$\text{ind}(D_{\geq}) = \int_M \omega_D(x)\, dx - \frac{1}{2} \left[\text{sign}(A) + \text{null}(A) \right]. \qquad (17.3)$$

Here $\text{null}(A)$ is the dimension of the kernel of A, and the signature $\text{sign}(A)$ is the "eta invariant," defined by analytic continuation to $s = 0$ of the "eta function"

$$\eta(A, s) = \text{tr}(|A|^{-s} \text{sign}\, A) = \text{tr}\left(A(A^2)^{-(s+1)/2} \right);$$

$$\text{sign}(A) = \eta(A, 0).$$

17.4   Expansions for APS Problems

The proof of the APS formula was not based on complete expansions of the traces of $e^{-t\Delta^{\pm}}$ for the Laplacians of $D_{\geq}$. However, it is natural to expect such expansions, and Gilkey raised the question explicitly a few years ago. Specifically, if D is an operator of Dirac type, and B an appropriate pseudodifferential boundary operator, is there a complete expansion for the heat kernel $e^{-t\Delta_B}$ associated with the Laplacian $\Delta_B = D_B^* D_B$? In fact, there is:

$$\text{tr}\, e^{-t\Delta_B} \sim c_0 t^{-n/2} + \sum_1^{\infty} (c_j + b_j) t^{(j-n)/2} + \sum_0^{\infty} \left(b_j' \log t + b_j'' \right) t^{j/2}. \qquad (17.4)$$

The terms down to and including t^0 were obtained in [9], and the others in [12]. This expansion gives the singularities in the meromorphic continuation of the zeta function $\text{tr}(\Delta_B^{-s})$ to the complex plane. To extend the corresponding eta function, there is a similar expansion

$$\text{tr}\, D_B e^{-t\Delta_B} \sim c_0 t^{-(n+1)/2} + \sum_1^{\infty} (c_j + b_j) t^{(j-n-1)/2} + \sum_0^{\infty} \left(b_j' \log t + b_j'' \right) t^{(j-1)/2}.$$

In the case considered by Atiyah, Patodi, and Singer, we get [13]

$$\zeta(\Delta_{\geq}^{\pm}, 0) = \text{tr}((\Delta_{\geq}^{\pm})^{-s})|_{s=0}$$

$$= \int_M c_n^{\pm}(x) \, dx \mp \frac{1}{4}\eta(A, 0) \mp \frac{1}{4}\text{null}(A) - \text{null}(\Delta_{\geq}^{\pm}).$$

Here $\text{tr}(\Delta^{-s}) = \sum_{\lambda_j \neq 0} \lambda_j^{-s}$, so $\zeta(\Delta_{\geq}^+, s) = \zeta(\Delta_{\geq}^-, s)$, and this reproves the Atiyah-Patodi-Singer formula.

The method in [12] and [13] is to expand $(\Delta + \lambda)^{-k}$ for $2k > n$. In [13], this is done by a straightforward separation of variables in the cylindrical neighborhood of the boundary. The more general case of [12] is based on a theory of "weakly parametric ψdo's" on a manifold without boundary. The construction of $(\Delta_B + \lambda)^{-1}$ for the boundary problem is then reduced to the expansion of a "weakly parametric ψdo" on the boundary.

A motivating example for the weakly parametric class is the case of the resolvent of a first-order ψdo A on a manifold M without boundary. A has a symbol

$$\sigma(A)(x, \xi) = a_1(x, \xi) + a_0(x, \xi) + a_{-1}(x, \xi) + \cdots.$$

$$\det a_1 \neq 0, \quad (a_1 + \lambda)^{-1} \text{ homogeneous for } |\xi| > 1, \ \lambda > 0.$$

Then the resolvent has a symbol expansion

$$\sigma(A + \lambda)^{-1} = (a_1 + \lambda)^{-1} + b_{-2}(x, \xi, \lambda) + \cdots \qquad (17.5)$$

In the case of a differential operator, a_1 and a_0 are polynomials in ξ, so $(a_1 + \lambda)^{-1}$ is smooth and homogeneous for all $(\xi, \lambda) \neq 0$. This implies that derivatives with respect to ξ improve the decay as $\lambda \to \infty$, and so the expansion (17.5) gives the asymptotics of $(A + \lambda)^{-1}$ as $\lambda \to \infty$, with all terms "local."

If, on the other hand, A is pseudodifferential, it may have a symbol such as

$$a_1(\xi) = [\xi],$$

a positive C^∞ function coinciding with $|\xi|$ when $|\xi| > 1$. In this case, terms such as

$$b_{-1} = ([\xi] + \lambda)^{-1} = 0(\lambda^{-1}) \text{ and } \partial_\xi b_{-1} = O(\lambda^{-2})$$

behave nicely as $\lambda \to \infty$, but

$$\partial_\xi b_{-1} = -b_{-1}^2 \partial_\xi^2[\xi] + \cdots = O(\lambda^{-2})$$

is not of the desired order $O(\lambda^{-3})$, and higher derivatives are no better than $O(\lambda^{-2})$. This implies that the terms which are negligible in the usual ψdo calculus are not negligible as $\lambda \to \infty$, so that calculus alone does not give the desired asymptotic expansion. But each term in the symbol does have an expansion in λ; the expansion for b_{-1} is

$$b_{-1} = ([\xi] + \lambda)^{-1} \sim \lambda^{-1} - [\xi]\lambda^{-2} + [\xi]^2\lambda^{-3} - \cdots .$$

The terms grow progressively worse in ξ but better in λ. It turns out that, for a ψdo whose symbol terms have expansions such as this, the integral kernel has a complete expansion as $\lambda \to \infty$. This expansion contains log terms with local coefficients, and nonlocal terms such as those in (17.4). The reader interested in the detailed technical definitions and statement of results for the class of "weakly parametric ψdo's" that we have formulated will find them in the announcement [14] or the paper [12] already cited.

The most direct application of the theory is to the case of an elliptic ψdo A, of positive integer order m, on a manifold M without boundary. Assume that

$$\sigma_m(A) + \lambda \text{ is invertible for } \lambda > 0.$$

Let Q be any ψdo of order q. Choose k with $km > q + n$. Then the kernel of $Q(A + \lambda)^{-k}$ has an expansion on the diagonal

$$Q(A + \lambda)^{-k}(x, x) \sim \sum_0^\infty c_j(x)\lambda^{(n+q-j)/m-k} + \sum_0^\infty \left[c_j h'(x) \log \lambda + c_j''(x) \right] \lambda^{-k-j}$$

with global coefficients $c_j''(x)$. If Q is a differential operator then $c_0'' \equiv 0$. In the case that $Q = \text{sign}(A)$ then $\int_M c_0''$ is essentially the eta invariant of A.

This expansion for the resolvent has already been obtained by Agranovich [1], from the singularities of QA^{-s}. These can be obtained by rewriting QA^{-s} as $QA^k A^{-s-k}$ for large integer k, and analyzing A^{-s-k} by the usual ψdo calculus. However, there seems to be no similar trick to handle boundary problems; this was what drove us to the theory suggested here. In principle, it yields new global invariants for these boundary problems. It remains to be seen whether some of them are interesting.

References

[1] M. S. Agranovich, *Some asymptotic formulas for pseudodifferential operators*, J. Functional Anal. and Appl. 21 (1987), 63–65.

[2] M. F. Atiyah, V. K. Patodi, and I. M. Singer, *Spectral asymmetry and Riemannian geometry I*, Math. Proc. Camb. Phil. Soc. 77 (1975), 43–69.

[3] M. F. Atiyah and I. M. Singer, *The index of elliptic operators I*, Ann. Math. 87 (1968), 484–530.

[4] N. Berline, E. Getzler, and M. Vergne, *Heat Kernels and Dirac Operators* Springer-Verlag, Berlin, 1992.

[5] T. Branson and P. B. Gilkey, *The asymptotics of the Laplacian on a manifold with boundary*, Comm. Part. Diff. Eq. 15 (1990), 245–272.

[6] A. P. Calderón, *Boundary value problems for elliptic equations*, Outlines of the Joint Soviet-American Symposium on Partial Differential Equations, Novosibirsk, August, 1963, 303–304.

[7] T. Carleman, *Proprietés asymptotiques des fonctions fondamentales des membranes vibrantes*, C. R. Congr. Math. Scand. Stockholm (1934), 34–44.

[8] J. J. Duistermaat and V. W. Guillemin, *The spectrum of positive elliptic operators and periodic bicharacteristics*, Inventiones Math. 29 (1975), 3–79.

[9] P. B. Gilkey, *Invariance Theory, the Heat Equation, and the Atiyah-Singer Index Theorem*, CRC Press, Boca Raton, 1995.

[10] P. Greiner, *An asymptotic expansion for the heat equation*, Arch. Rat. Mech. Anal. 41 (1971), 163–218.

[11] G. Grubb, *Heat operator trace expansions and index for general Atiyah-Patodi-Singer boundary problems*, Comm. P. D. E. 17 (1992), 2031–2077.

[12] G. Grubb and R. Seeley, *Weakly parametric pseudo-differential operators and Atiyah-Patodi-Singer problems*, Inventiones Math. 121 (1995), 481–529.

[13] ———, *Zeta and eta functions for Atiyah-Patodi-Singer operators*, Journal of Geometric Analysis 6 (1996), 31–77.

[14] ———, *Developpements asymptotiques pour l'operateur d'Atiyah-Patodi-Singer*, C. R. Acad. Sci. Paris 317 (1993), 1123–1126.

[15] S. Minakshisundaram and A. Pleijel, *Some properties of the eigenfunctions of the Laplace operator on Riemannian manifolds*, Can. J. Math. 1 (1949), 242–256.

[16] V. K. Patodi, *Curvature and the fundamental solution of the heat operator*, J. Indian Math. Soc. 34 (1970), 269–285; *Curvature and the eigenforms of the Laplace operators*, J. Differential Geom. 5 (1971), 233–249; *An analytic proof of the Riemann-Roch-Hirzebruch theorem for Kähler*

manifolds, J. Differential Geom. 5 (1971), 251–283; *Holomorphic Lefschetz fixed-point formula*, Bull. Amer. Math. Soc. 79 (1973), 825–828.

[17] R. Seeley, *Complex powers of an elliptic operator*, Proc. Sympos. Pure Math. 10, 28–307, Amer. Math. Soc. Providence, 1968.

[18] ———, *Analytic extension of the trace associated with elliptic boundary problems*, Amer. J. Math. 91 (1969), 963–983.

[19] M. A. Shubin, *Pseudodifferential Operators and Spectral Theory*, Springer-Verlag, Berlin, 1987.

[20] H. Weyl, *Das asymptotische Verteilungsgesetz der Eigenwerte linearer partieller Differentialgleichungen*, Math. Anal. 71 (1912), 441–479.

Chapter 18

Analysis on Metric Spaces

Stephen Semmes

It is well known that a lot of standard analysis works in the very general setting of "spaces of homogeneous type." There is a lot of analysis, related to Sobolev spaces and differentiability properties of functions, which does not work at this level of generality. There are nontrivial results about when it does work.

We understand very little about geometry, just a few special cases.

18.1   Introduction

Let $(M, d(x, y))$ be a metric space. For the record, this means that M is a nonempty set, $d(x, y)$ is a nonnegative symmetric function on $M \times M$, which vanishes exactly on the diagonal and which satisfies the triangle inequality

$$d(x, z) \leq d(x, y) + d(y, z) \tag{18.1}$$

for all $x, y, z \in M$.

We all know that a lot of analysis, of continuity and convergence and so forth, makes sense on abstract metric spaces. On the other hand, a lot of analysis makes sense on abstract measure spaces, spaces with a measure but no metric, analysis of integration and L^p spaces and so forth. And then there is a lot of analysis that makes sense on Euclidean spaces which does not carry over to either of these abstract contexts because they do not have enough structure. Coifman and Weiss [1], [2] had the marvelous realization that one could go a long way with only a modest compatibility between measure and

---

Research partially supported by an NSF grant.

metric. If $(M, d(x, y))$ is a metric space and μ is a nonnegative Borel measure
on M, not identically zero, then we say that μ is a *doubling measure* if there
is a constant $C > 0$ so that

$$\mu(B(x, 2r)) \leq C\,\mu(B(x, r)) \qquad (18.2)$$

for all $x \in M$ and $r > 0$, where $B(x, r)$ denotes the (open) ball in M with
center x and radius r. In this case we call the triple $(M, d(x, y), \mu)$ a *space
of homogeneous type*. We have both a metric and a measure, and we can
formulate many ideas in analysis in this setting. Examples include Euclidean
spaces as well as many familiar fractals, like Cantor sets and snowflakes.

For example, if f is a locally integrable function on M, then we say that
x is a Lebesgue point for f if

$$\lim_{r \to 0} \frac{1}{\mu(B(x, r))} \int_{B(x, r)} |f(y) - f(x)|\, d\mu(y) = 0. \qquad (18.3)$$

This concept makes sense as soon as we have both a measure and a metric,
and it turns out that the doubling condition is strong enough to ensure that
almost every point in M is a Lebesgue point for any given locally integrable
function. This is true for practically the same reason that it is true on
Euclidean spaces; a suitable covering lemma of the Vitali variety works on
spaces of homogeneous type, and then the rest of the argument goes through.

To quiet the nervous voices in the back of my mind, let me assume also
that $(M, d(x, y))$ is complete. This ensures that closed and bounded subsets
of M are compact (in the presence of a doubling measure), and that obviates
some technical issues (like the Borel regularity of the measure).

Not only does one have Lebesgue points on spaces of homogeneous type,
but one also has working theories of Hardy spaces, BMO, and singular integral operators. Specific examples of interesting singular integral operators
on general spaces of homogeneous type are harder to come by. There do
not seem to be anything like Riesz transforms in general; they require more
geometric structure. One can build analogues of imaginary-order fractional
integral operators, but they are less interesting than Riesz transforms.

What about Lipschitz functions and Sobolev spaces? One of the great
features of analysis on Euclidean spaces is the special structure of Lipschitz
functions and Sobolev spaces, e.g., differentiability almost everywhere, various mechanisms to control a function in terms of its gradient (Sobolev embeddings, Poincaré inequalities, isoperimetric inequalities). Are there reasonable
versions of these results for general spaces of homogeneous type? The short
answer is no. One can see this with Cantor sets and snowflakes, for instance.

For the record, a Lipschitz function on a metric space $(M, d(x, y))$ means
a (real-valued) function f such that

$$|f(x) - f(y)| \leq C\, d(x, y) \qquad (18.4)$$

for all $x, y \in M$. If we replace $d(x, y)$ by $d(x, y)^\alpha$ for some $\alpha \in (0, 1)$, then we say that f is Hölder continuous of order α. In Euclidean analysis there is an enormous difference between Lipschitz functions and Hölder continuous functions, and more generally between integer orders of smoothness and fractional orders of smoothness. At the integers there is much more rigidity and structure. At the level of spaces of homogeneous type one does not see this distinction. Indeed, one is often feeling free on spaces of homogeneous type to replace a metric $d(x, y)$ with $d(x, y)^s$ for any $s > 0$, which will not even be a metric when $s > 1$ (only a quasimetric).

Sometimes in analysis the special nature of the integer orders of smoothness is a nuisance, but there is a lot of geometry there that we should not forget. The concept of Lipschitz functions has the wonderful property that it makes sense on any metric space and enjoys interesting rigidity properties on some metric spaces, like Euclidean spaces. Lipschitz functions are nicer than Riesz transforms in that one always has them even if they do not always have interesting structure.

Here is a special feature of analysis on Euclidean spaces. Let f be a compactly supported function on $\mathbb{R}^n$, or at least a function which tends to 0 at ∞. Then

$$|f(x)| \le C \int_{\mathbb{R}^n} \frac{1}{|x - y|^{n-1}} |\nabla f(y)| \, dy \qquad (18.5)$$

for all $x \in \mathbb{R}^n$. This is true, for instance, if f is C^1, and follows from a well-known identity. It can be proved by computing $f(x)$ in terms of an integral of ∇f over each ray emanating from x, and then averaging over the rays. (See [15].) It holds more generally for Lipschitz functions, or functions in Sobolev spaces, by approximation arguments. (In this case we should take ∇f in the sense of distributions.) If f is not continuous we may have to settle for (18.5) holding almost everywhere, etc.

This inequality captures a nontrivial amount of Euclidean geometry. Once we have it we can derive Sobolev embeddings through estimates on the potential operator

$$I_1(g)(x) = \int_{\mathbb{R}^n} \frac{1}{|x - y|^{n-1}} g(y) \, dy. \qquad (18.6)$$

The usual estimates for this potential operator (as in [15]) work much more generally than simply on Euclidean spaces; they depend only on fairly crude considerations of metric and measure. It is (18.5) which contains subtle information about Euclidean geometry, and there is nothing like it for spaces of homogeneous type in general.

Let us try to formulate the idea of (18.5) in a general setting. The main point is to have a generalization of $|\nabla f(y)|$, some quantity which controls the infinitesimal oscillation of f. There are a number of ways to do this, or one can make discrete versions that avoid some technicalities. For the present purposes the convenient choice is the following notion of generalized gradient, for which we follow Hajlasz, Heinonen, and Koskela.

Definition 18.1. If $(M, d(x, y))$ is a metric space and f and g are two Borel measurable functions on M, with f real-valued and g taking values in $[0, \infty]$, then we say that g is a *generalized gradient* of f if

$$|f(\gamma(a)) - f(\gamma(b))| \leq \int_a^b g(\gamma(t)) \, dt \tag{18.7}$$

whenever $\gamma : [a, b] \to M$ is 1-Lipschitz, i.e., $d(\gamma(s), \gamma(t)) \leq |s - t|$ for all $s, t \in [a, b]$.

For instance, if f is a Lipschitz function on M, then

$$g(x) = \liminf_{r \to 0} \sup_{y \in B(x,r)} \frac{|f(y) - f(x)|}{r} \tag{18.8}$$

is always a generalized gradient for f.

To formulate a general version of (18.5) we also need a measure, for which we shall use Hausdorff measure. Let $(M, d(x, y))$ and $n > 0$ be given, and let A be a subset of M. For each $\delta > 0$ set

$$H^n_\delta(A) = \inf \{ \sum_j (\operatorname{diam} E_j)^n : \{E_j\} \text{ is a sequence of sets in } M$$

$$\text{which covers } A \text{ and satisfies } \operatorname{diam} E_j < \delta \text{ for all } j\},$$

and then define the n-dimensional Hausdorff measure of A by

$$H^n(A) = \lim_{\delta \to 0} H^n_\delta(A). \tag{18.9}$$

The limit exists because of monotonicity in δ. The natural generalization of (18.5) to a general metric space $(M, d(x, y))$ is then

$$|f(x)| \leq C \int_M \frac{1}{d(x, y)^{n-1}} |g(y)| \, dH^n(y) \tag{18.10}$$

for all $x \in M$ whenever f is a function on M with compact support and g is a generalized gradient of f in the sense of Definition 18.1. (For this formulation we need M to be unbounded and there are other formulations for the bounded case.)

This is not to say that such an estimate holds in general. When it is true it contains real information about the geometry of M. To avoid trivialities one should require something about the Hausdorff dimension of M though; e.g., one should not take n stupidly too small so that the right-hand side is always infinite.

When does (18.10) hold? I can prove the following.

Theorem 18.2. *Let $(M, d(x, y))$ be a metric space and n be a positive integer. The estimate (18.10) holds for some C and all x, f, g as above if M satisfies the following five conditions:*

(i) *M is complete and unbounded;*

(ii) *M is doubling, which means that every ball in M can be covered by a bounded number of balls of half the radius;*

(iii) *every bounded subset of M has finite H^n-measure;*

(iv) *M is a topological manifold of dimension n;*

(v) *M is locally linearly contractible, which means that there is a constant $k > 1$ so that if B is any ball in M then B can be contracted to a point inside of kB (the ball with the same center as B but k times the radius).*

One could say that the preceding assumptions on M are sufficient to imply that M has some Euclidean behavior. Note that $M = \mathbb{R}^n$ satisfies all these conditions, and indeed much better. Condition (v) is probably the trickiest of the conditions to understand. It says that M is locally contractible, with a scale-invariant estimate. The requirement that M be a topological manifold implies already that M is locally contractible, but (iv) is not quantitative; it is (v) which provides the quantitative link between the metric and the topology. It prevents cusps and long thin tubes, for instance. The doubling condition is the only other part of the hypotheses which comes with a bound.

Condition (iv) implies that M has topological dimension n, and hence Hausdorff dimension $\geq n$. From (iii) we conclude that the Hausdorff dimension is exactly n. We do not demand good control on the Hausdorff measure here, but bad behavior would mean that the Hausdorff measure is too large and we would pay for that in (18.10) (in the right side being large).

Theorem 18.2 is proved in [13]. In fact, one has approximately the same structure as for Euclidean spaces, namely a large family of curves that emanate from a given point x and that are fairly well distributed in mass. These curves play the role of the rays which emanate from a given point in a Euclidean space. It is not so easy to produce the curves directly; instead they are found in the fibers of some special mappings with controlled Lipschitz behavior and topological nondegeneracy. In this outline form the methods are the same as in [4], which treated the case of geometries obtained by perturbing Euclidean spaces via "metric doubling measures." In the present situation the construction of the special mappings is more involved because of the absence of a background Euclidean structure. See also [14] for a version of [4] in which the basic principles are implemented more clearly and with more room for maneuver.

Part of the point here is that there are plenty of examples of spaces with good geometric properties but no good parameterization. (See [11], [12].) Theorem 18.2 and its proof show that many natural constructions that would be easy if there were good parameterizations can still be carried out when such parameterizations fail to exist.

There are many variations on this theme in [13], e.g., localized control on the oscillation of a function in terms of its generalized gradient, finding plenty of curves that are not too long, concluding Sobolev or Poincaré inequalities under suitable measure-theoretic assumptions. Let us forget all of that and instead play with the conceptual meaning of (18.10).

Let us put ourselves more firmly in the mode of spaces of homogeneous type and require that $(M, d(x,y))$ be *Ahlfors regular* of dimension n, which means that it is complete and that there is a constant $C > 0$ so that

$$C^{-1} r^n \leq H^n(B(x,r)) \leq C\, r^n \qquad (18.11)$$

for all $x \in M$ and $r > 0$. This implies that M is doubling and unbounded, and that H^n is a doubling measure on M. Under this assumption the right side of (18.10) behaves (as an integral operator) in practically the same manner as on $\mathbb{R}^n$.

Notice that we really need to have some concept of dimension in this story of (18.10), because we need to know what number to put in the denominator of the integrand in (18.10). This is not true of the usual theory of spaces of homogeneous type.

Under what conditions can we hope that something like (18.10) holds? Let us consider the space $(\mathbb{R}^n, |x - y|^s)$ as an example, $s > 0$, $s \neq 1$. If $s > 1$ then this is not a metric space, and it is not even close, because there are no nonconstant Lipschitz functions on $(\mathbb{R}^n, |x - y|^s)$ when $s > 1$. This reflects an important point about metric spaces and the triangle inequality: there are plenty of Lipschitz functions on metric spaces. When we ask for versions of (18.10), it is reasonable to start by making the a priori assumption that our function f is Lipschitz, in order to have some reasonable behavior. If $s < 1$ then $(\mathbb{R}^n, |x - y|^s)$ is a kind of snowflake, and there are no nonconstant Lipschitz mappings from an interval into it. The concept of generalized gradient is then vacuous. Thus the power $s = 1$ is determined uniquely by the existence of nontrivial Lipschitz functions on $\mathbb{R}^n$ and nontrivial rectifiable curves in $\mathbb{R}^n$.

Is anything like (18.10) ever true for spaces whose geometry is not somehow very close to Euclidean? Some kind of fractals perhaps? The ones ‚with the best hope are the ones with the most rectifiable curves inside of them, but the usual examples (like the Sierpiński carpet) do not have enough rectifiable curves in them. For metric spaces which can be realized as subsets of Euclidean spaces it might be that something like (18.10) would force approximately-Euclidean behavior, but for general metric spaces this is not true, because there are estimates like (18.10) for the Heisenberg group equipped with its Carnot metric. These metric spaces are Ahlfors-regular, but with a dimension one larger than the topological dimension.

So under what conditions is something like (18.10) true? One can ask similar questions about Sobolev inequalities, isoperimetric inequalities, etc., but we may as well stick to (18.10). It is a nice property that we can formulate very generally, and we know that once we have it we can do a lot through arguments typical for spaces of homogeneous type.

It seems reasonable to me to conjecture that if one has something like (18.10) on an Ahlfors-regular space of dimension n, then n has to be an integer.

Theorem 18.2 says that we can see a phenomenon like (18.10) through quantitative topology. The proof is very much for spaces approximately Euclidean; it was very important that the topological and Hausdorff dimensions coincide. The method is completely ineffective for the Heisenberg group.

There are a lot of aspects of analysis for which Euclidean spaces and the Heisenberg group are very similar. This includes (18.10) and the differentiability almost everywhere of Lipschitz mappings. (See [9].) It is not so clear how to distinguish them in simple language; I mean language that makes sense for Ahlfors-regular metric spaces for instance. One possibility is that, while both Euclidean spaces and Heisenberg groups satisfy (18.10), for Euclidean spaces this might be true in a much stabler way; i.e., so that there is a much more generous notion of "approximately Euclidean" for which (18.10) is preserved. This idea is illustrated by Theorem 18.2, where we see that (18.10) can be captured by relatively loose conditions. In Theorem 18.2 one need not specify much structure that is approximately Euclidean, even though in the end these conditions imply that the structure is more like Euclidean than is obvious at first. It is not clear that Heisenberg geometry is so stable, that relatively loose conditions can still capture some nontrivial aspect of the geometry, as in (18.10).

In fact I suspect that Heisenberg geometry is not so stable, precisely because it is not so closely connected to topology the way that Euclidean geometry is. Roughly speaking, one expects something like Euclidean geometry when the topological dimension and the Hausdorff dimension coincide. One can make precise statements of this nature, as discussed in [10]. Note, however, that Heisenberg geometry is much more stable than most "fractal" geometries, because of results like the differentiability almost everywhere of Lipschitz mappings.

For what other geometries is there good behavior for analysis almost like there is on Euclidean spaces, like (18.10) for instance? This question makes me a little nervous; maybe we are a little bit lucky to have come across Heisenberg geometry, especially since it cannot appear inside a Euclidean space even up to bilipschitz equivalence, because of [9]. (See [11].) Maybe there are many other geometries with similar properties but which are very different from Euclidean and Heisenberg geometries. They might exist and we just do not know where to look for them. Or maybe there are not so many geometries that are truly different and for which (18.10) works, but then how do we extract some common principle which includes both Euclidean geometry and Heisenberg geometry and excludes the rest?

Incidentally, the only examples that I know of metric spaces which are doubling but which do not admit bilipschitz parameterizations into some $\mathbb{R}^n$ are based on the Heisenberg group (or some Carnot group). The argument uses the differentiability almost everywhere of real-valued Lipschitz functions on the Heisenberg group, as in [9]. It would be very nice to have other examples, preferably more elementary or based on more direct principles.

What can we say about geometries as a family? How to deform geometries, when are geometries very stable, so that relatively loose conditions imply that two spaces have similar geometry? Or how to understand the variety in geometry, the ways in which geometries are truly different, in terms comprehensible to human beings? I know some examples, but I think that we are blind to what really happens in the variety of geometry.

As to the stability of geometry, of Euclidean geometry, the story of uniform rectifiability of David and myself (as in [5], [6], [3], [10]) provides a language in which we can say that relatively loose conditions on a space imply that the geometry is approximately Euclidean. The older story of ordinary rectifiability also does this in a less structured way.

Issues like these show up in the recent work of Heinonen and Koskela [7], [8] too. There one asks for certain Sobolev-Poincaré inequalities and one makes conclusions about the geometry of mappings. More precisely, one asks for bounds on functions whose gradient lies in L^n, where n is the dimension as in (18.11). This is the conformally invariant case, and such bounds are weaker than the usual range of Sobolev-Poincaré inequalities, where one wants to control functions even when the gradient lies in L^p for $p < n$. Again one can ask for which spaces such inequalities hold, how stable are the conditions under which they hold, and again the answer is not so clear. The conditions arising in [7], [8] are much closer to the edge of invariance; it is basically a matter of knowing when nontrivial continua can be seen by the invariant Sobolev space of functions whose gradient lies in L^n. Perhaps they can be tied to geometry in a more delicate way.

References

[1] R. Coifman and G. Weiss, *Analyse Harmonique Non-commutative Certains Espaces Homogènes*, Lecture Notes in Math. 242, Springer-Verlag, 1971.

[2] ———, *Extensions of Hardy spaces and their use in analysis*, Bull. Amer. Math. Soc. 83 (1977), 569–645.

[3] G. David, *Wavelets and Singular Integrals on Curves and Surfaces*, Lecture Notes in Math. 1465, Springer-Verlag, 1991.

[4] G. David and S. Semmes, *Strong A_∞-weights, Sobolev inequalities, and quasiconformal mappings*, in *Analysis and Partial Differential Equations*, edited by C. Sadosky, Lecture Notes in Pure and Applied Mathematics 122, Marcel Dekker, 1990.

[5] ———, *Singular Integrals and Rectifiable Sets in $\mathbb{R}^n$: au-delà des graphes lipschitziens*, Astérisque 193, Société Mathématique de France, 1991.

[6] ———, *Analysis of and on Uniformly Rectifiable Sets*, Mathematical Surveys and Monographs 38, 1993, American Mathematical Society.

[7] J. Heinonen and P. Koskela, *Definitions of quasiconformality*, Inv. Math. 120 (1995), 61–79.

[8] ———, *Quasiconformal maps in metric spaces with controlled geometry*, Acta Mathematica 181 (1998), 1–61.

[9] P. Pansu, *Métriques de Carnot-Carathéodory et quasiisométries des espaces symmétriques de rang un*, Ann. Math. 129 (1989), 1–60.

[10] S. Semmes, *Finding structure in sets with little smoothness*, Proc. I.C.M. (Zürich, 1994), 875–885, Birkhäuser, 1995.

[11] ———, *On the nonexistence of bilipschitz parameterizations and geometric problems about A_∞ weights*, Revista Matemática Iberoamericana 12 (1996), 337–410.

[12] ———, *Good metric spaces without good parameterizations*, Revista Matemática Iberoamericana 12 (1996), 187–275.

[13] ———, *Finding Curves on General Spaces through Quantitative Topology with Applications for Sobolev and Poincaré Inequalities*, Selecta Math. (N.S.) 2 (1996), 155–295.

[14] ————, *Some remarks about metric spaces, spherical mappings, functions and their derivatives*, Publications Matématiques 40 (1996), 411–430.

[15] E. M. Stein, *Singular Integrals and Differentiability Properties of Functions*, Princeton University Press, 1970.

Chapter 19

Developments in Inverse Problems since Calderón's Foundational Paper

Gunther Uhlmann

19.1 Introduction

In 1980 A. P. Calderón published a short paper entitled "On an inverse boundary-value problem" [13]. This pioneering contribution motivated many developments in inverse problems, in particular in the construction of "complex geometrical optics" solutions of partial differential equations to solve several inverse problems. We survey these developments in this paper. We emphasize the new results since the survey paper [68] was written.

The problem that Calderón proposed in [13] is whether it is possible to determine the conductivity of a body by making current and voltage measurements at the boundary. This problem arises in geophysical prospection [70]. Apparently Calderón thought of this problem while working as an engineer in Argentina, but he did not publish his results until several decades later. More recently this noninvasive inverse method, also referred in the literature as *Electrical Impedance Tomography*, has been proposed as a possible diagnostic tool and in medical imaging ([5], [69]). One concrete clinical application, which seems to be very promising, is in the monitoring of pulmonary edema ([23], [43]).

We now describe more precisely the mathematical problem.

Research partially supported by NSF and ONR grant N00014-93-1-0295.

Let $\Omega \subseteq \mathbb{R}^n$ be a bounded domain with smooth boundary (many of the results we will describe are valid for domains with Lipschitz boundaries). The electrical conductivity of Ω is represented by a bounded and positive function $\gamma(x)$. In the absence of sinks or sources of current the equation for the potential is given by

$$\operatorname{div}(\gamma \nabla u) = 0 \text{ in } \Omega \tag{19.1}$$

since, by Ohm's law, $\gamma \nabla u$ represents the current flux.

Given a potential $f \in H^{\frac{1}{2}}(\partial \Omega)$ on the boundary the induced potential $u \in H^1(\Omega)$ solves the Dirichlet problem

$$\begin{aligned} \operatorname{div}(\gamma \nabla u) &= 0 \text{ in } \Omega, \\ u\big|_{\partial \Omega} &= f. \end{aligned} \tag{19.2}$$

The Dirichlet to Neumann map, or voltage to current map, is given by

$$\Lambda_\gamma(f) = \left(\gamma \frac{\partial u}{\partial \nu} \right) \bigg|_{\partial \Omega}, \tag{19.3}$$

where ν denotes the unit outer normal to $\partial \Omega$.

The inverse problem is to determine γ knowing Λ_γ. More precisely we want to study properties of the map

$$\gamma \overset{\Lambda}{\longrightarrow} \Lambda_\gamma. \tag{19.4}$$

Note that $\Lambda_\gamma : H^{\frac{1}{2}}(\partial \Omega) \to H^{-\frac{1}{2}}(\partial \Omega)$ is bounded. We can divide this problem into several parts.

(a) Injectivity of Λ (identifiability)

(b) Continuity of Λ and its inverse if it exists (stability)

(c) The range of Λ (characterization problem)

(d) Formula to recover γ from Λ_γ (reconstruction)

(e) An approximate numerical algorithm to find an approximation of the conductivity given a finite number of voltage and current measurements at the boundary (numerical reconstruction).

It is difficult to find a systematic way of prescribing voltage measurements at the boundary to be able to find the conductivity. Calderón took instead a different route.

Using the divergence theorem we have

$$Q_\gamma(f) := \int_\Omega \gamma |\nabla u|^2 dx = \int_{\partial\Omega} \Lambda_\gamma(f) f \, dS, \qquad (19.5)$$

where dS denotes surface measure and u is the solution of (19.2). In other words, $Q_\gamma(f)$ is the quadratic form associated to the linear map $\Lambda_\gamma(f)$, i.e., to know $\Lambda_\gamma(f)$ or $Q_\gamma(f)$ for all $f \in H^{\frac{1}{2}}(\partial\Omega)$ is equivalent. $Q_\gamma(f)$ measures the energy needed to maintain the potential f at the boundary. Calderón's point of view is that if one looks at $Q_\gamma(f)$ the problem is changed into finding enough solutions $u \in H^1(\Omega)$ of the equation (19.1) in order to find γ in the interior. We will explain this approach further in the next section where we study the linearization of the map

$$\gamma \xrightarrow{\ Q\ } Q_\gamma. \qquad (19.6)$$

Here we consider Q_γ as the bilinear form associated to the quadratic form (19.5).

In §19.2 we describe Calderón's paper and how he used complex exponentials to prove that the linearization of (19.6) is injective at constant conductivities. He also gave an approximation formula to reconstruct a conductivity which is a priori close to a constant conductivity.

In §19.3 we describe the construction by Sylvester and Uhlmann [61, 62] of complex geometrical optics solutions for the Schrödinger equation associated to a bounded potential. These solutions behave like Calderón's complex exponential solutions for large complex frequencies. In §19.4 we use these solutions to prove in dimension $n \geq 3$ a global identifiability result [61], stability estimates [3] and a reconstruction method for the inverse problem [35], [45]. We also describe an extension of the identifiability result to nonlinear conductivities [53].

In §19.5 we consider the two-dimensional case. In particular, we follow recent work of Brown and Uhlmann [12] to improve the regularity result in A. Nachman's result [36]. In turn the [12] paper relies in work of Beals and Coifman [8] and L. Sung [59] in inverse scattering for a class of first-order systems in two dimensions.

In §19.6 we consider other inverse boundary-value problems arising in applications. A common feature of these problems is that they can be reduced to consider first-order scalar and systems perturbations of the Laplacian. In the scalar case we consider an inverse boundary-value problem for the Schrödinger equation in the presence of a magnetic potential. We also consider an inverse boundary-value problem for the elasticity system. The problem is to determine the elastic parameters of an elastic body by making displacements and traction measurements at the boundary. In §19.7 we give a general method, due to Nakamura and Uhlmann [39], to construct the complex geometrical optics solutions in this case and a method, due to Tolmasky [66], to construct these solutions for first-order perturbations of the Laplacian with less regular conductivities.

Finally we consider in §19.8 the case of anisotropic conductivities, i.e., the conductivity depends also of direction. In particular, we outline recent progress in the study of the quasilinear case [57].

19.2   Calderón's Paper

Calderón proved in [13] that the map Q is analytic. The Fréchet derivative of Q at $\gamma = \gamma_0$ in the direction h is given by

$$dQ|_{\gamma=\gamma_0}(h)(f,g) = \int_\Omega h\nabla u \cdot \nabla v\,dx, \qquad (19.7)$$

where $u, v \in H^1(\Omega)$ solve

$$\begin{cases} \operatorname{div}(\gamma_0\nabla u) = \operatorname{div}(\gamma_0\nabla v) = 0 \text{ in } \Omega, \\ u\big|_{\partial\Omega} = f \in H^{\frac{1}{2}}(\partial\Omega), \quad v\big|_{\partial\Omega} = g \in H^{\frac{1}{2}}(\partial\Omega). \end{cases} \qquad (19.8)$$

So the linearized map is injective if the products of $H^1(\Omega)$ solutions of $\operatorname{div}(\gamma_0\nabla u) = 0$ is dense in, say, $L^2(\Omega)$.

Calderón proved injectivity of the linearized map in the case $\gamma_0 = \text{con-}$ stant, which we assume for simplicity to be the constant function 1. The question is reduced to whether the product of gradients of harmonic functions is dense in, say, $L^2(\Omega)$.

Calderón took the following harmonic functions

$$u = e^{x\cdot\rho}, \quad v = e^{-x\cdot\bar\rho}, \qquad (19.9)$$

where $\rho \in \mathbb{C}^n$ with

$$\rho \cdot \rho = 0. \qquad (19.10)$$

We remark that the condition (19.10) is equivalent to the following:

$$\rho = \frac{\eta + ik}{2}, \ \eta, k \in \mathbb{R}^n, \qquad (19.11)$$

$$|\eta| = |k|, \ \eta \cdot k = 0.$$

Then plugging the solutions (19.9) into (19.7) we obtain if $dQ|_{\gamma_0=1}(h) = 0$

$$|k|^2(\chi_\Omega h)^\wedge(k) = 0, \qquad \forall\, k \in \mathbb{R}^n,$$

where χ_Ω denotes the characteristic function of Ω. Then we easily conclude that $h = 0$. However, one cannot apply the implicit function theorem to conclude that γ is invertible near a constant since conditions on the range of Q that would allow use of the implicit function theorem are either false or not known.

Calderón also observed that using the solutions (19.9) one can find an approximation for the conductivity γ if

$$\gamma = 1 + h \tag{19.12}$$

and h is small enough in L^∞ norm.

We are given

$$G_\gamma = Q_\gamma \left(e^{x\cdot\rho}\Big|_{\partial\Omega}, e^{-x\cdot\bar\rho}\Big|_{\partial\Omega} \right)$$

with $\rho \in \mathbb{C}^n$ as in (19.12). Now

$$\begin{aligned}
G_\gamma &= \int_\Omega (1+h)\nabla u \cdot \nabla v \, dx \\
&+ \int_\Omega h(\nabla\delta u \cdot \nabla v + \nabla u \cdot \nabla\delta v) \, dx \\
&+ \int_\Omega (1+h)\nabla\delta u \cdot \nabla\delta v \, dx
\end{aligned} \tag{19.13}$$

with u, v as in (19.9) and

$$\begin{aligned}
\operatorname{div}(\gamma\nabla(u + \delta u)) &= \operatorname{div}(\gamma\nabla(v + \delta v)) = 0 \text{ in } \Omega, \\
\delta u\Big|_{\partial\Omega} &= \delta v\Big|_{\partial\Omega} = 0.
\end{aligned} \tag{19.14}$$

Now standard elliptic estimates applied to (19.14) show that

$$\|\nabla\delta u\|_{L^2(\Omega)}, \quad \|\nabla\delta v\|_{L^2(\Omega)} \le C\|h\|_{L^\infty(\Omega)}|k|e^{\frac{1}{2}r|k|} \tag{19.15}$$

for some $C > 0$ where r denotes the radius of the smallest ball containing Ω.

Now plugging u, v into (19.13) we obtain

$$\widehat{\chi_\Omega\gamma}(k) = -2\frac{G_\gamma}{|k|^2} + R(k) = \widehat{F}(k) + R(k), \tag{19.16}$$

where F is determined by G_γ and therefore known. Using (19.15), we can show that $R(k)$ satisfies the estimate

$$|R(k)| \le C\|h\|^2_{L^\infty(\Omega)}e^{r|k|}. \tag{19.17}$$

In other words, we know $\widehat{\chi_\Omega\gamma}(k)$ up to a term that is small for k small enough. More precisely, let $1 < \alpha < 2$. Then for

$$|k| \le \frac{2-\alpha}{r}\log\frac{1}{\|h\|_{L^\infty}} =: \sigma \tag{19.18}$$

we have that

$$|R(k)| \leq C\|h\|_{L^\infty(\Omega)}^\alpha \qquad (19.19)$$

for some $C > 0$.

We take $\hat{\eta}$ a C^∞ cutoff so that $\hat{\eta}(0) = 1$, $\mathrm{supp}\,\hat{\eta}(k) \subset \{k \in \mathbb{R}^n, |k| \leq 1\}$ and $\eta_\sigma(\chi) = \sigma^n \eta(\sigma x)$. Then we obtain

$$\widehat{\chi_\Omega \gamma}(k)\hat{\eta}\left(\frac{k}{\sigma}\right) = \frac{-2G_\gamma \gamma}{|k|^2}\hat{\eta}\left(\frac{k}{\sigma}\right) + R(k)\hat{\eta}\left(\frac{k}{\sigma}\right).$$

Using this we get the following estimate,

$$|\rho(x)| \leq C\|h\|_{L^\infty(\Omega)}^\alpha \left[\log \frac{1}{\|h\|_{L^\infty(\Omega)}}\right]^n, \qquad (19.20)$$

where $\rho(x) = (\chi_\Omega \gamma * \eta_\sigma)(x) - (F * \eta_\sigma)(x)$. Formula (19.20) gives then an approximation to the smoothed out conductivity $\chi_\Omega \gamma * \eta_\sigma$ for h sufficiently small.

This approximation estimate of Calderón and modifications of it have been tried out numerically [24].

This estimate uses the harmonic exponentials for low frequencies. In the next section we consider high (complex) frequency solutions of the conductivity equation

$$L_\gamma u = \mathrm{div}(\gamma \nabla u) = 0.$$

19.3 Complex Geometrical Optics for the Schrödinger Equation

Let $\gamma \in C^2(\mathbb{R}^n)$, γ strictly positive in $\mathbb{R}^n$ and $\gamma = 1$ for $|x| \geq R$, some $R > 0$. Let $L_\gamma u = \mathrm{div}(\gamma \nabla u)$. Then we have

$$\gamma^{-\frac{1}{2}} L_\gamma(\gamma^{-\frac{1}{2}} u) = (\Delta - q)u, \qquad (19.21)$$

where

$$q = \frac{\Delta\sqrt{\gamma}}{\sqrt{\gamma}}. \qquad (19.22)$$

Therefore, to construct solutions of $L_\gamma u = 0$ in $\mathbb{R}^n$ it is enough to construct solutions of the Schrödinger equation $(\Delta - q)u = 0$ with q of the form (19.22). The next result proven in [61, 62] states the existence of complex geometrical optics solutions for the Schrödinger equation associated to any bounded and compactly supported potential.

Theorem 19.1. *Let $q \in L^\infty(\mathbb{R}^n)$, $n \geq 2$, with $q(x) = 0$ for $|x| \geq R > 0$. Let $-1 < \delta < 0$. There exists $\epsilon(\delta)$ such that, for every $\rho \in \mathbb{C}^n$ satisfying*

$$\rho \cdot \rho = 0$$

and

$$\frac{\|(1 + |x|^2)^{1/2}q\|_{L^\infty(\mathbb{R}^n)} + 1}{|\rho|} \leq \epsilon,$$

there exists a unique solution to

$$(\Delta - q)u = 0$$

of the form

$$u = e^{x \cdot \rho}(1 + \psi_q(x, \rho)) \tag{19.23}$$

with $\psi_q(\cdot, \rho) \in L^2_\delta(\mathbb{R}^n)$. Moreover $\psi_q(\cdot, \rho) \in H^2_\delta(\mathbb{R}^n)$ and for $0 \leq s \leq 1$ there exists $C = C(n, s, \delta) > 0$ such that

$$\|\psi_q(\cdot, \rho)\|_{H^s_\delta} \leq \frac{C}{|\rho|^{1-s}}. \tag{19.24}$$

Here

$$L^2_\delta(\mathbb{R}^n) = \{f : \int (1 + |x|^2)^\delta |f(x)|^2 dx < \infty\}$$

with the norm given by $\|f\|^2_{L^2_\delta} = \int (1 + |x|^2)^\delta |f(x)|^2 dx$ and $H^m_\delta(\mathbb{R}^n)$ denotes the corresponding Sobolev space. Note that for large $|\rho|$ these solutions behave like Calderón's exponential solutions. Equation for ψ_q is given by

$$(\Delta + 2\rho \cdot \nabla)\psi_q = q(1 + \psi_q). \tag{19.25}$$

The equation (19.25) is solved by constructing an inverse for $(\Delta + 2\rho \cdot \nabla)$ and solving the integral equation

$$\psi_q = (\Delta + 2\rho \cdot \nabla)^{-1}(q(1 + \psi_q)). \tag{19.26}$$

Lemma 19.2. *Let $-1 < \delta < 0$, $0 \leq s \leq 1$. Let $\rho \in \mathbb{C}^n - 0$, $\rho \cdot \rho = 0$. Let $f \in L^2_{\delta+1}(\mathbb{R}^n)$. Then there exists a unique solution $u_\rho \in L^2_\delta(\mathbb{R}^n)$ of the equation*

$$\Delta_\rho u_\rho := (\Delta + 2\rho \cdot \nabla)u_\rho = f. \tag{19.27}$$

Moreover, $u_\rho \in H^2_\delta(\mathbb{R}^n)$ and

$$\|u_\rho\|_{H^s_\delta(\mathbb{R}^n)} \leq \frac{C_{s,\delta}\|f\|_{L^2_{\delta+1}}}{|\rho|^{s-1}}$$

for $0 \leq s \leq 1$ and for some constant $C_{s,\delta} > 0$.

The integral equation (19.26) can then be solved in $L^2_\delta(\mathbb{R}^n)$ for large $|\rho|$ since

$$(I - (\Delta + 2\rho \cdot \nabla)^{-1}q)\psi_q = (\Delta + 2\rho \cdot \nabla)^{-1}q$$

and $\|(\Delta + 2\rho \cdot \nabla)^{-1}q\|_{L^2_\delta \to L^2_\delta} \leq \frac{C}{|\rho|}$ for some $C > 0$ where $\| \cdot \|_{L^2_\delta \to L^2_\delta}$ denotes the operator norm between $L^2_\delta(\mathbb{R}^n)$ and $L^2_\delta(\mathbb{R}^n)$. We will not give details of the proof of Lemma 19.2 here. We refer to the papers [61, 62]. We describe the underlying ideas in the case $n \geq 3$.

The point is that the operator $\Delta_\rho = \Delta + 2\rho \cdot \nabla$ has a symbol $-|\xi|^2 + 2i\rho \cdot \xi$ which is jointly homogeneous of degree 2 in (ξ, ρ). Since we want to look at the behavior of Δ_ρ in ρ we consider ρ as another dual variable (this will be made more precise in §19.7).

Now the characteristic variety of Δ_ρ in ξ-space for every ρ is a codimension two real submanifold. One simple example that exhibits both behaviors is the equation $|\rho|(\partial x_1 + i\partial x_2)$ in $\mathbb{R}^n$. We have that the "principal symbol" of $|\rho|(\partial x_1 + i\partial x_2)$ is homogeneous of degree two in (ξ, ρ) and its characteristic variety has codimension two. The point then is that Δ_ρ is microlocally equivalent to $|\rho|(\partial x_1 + i\partial x_2)$ and the estimates follow from the Nirenberg-Walker [44] estimates for the $\overline{\partial}$ equation in two dimensions. Namely, in [44] is proved the following:

Lemma 19.3. *Let $n = 2$. Let $-1 < \delta < 0$. Let $L = \partial$ or $\overline{\partial}$. Then given $f \in L^2_{\delta+1}(\mathbb{R}^2)$ there exists a unique $u \in L^2_\delta(\mathbb{R}^2)$ so that*

$$Lu = f.$$

Moreover, $\|u\|_{L^2_\delta} \leq C\|f\|_{L^2_{\delta+1}}$ for some $C = C(\delta) > 0$.

Now if we apply this result to $|\rho|\frac{(\partial x_1 + i\partial x_2)}{2}$ with the variables $x_3, \dots, x_n$ as parameters we get Lemma 19.2 for $s = 0$ since

$$
\begin{aligned}
\|u\|^2_{L^2_\delta} &= \int_{\mathbb{R}^n}(1 + |x|^2)^\delta|u(x)|^2 dx \leq \int_{\mathbb{R}^n}(1 + |x_1|^2 + |x_2|^2)^\delta|u(x)|^2 dx \\
&\leq \frac{C}{|\rho|^2}\int_{\mathbb{R}^n}(1 + |x_1|^2 + |x_2|^2)^{\delta+1}|f(x)|^2 dx \\
&\leq \frac{C}{|\rho|^2}\int_{\mathbb{R}^n}(1 + |x|^2)^{\delta+1}|f(x)|^2 dx.
\end{aligned}
$$

We mention here the following extension due to R. Brown [10] of Lemma 19.2 to Besov spaces.

Lemma 19.4. *Let $\rho \in \mathbb{C}^n - 0$, $n \geq 3$, satisfying $\rho \cdot \rho = 0$. Then for $-1 < \delta < 0$ and $0 \leq s \leq \frac{1}{2}$ we have that*

$$\|\Delta_\rho^{-1}f\|_{B^{s,\delta}_{2,2}} \leq \frac{C}{|\rho|^{1-2s}}\|f\|_{B^{-s,\delta+1}_{2,2}},$$

where $C = C(n, s, \delta)$.

Here $B_{p,q}^{s,\delta} = \{f : (1 + |x|^2)^{\frac{\delta}{2}} f \in B_{p,q}^s\}$ with the norm

$$\|f\|_{B_{p,q}^{s,\delta}} = \|(1 + |x|^2)^{\frac{\delta}{2}} f\|_{B_{p,q}^s}$$

and $B_{p,q}^s$ denotes the standard Besov space, i.e., $f \in B_{p,q}^s$ if and only if

$$\|f\|_{L^p} + \left(\int_{\mathbb{R}^n} \left(\int_{\mathbb{R}^n} |f(x + h) - f(x)|^p dx \right)^{q/p} h^{-n-sq} dh \right)^{\frac{1}{q}} \quad (19.28)$$

is finite and (19.28) gives a norm.

We shall discuss the proof of Lemma 19.2 in the two-dimensional case in §19.5 together with an extension of the estimates to weighted L^p spaces.

We assume that $\gamma \in C^2(\overline{\Omega})$ in order to have $q \in L^\infty(\Omega)$. Brown [10] showed that one can relax the smoothness assumption on the conductivity further. Let γ be a bounded function on $\mathbb{R}^n$ strictly positive and γ equal 1 for $|x| > M$ and for some $0 < s < 1$

$$\|\nabla \gamma\|_{B_{\infty,2}^{1-s}} \leq M. \quad (19.29)$$

Let $u \in C^\infty(\mathbb{R}^n)$. We denote by $m_q(u)$ the distribution defined by

$$m_q(u)(\varphi) = -\int_{\mathbb{R}^n} \nabla\sqrt{\gamma} \cdot \nabla \left(\frac{1}{\sqrt{\gamma}} u\varphi \right) dx, \quad \forall \, \varphi \in C_0^\infty(\mathbb{R}^n).$$

Note that if $\gamma \in C^2(\mathbb{R}^n)$ then

$$m_q(u) = qu \text{ with } q = \frac{\Delta\sqrt{\gamma}}{\sqrt{\gamma}}.$$

In [10] it was proven that the map m_q is bounded between certain Besov spaces. More precisely we have

$$\|m_q(u)\|_{B_{2,2}^{-s,\delta+1}} \leq C\|u\|_{B_{2,2}^{s,\delta}} \quad (19.30)$$

for $-1 < \delta < 0$, $0 < s < 1$. Combining Lemma 19.4 and (19.30) one concludes

Theorem 19.5 ([10]). *Let γ be a bounded function in $\mathbb{R}^n$ strictly positive and one outside a large ball. Let $\rho \in \mathbb{C}^n$ satisfy $\rho \cdot \rho = 0$. Let $0 < s < \frac{1}{2}$ and $-1 < \delta < 0$. Let $f \in B_{2,2}^{-s,\delta+1}$. Then $\exists \, R > 0$ such that for $|\rho| > R$ there exists a unique solution $\psi \in B_{2,2}^{s,\delta}$ to*

$$\Delta\psi + 2\rho \cdot \nabla\psi - m_q(\psi) = f.$$

Furthermore, we have for some $C = C(n, s, \delta, M) > 0$

$$\|\psi\|_{B_{2,2}^{s,\delta}} \leq \frac{C}{|\rho|^{1-2s}} \|f\|_{B_{2,2}^{-s,\delta+1}}.$$

Local constructions of complex geometrical optics solutions can be found in [22] and [23].

19.4   The Inverse Conductivity Problem in $n \geq 3$

The identifiability question was resolved in [61] for smooth enough conductivities. The result is

Theorem 19.6. *Let* $\gamma_i \in C^2(\overline{\Omega})$, γ_i *strictly positive,* $i = 1, 2$. *If* $\Lambda_{\gamma_1} = \Lambda_{\gamma_2}$ *then* $\gamma_1 = \gamma_2$ *in* $\overline{\Omega}$.

In dimension $n \geq 3$ this result is a consequence of a more general result. Let $q \in L^\infty(\Omega)$. We define the Cauchy data as the set

$$\mathcal{C}_q = \left\{ \left(u\big|_{\partial\Omega}, \frac{\partial u}{\partial\nu}\Big|_{\partial\Omega} \right) \right\}, \quad \text{where } u \in H^1(\Omega) \tag{19.31}$$

is a solution of

$$(\Delta - q)u = 0 \text{ in } \Omega. \tag{19.32}$$

We have that $\mathcal{C}_q \subseteq H^{\frac{1}{2}}(\partial\Omega) \times H^{-\frac{1}{2}}(\partial\Omega)$. If 0 is not a Dirichlet eigenvalue of $\Delta - q$, then in fact $\mathcal{C}_q$ is a graph, namely

$$\mathcal{C}_q = \{(f, \Lambda_q(f)) \in H^{\frac{1}{2}}(\partial\Omega) \times H^{-\frac{1}{2}}(\partial\Omega)\},$$

where $\Lambda_q(f) = \frac{\partial u}{\partial\nu}\big|_{\partial\Omega}$ with $u \in H^1(\Omega)$ the solution of

$$\begin{aligned}
(\Delta - q)u &= 0 \text{ in } \Omega, \\
u\big|_{\partial\Omega} &= f.
\end{aligned}$$

Λ_q is the *Dirichlet to Neumann* map in this case.

Theorem 19.7. *Let* $q_i \in L^\infty(\Omega)$, $i = 1, 2$. *Assume* $\mathcal{C}_{q_1} = \mathcal{C}_{q_2}$, *then* $q_1 = q_2$.

We now show that Theorem 19.7 implies Theorem 19.6.
Using (19.21) we have that

$$\mathcal{C}_{q_i} = \left\{ \left(f, \left(\frac{1}{2}\gamma_i^{-\frac{1}{2}}\Big|_{\partial\Omega} \frac{\partial\gamma_i}{\partial\nu}\Big|_{\partial\Omega} \right) f + \gamma_i^{-\frac{1}{2}}\Big|_{\partial\Omega} \Lambda_{\gamma_i} \left(\gamma_i^{-\frac{1}{2}}\Big|_{\partial\Omega} f \right) \right), \quad f \in H^{\frac{1}{2}}(\partial\Omega) \right\}.$$

Then we conclude $\mathcal{C}_{q_1} = \mathcal{C}_{q_2}$ since we have the following result due to Kohn and Vogelius [29].

Theorem 19.8. *Let* $\gamma_i \in C^1(\overline{\Omega})$ *and strictly positive. Assume* $\Lambda_{\gamma_1} = \Lambda_{\gamma_2}$. *Then*

$$\partial^\alpha \gamma_1\big|_{\partial\Omega} = \partial^\alpha \gamma_2\big|_{\partial\Omega}, \quad |\alpha| \leq 1.$$

Remark. In fact, Kohn and Vogelius proved that if $\gamma_i \in C^\infty(\overline{\Omega})$, γ_i strictly positive, then $\Lambda_{\gamma_1} = \Lambda_{\gamma_2}$ implies that

$$\partial^\alpha \gamma_1 \Big|_{\partial\Omega} = \partial^\alpha \gamma_2 \Big|_{\partial\Omega} \qquad \forall\, \alpha.$$

This settled the identifiability question in the real-analytic category. They extended the identifiability result to piecewise real-analytic conductivities in [30]. For related results see [18].

Proof of 19.7. Let $u_i \in H^1(\Omega)$ be a solution of

$$(\Delta - q_i)u_i = 0 \text{ in } \Omega, \qquad i = 1, 2.$$

Then using the divergence theorem we have that

$$\int_\Omega (q_1 - q_2)u_1 u_2 dx = \int_{\partial\Omega} \left(\frac{\partial u_1}{\partial \nu} u_2 - u_1 \frac{\partial u_2}{\partial \nu} \right) dS. \qquad (19.33)$$

Now it is easy to prove that if $C_{q_1} = C_{q_2}$ then the LHS of (19.33) is zero.

Now we extend $q_i = 0$ in Ω^c. We take solutions of $(\Delta - q_i)u_i = 0$ in $\mathbb{R}^n$ of the form

$$u_i = e^{x \cdot \rho_i}(1 + \psi_{q_i}(x, \rho_i)), \qquad i = 1, 2 \qquad (19.34)$$

with $|\rho_i|$ large, $i = 1, 2$, with

$$\rho_1 = \frac{\eta}{2} + i\left(\frac{k+l}{2}\right), \qquad (19.35)$$

$$\rho_2 = -\frac{\eta}{2} + i\left(\frac{k-l}{2}\right),$$

and $\eta, k, l \in \mathbb{R}^n$ such that

$$\eta \cdot k = k \cdot l = \eta \cdot l = 0, \qquad (19.36)$$
$$|\eta|^2 = |k|^2 + |l|^2.$$

Condition (19.36) guarantees that $\rho_i \cdot \rho_i = 0$, $i = 2$. Replacing (19.34) into

$$\int_\Omega (q_1 - q_2)u_1 u_2 dx = 0 \qquad (19.37)$$

we conclude

$$\widehat{(q_1 - q_2)}(-k) = -\int_\Omega e^{ix \cdot k}(q_1 - q_2)(\psi_{q_1} + \psi_{q_2} + \psi_{q_1}\psi_{q_2})dx. \qquad (19.38)$$

Now $\|\psi_{q_i}\|_{L^2(\Omega)} \le \frac{C}{|\rho_i|}$. Therefore by taking $|l| \to \infty$ we get that

$$\widehat{q_1 - q_2}(k) = 0, \qquad \forall\, k \in \mathbb{R}^n,$$

concluding the proof.

We now discuss Theorem 19.8.

Sketch of proof of Theorem 19.8. We outline an alternative proof to the one given by Kohn and Vogelius of Theorem 19.8. In the case $\gamma \in C^\infty(\overline{\Omega})$ we know, by another result of Calderón [14], that Λ_γ is a classical pseudodifferential operator of order 1. Let (x', x^n) be coordinates near a point $x_0 \in \partial\Omega$ so that the boundary is given by $x^n = 0$. If $\lambda_\gamma(x', \xi')$ denotes the full symbol of Λ_γ in these coordinates. It was proved in [63] that

$$\lambda_\gamma(x', \xi') = \gamma(x', 0)|\xi'| + a_0(x', \xi') + r(x', \xi'), \qquad (19.39)$$

where $a_0(x', \xi')$ is homogeneous of degree 0 in ξ' and is determined by the normal derivative of γ at the boundary and tangential derivatives of γ at the boundary. The term $r(x', \xi')$ is a classical symbol of order -1. Then $\gamma\big|_{\partial\Omega}$ is determined by the principal symbol of Λ_γ and $\frac{\partial\gamma}{\partial x^n}\big|_{\partial\Omega}$ is determined by the principal symbol and the term homogeneous of degree 0 in the expansion of the full symbol of Λ_γ. More generally, the higher order normal derivatives of the conductivity at the boundary can be determined recursively. In [34] one can find a more general approach to the calculation of the full symbol of the Dirichlet to Neumann map.

The case $\gamma \in C^1(\overline{\Omega})$ of Theorem 19.8 follows using an approximation argument [63]. For other results and approaches to boundary determination of the conductivity, see [4], [11], [35].

It is not clear at present what is the optimal regularity on the conductivity for Theorem 19.6 to hold. Chanillo proves in [15] that Theorem 19.6 is valid under the assumption that $\Delta\gamma \in F^p$, $p > \frac{n-1}{2}$, where F^p is the Fefferman-Phong class and it is also small in this class. He also presents an argument of Jerison and Kenig that shows that if one assumes $\gamma_i \in W^{2,p}(\Omega)$ with $p > \frac{n}{2}$ then Theorem 19.6 holds. An identifiability result was proven by Isakov [26] for conductivities having jump-type singularities across a submanifold.

R. Brown [10] has shown that Theorem 19.6 is valid if one assumes $\gamma_i \in C^{\frac{3}{2}+\epsilon}(\overline{\Omega})$ by using the arguments in [61] combined with Theorem 19.5.

The arguments used in the proofs of Theorems 19.6, 19.7, 19.8 can be pushed further to prove the following stability estimates. For stability estimates for the inverse scattering problem at a fixed energy see [52].

Theorem 19.9 ([3]). *Suppose that $s > \frac{n}{2}$ and that γ_1 and γ_2 are C^∞ conductivities on $\overline{\Omega} \subseteq \mathbb{R}^n$ satisfying* *(i)* $0 < \frac{1}{E} \le \gamma_j \le E,$

(ii) $\|\gamma_j\|_{H^{s+2}(\Omega)} \le E.$

Then there exists $C = C(\Omega, E, n, s)$ *and* $0 < \sigma < 1$ *(*$\sigma = \sigma(n, s)$*) such that*

$$\|\gamma_1 - \gamma_2\|_{L^\infty(\Omega)} \le C\{|\log\|\Lambda_{\gamma_1} - \Lambda_{\gamma_2}\|_{\frac{1}{2}, -\frac{1}{2}}|^{-\sigma} + \|\Lambda_{\gamma_1} - \Lambda_{\gamma_2}\|_{\frac{1}{2}, -\frac{1}{2}}\} \qquad (19.40)$$

where $\| \cdot \|_{\frac{1}{2}, -\frac{1}{2}}$ *denotes the operator norm as operators from* $H^{\frac{1}{2}}(\partial\Omega)$ *to* $H^{-\frac{1}{2}}(\partial\Omega)$.

This result is a consequence of the next two results.

Theorem 19.10 ([3]). *Assume* 0 *is not a Dirichlet eigenvalue of* $\Delta - q_i$, $i = 1, 2$. *Let* $s > \frac{n}{2}$, $n \ge 3$ *and*

$$\|q_j\|_{H^s(\Omega)} \le M.$$

Then there exists $C = C(\Omega, M, n, s)$ *and* $0 < \sigma < 1$ *(* $\sigma = \sigma(n, s)$*) such that*

$$\|q_1 - q_2\|_{H^{-1}(\Omega)} \le C(|\log\|\Lambda_{q_1} - \Lambda_{q_2}\|_{\frac{1}{2}, -\frac{1}{2}}|^{-\sigma} + \|\Lambda_{q_1} - \Lambda_{q_2}\|_{\frac{1}{2}, -\frac{1}{2}}. \qquad (19.41)$$

The stability estimate at the boundary is of Hölder type.

Theorem 19.11 ([63]). *Suppose that* γ_1 *and* γ_2 *are* C^∞ *functions on* $\overline{\Omega} \subseteq \mathbb{R}^n$ *satisfying*

(i) $0 < \frac{1}{E} \le \gamma_i \le E,$

(ii) $\|\gamma_i\|_{C^2(\overline{\Omega})} \le E.$

Given any $0 < \sigma < \frac{1}{n+1}$, *there exists* $C = C(\Omega, E, n, \sigma)$ *such that*

$$\|\gamma_1 - \gamma_2\|_{L^\infty(\partial\Omega)} \le C\|\Lambda_{\gamma_1} - \Lambda_{\gamma_2}\|_{\frac{1}{2}, -\frac{1}{2}} \qquad (19.42)$$

and

$$\left\|\frac{\partial\gamma_1}{\partial\nu} - \frac{\partial\gamma_2}{\partial\nu}\right\|_{L^\infty(\partial\Omega)} \le C\|\Lambda_{\gamma_1} - \Lambda_{\gamma_2}\|_{\frac{1}{2}, -\frac{1}{2}}^\sigma. \qquad (19.43)$$

The complex geometrical optics solution of Theorems 19.6 and 19.7 were also used by A. Nachman [35] and R. Novikov [45] to give a *reconstruction* procedure of the conductivity from Λ_γ.

We first can reconstruct γ at the boundary since $\gamma\big|_{\partial\Omega}|\xi'|$ is the principal symbol of Λ_γ (see (19.39)). In other words, in coordinates (x', x^n) so that $\partial\Omega$ is locally given by $x^n = 0$ we have

$$\gamma(x', 0) = \lim_{s \to \infty} e^{-is<x', \omega'>}\frac{1}{s}\Lambda_\gamma(e^{is<x', \omega'>})$$

with $\omega' \in \mathbb{R}^{n-1}$ and $|\omega'| = 1$.

In a similar fashion, using (19.39), one can find $\frac{\partial \gamma}{\partial \nu}\big|_{\partial\Omega}$ by computing the principal symbol of $(\Lambda_\gamma - \gamma|_{\partial\Omega}\Lambda_1)$ where Λ_1 denotes the Dirichlet to Neumann map associated to the conductivity 1.

Therefore if we know Λ_γ we can determine Λ_q. We will then show how to reconstruct q from Λ_q. Once this is done, to find $\sqrt{\gamma}$, we solve the problem

$$\Delta u - qu = 0 \text{ in } \Omega, \tag{19.44}$$

$$u\big|_{\partial\Omega} = \sqrt{\gamma}\big|_{\partial\Omega}.$$

Let $q_1 = q$, $q_2 = 0$ in formula (19.33). Then we have

$$\int_\Omega quv = \int_{\partial\Omega} (\Lambda_q - \Lambda_0) \left(v\big|_{\partial\Omega}\right) u\big|_{\partial\Omega} dS, \tag{19.45}$$

where $u, v \in H^1(\Omega)$ solve $\Delta u - qu = 0$, $\Delta v = 0$ in Ω. Here Λ_0 denotes the Dirichlet to Neumann map associated to the potential $q = 0$. We choose $\rho_i, i = 1, 2$ as in (19.35) and (19.36).

Take $v = e^{x\cdot\rho_1}$, $u := u_\rho = e^{x\cdot\rho_2}(1 + \psi_q(x, \rho_2))$ as in Theorem 19.1. By taking $\lim\limits_{|l|\to\infty}$ in (19.45) we conclude

$$\hat{q}(-k) = \lim_{|l|\to\infty} \int_{\partial\Omega} (\Lambda_q - \Lambda_0) \left(e^{x\cdot\rho_1}\big|_{\partial\Omega}\right) u_\rho\big|_{\partial\Omega} dS.$$

So the problem is then to recover the boundary values of the solutions u_ρ from Λ_q.

The idea is to find $u_\rho\big|_{\partial\Omega}$ by looking at the exterior problem. Namely by extending $q = 0$ outside Ω, u_ρ solves

$$\Delta u_\rho = 0 \text{ in } \mathbb{R}^n - \Omega, \tag{19.46}$$

$$\frac{\partial u_\rho}{\partial \nu}\bigg|_{\partial\Omega} = \Lambda_q \left(u_\rho\big|_{\partial\Omega}\right).$$

Also note that

$$e^{-x\cdot\rho} u_\rho - 1 \in L_\delta^2(\mathbb{R}^n). \tag{19.47}$$

Let $\rho \in \mathbb{C}^n - 0$ with $\rho\cdot\rho = 0$. Let $G_\rho(x, y) \in \mathcal{D}'(\mathbb{R}^n \times \mathbb{R}^n)$ denote the Schwartz kernel of the operator Δ_ρ^{-1}. Then we have that

$$g_\rho(x) = e^{x\cdot\rho} G_\rho(x) \tag{19.48}$$

is a Green's kernel for Δ, namely

$$\Delta g_\rho = \delta_0. \tag{19.49}$$

We shall write the solution of (19.46) and (19.47) in terms of single- and double-layer potentials using this Green's kernel (also called Faddeev's Green's kernel [20]. For applications to inverse scattering at fixed energy of this Green's kernel see [19], [35], [45], [46], [47].)

We define the single- and double-layer potentials

$$S_\rho f(x) = \int_{\partial\Omega} g_\rho(x-y)f(y)dS_y, \qquad x \in \mathbb{R}^n - \Omega, \tag{19.50}$$

$$D_\rho f(x) = \int_{\partial\Omega} \frac{\partial g_\rho}{\partial\nu}(x-y)f(y)dS_y, \qquad x \in \mathbb{R}^n - \Omega, \tag{19.51}$$

$$B_\rho f(x) = \text{p.v.} \int_{\partial\Omega} \frac{\partial g_\rho}{\partial\nu}(x-y)f(y)dS_y, \qquad x \in \partial\Omega. \tag{19.52}$$

Nachman showed that $f_\rho = u_\rho\big|_{\partial\Omega}$ is a solution of the integral equation

$$f_\rho = e^{x\cdot\rho} - \left(S_\rho\Lambda_q - B_\rho - \frac{1}{2}I\right)f_\rho. \tag{19.53}$$

Moreover, (19.53) is an inhomogeneous integral equation of Fredholm type for f_ρ, and it has a unique solution in $H^{\frac{3}{2}}(\partial\Omega)$. The uniqueness of the homogeneous equation follows from the uniqueness of the solutions in Theorem 19.7.

We end this section by considering an extension of Theorem 19.6 to quasilinear conductivities.

Let $\gamma(x,t)$ be a function with domain $\overline{\Omega}\times\mathbb{R}$. Let α be such that $0 < \alpha < 1$. We assume

$$\gamma \in C^{1,\alpha}(\overline{\Omega} \times [-T,T]) \quad \forall\, T, \tag{19.54}$$

$$\gamma(x,t) > 0 \,\forall\, (x,t) \in \overline{\Omega} \times \mathbb{R}. \tag{19.55}$$

If $f \in C^{2,\alpha}(\partial\Omega)$, there exists a unique solution of the Dirichlet problem (see e.g., [19])

$$\begin{cases} \text{div}(\gamma(x,u)\nabla u) &= 0 \quad \text{in } \Omega, \\ u\big|_{\partial\Omega} &= f. \end{cases} \tag{19.56}$$

Then the Dirichlet to Neumann map is defined by

$$\Lambda_\gamma(f) = \gamma(x,f)\big|_{\partial\Omega}\frac{\partial u}{\partial\nu}\big|_{\partial\Omega}, \tag{19.57}$$

where u is a solution to (19.56). Sun [53] proved the following result.

Theorem 19.12 ([53]). *Let $n \geq 3$. Assume $\gamma_i \in C^{1,1}(\overline{\Omega} \times [-T, T]) \; \forall \, T > 0$ and $\Lambda_{\gamma_1} = \Lambda_{\gamma_2}$. Then $\gamma_1(x, t) = \gamma_2(x, t)$ on $\overline{\Omega} \times \mathbb{R}$.*

The main idea is to linearize the Dirichlet to Neumann map at constant boundary data equal to t (then the solution of (19.56) is equal to t). Isakov [27] was the first to use a linearization technique to study an inverse parabolic problems associated to nonlinear equations. The case of the Dirichlet to Neumann map associated to the Schrödinger equation with a nonlinear potential was considered in [29], [30] under some assumptions on the potential. We note that, in contrast to the linear case, one cannot reduce the study of the inverse problem of the conductivity equation (19.56) to the Schrödinger equation with a nonlinear potential since a reduction similar to (19.21) is not possible in this case. The main technical lemma in the proof of Theorem 19.12 is

Lemma 19.13. *Let $\gamma(x, t)$ be as in (19.54) and (19.55). Let $1 < p < \infty$, $0 < \alpha < 1$. Let us define*

$$\gamma^t(x) = \gamma(x, t). \tag{19.58}$$

Then for any $f \in C^{2,\alpha}(\partial\Omega)$, $t \in \mathbb{R}$,

$$\lim_{s \to 0} \left\| \frac{1}{s}\Lambda_\gamma(t + sf) - \Lambda_{\gamma^t}(f) \right\|_{W^{1 - \frac{1}{p}, p}(\partial\Omega)} = 0. \tag{19.59}$$

The proof of Theorem 19.12 follows immediately from the lemma. Namely, (19.59) and the hypotheses $\Lambda_{\gamma_1} = \Lambda_{\gamma_2} \Rightarrow \Lambda_{\gamma_1^t} = \Lambda_{\gamma_2^t}$ for all $t \in \mathbb{R}$. Then using the linear result Theorem 19.6, we conclude that $\gamma_1^t = \gamma_2^t$ proving the result.

19.5   The Two-Dimensional Case

A. Nachman proved in [34] that, in the two-dimensional case, one can uniquely determine conductivities in $W^{2,p}(\Omega)$ for some $p > 1$ from Λ_γ. An essential part of Nachman's argument is the construction of the complex geometrical optics solutions (19.23) for all complex frequencies $\rho \in \mathbb{C}^2 - 0$, $\rho \cdot \rho = 0$ for potentials of the form (19.22). Then he applies the $\overline{\partial}$-method in inverse scattering, pioneered in one dimension by Beals and Coifman [6] and extended to higher dimensions by several authors [7], [1], [37], [47], [67]. We note one cannot construct these solutions for a general potential for all non-zero complex frequencies as observed by Tsai [67].

In fact, the analogue of Theorem 19.7 is open, in two dimensions, for a general potential $q \in L^\infty(\Omega)$. We describe later in 19.5.2 of this section progress made in the identifiability problem in this case. In 19.5.1 we outline a different approach to Nachman's result that allows less regular conductivities.

19.5.1   The Inverse Conductivity Problem

In this section we describe an extension of Nachman's result to $W^{1,p}(\Omega)$, $p > 2$, conductivities proved by Brown and the author [12]. We follow an earlier approach of Beals and Coifman [8] and L. Sung [59], who studied scattering for a first-order system whose principal part is $\begin{pmatrix} \bar{\partial} & 0 \\ 0 & \partial \end{pmatrix}$.

Theorem 19.14. *Let $n = 2$. Let $\gamma \in W^{1,p}(\Omega), p > 2$, γ strictly positive. Assume $\Lambda_{\gamma_1} = \Lambda_{\gamma_2}$. Then*

$$\gamma_1 = \gamma_2 \ in \ \overline{\Omega}.$$

We first reduce the conductivity equation to a first-order system. We define

$$q = -\frac{1}{2}\partial \log \gamma \tag{19.60}$$

and define a matrix potential Q by

$$Q = \begin{pmatrix} 0 & q \\ \bar{q} & 0 \end{pmatrix}. \tag{19.61}$$

We let D be the operator

$$D = \begin{pmatrix} \bar{\partial} & 0 \\ 0 & \partial \end{pmatrix}. \tag{19.62}$$

An easy calculation will show that, if u satisfies the conductivity equation $\mathrm{div}(\gamma \nabla u) = 0$, then

$$\begin{pmatrix} v \\ w \end{pmatrix} = \gamma^{\frac{1}{2}} \begin{pmatrix} \partial u \\ \bar{\partial} u \end{pmatrix} \tag{19.63}$$

solves the system

$$D \begin{pmatrix} v \\ w \end{pmatrix} - Q \begin{pmatrix} v \\ w \end{pmatrix} = 0. \tag{19.64}$$

In [12] are constructed matrix solutions of (19.64) of the form

$$\psi = m(z,k) \begin{pmatrix} e^{izk} & 0 \\ 0 & e^{-i\bar{z}k} \end{pmatrix}. \tag{19.65}$$

where $z = x_1 + ix_2$, $k \in \mathbb{C}$, with $m \to 1$ as $|z| \to \infty$ in a sense to be described below. To construct m we solve the integral equation

$$m - D_k^{-1}Qm = 1, \tag{19.66}$$

where, for a matrix-valued function A,

$$D_k A = E_k^{-1} D E_k A \tag{19.67}$$

with

$$E_k A = A^d + \Lambda_k^{-1} A^{\text{off}} \tag{19.68}$$

and

$$\Lambda_k(z) = \begin{pmatrix} e^{i(z\bar{k}+\bar{z}k)} & 0 \\ 0 & e^{-i(zk+\bar{z}\bar{k})} \end{pmatrix}. \tag{19.69}$$

Here A^d denotes the diagonal part of A and A^{off} the antidiagonal part.

Let

$$J = \frac{1}{2} \begin{pmatrix} -i & 0 \\ 0 & i \end{pmatrix}. \tag{19.70}$$

Then we have

$$\mathcal{J} A =: [J, A] = 2J A^{\text{off}} = -2A^{\text{off}} J. \tag{19.71}$$

where $[\ ,\ ]$ denotes the commutator.

To end with the preliminary notation, we recall the definition of the weighted L^p space

$$L_\alpha^p(\mathbb{R}^2) = \left\{ f : \int (1 + |x|^2)^\alpha |f(x)|^p dx < \infty \right\}.$$

The next result gives the solvability of (19.66) in an appropriate space.

Theorem 19.15. *Let $Q \in L^p(\mathbb{R}^2), p > 2$, and compactly supported. Assume that Q is a Hermitian matrix. Choose r so that $\frac{1}{p} + \frac{1}{r} > \frac{1}{2}$ and then β so that $\beta r > 2$. Then the operator $(I - D_k^{-1}Q)$ is invertible in $L_{-\beta}^r$. Moreover, the inverse is differentiable in k in the strong operator topology.*

Theorem 19.15 implies the existence of solutions of the form (19.65) with $m - 1 \in L_{-\beta}^r(\mathbb{R}^2)$ with β, r as in Theorem 19.15. Next we compute $\frac{\partial}{\partial k}m(z,k)$.

Theorem 19.16. *Let m be the solution of (19.66) with $m - 1 \in L^r_{-\beta}(\mathbb{R}^2)$. Then*

$$\frac{\partial}{\partial \overline{k}} m(z, k) = m(z, \overline{k}) \Lambda_k(z) S_Q(k), \tag{19.72}$$

where the scattering data S_Q is given by (see [8])

$$S_Q(k) = -\frac{1}{\pi} \mathcal{J} \int_{\mathbb{R}^2} E_k Q m d\mu, \tag{19.73}$$

where $d\mu$ denotes Lebesgue measure in $\mathbb{R}^2$.

Finally, we need an estimate for the growth of m in the variable k. The following result is a straightforward generalization of Proposition 2.23 in [59]. (We remark that the proof in [59] is incorrect. A corrected proof, kindly provided by L. Sung, appears in [12].)

Theorem 19.17. *Let $Q \in L^p(\mathbb{R}^2), p > 2$, and compactly supported. Then there exists $R = R(Q)$ so that, for all $q > \frac{2p}{p-2}$,*

$$sup_z \| m(z, \cdot) - 1 \|_{L^q\{k; |k| > R\}} \leq C,$$

where the constants depend on p, q and Q.

Outline of proof of Theorem 19.14. We recall (see Theorem 19.8) that if $\gamma_i \in W^{1,p}(\Omega)$ and $\Lambda_{\gamma_1} = \Lambda_{\gamma_2}$, then $\partial^\alpha \gamma_1 \big|_{\partial \Omega} = \partial^\alpha \gamma_2 \big|_{\partial \Omega} \; \forall \; |\alpha| \leq 1$. Therefore we can extend $\gamma_i \in W^{1,p}(\mathbb{R}^2)$, $\gamma_1 = \gamma_2$ in $\mathbb{R}^2 - \Omega$ and $\gamma_i = 1$ outside a large ball. Thus $Q \in W^{1,p}(\mathbb{R}^2)$. Then Theorems 19.15 and 19.16 apply. Now we follow the following steps.

Step 1: $\Lambda_{\gamma_1} = \Lambda_{\gamma_2} \Rightarrow S_{Q_1} = S_{Q_2} := S$.
This just follows using that

$$\overline{\partial} m_i - Q_i m_i = 1, \; i = 1, 2$$

and integrating by parts in (19.73) (in a ball containing the support of Q_i).

Step 2: Using the $\overline{\partial}$-equation (19.72) and Step 1 we conclude that

$$\frac{\partial}{\partial \overline{k}} (m_1 - m_2) - (m_1 - m_2)(z, \overline{k}) \Lambda_k(z) S_Q(k) = 0. \tag{19.74}$$

Therefore $m_1 - m_2 \in L^{\tilde{p}}_{-\beta}(\mathbb{R}^2)$ satisfies the pseudoanalytic equation

$$\overline{\partial}_k(m_1 - m_2) = r(z, k)(m_1 - m_2)(z, \overline{k}), \tag{19.75}$$

where $r(z, k) = S_Q(k) \Lambda_k(z)$.

Step 3: We define

$$(\widetilde{m}_1 - \widetilde{m}_2) = (m_1 - m_2)(z, \overline{k})e^{\overline{\partial}^{-1}r}. \qquad (19.76)$$

It is easy to check that

$$\overline{\partial}(\widetilde{m}_1 - \widetilde{m}_2) = 0.$$

Then we can conclude that $\widetilde{m}_1 = \widetilde{m}_2$ and therefore $m_1 = m_2$, which in turn easily implies that $Q_1 = Q_2$ and therefore $\gamma_1 = \gamma_2$ by using the following result, combined with Theorems 19.16 and 19.17.

Lemma 19.18. *Let $f \in L^2(\mathbb{R}^2)$ and $w \in L^p(\mathbb{R}^2)$ for some finite p. Assume that $we^{\overline{\partial}^{-1}f}$ is analytic. Then $w = 0$.*

The idea of the proof of Lemma 19.18 is the observation that since $r \in L^2(\mathbb{R}^2)$, $u = \overline{\partial}^{-1}r$ is in $VMO(\mathbb{R}^2)$ (the space of functions with vanishing mean oscillation) and thus is $O(\log|z|)$ as $|z| \to \infty$. Hence $e^u w \in L^{\widetilde{p}}$ for $\widetilde{p} > p$. By Liouville's theorem it follows that $e^u w = 0$. The details can be found in [12].

We remark that Theorem 19.12 is also valid in the two-dimensional case [54].

19.5.2   The Potential Case

As we mentioned at the beginning of this section, the analogue of Theorem 19.7 is unknown at present for a general potential $q \in L^\infty(\Omega)$. By Nachman's result it is true for potentials of the form $q = \frac{\Delta u}{u}$ with $u \in W^{2,p}(\Omega), u > 0$ for some p, $p > 1$. Sun and Uhlmann proved generic uniqueness for pairs of potentials in [56]. In [55] it is shown that one can determine the singularities of an L^∞ potentials from the Dirichlet to Neumann map. Namely we have

Theorem 19.19. *Let $\Omega \subset \mathbb{R}^2$, be a bounded open set with smooth boundary. Let $q_i \in L^\infty(\Omega)$ satisfying*

$$\mathcal{C}_{q_1} = \mathcal{C}_{q_2}.$$

Then

$$q_1 - q_2 \in C^\alpha(\Omega)$$

for all $0 \le \alpha < 1$.

We shall outline in the remaining of this section the proof of an identifiability result near the 0 potential. This proof exhibits some of the features of the proof of Theorem 19.19. We also use directly Calderón's result that the product of harmonic functions is dense in $L^2(\Omega)$.

Theorem 19.20. *Let $\Omega \subseteq \mathbb{R}^2$ be a bounded open set with smooth boundary. Let $q_i \in W^{1,\infty}(\Omega)$, $i = 1, 2$. Then there is $\epsilon(\Omega) > 0$ such that if $\|q_i\|_{W^{1,\infty}(\Omega)} < \epsilon(\Omega)$ and*

$$\Lambda_{q_1} = \Lambda_{q_2}.$$

Then

$$q_1 = q_2.$$

We first state an extension of Theorem 19.7 to potentials in some weighted L^p spaces. We also find an asymptotic expansion for the remainder term ψ.

Theorem 19.21. *Let $1 < p < \infty$ and*

$$-\frac{2}{p} < \delta < -1 + \frac{2}{p'} \text{ with } \frac{1}{p} + \frac{1}{p'} = 1. \tag{19.77}$$

There exists a constant $\epsilon(\delta, p)$ and a constant $C > 0$ such that for every $q \in L^p_{\delta+1}(\mathbb{R}^2) \cap L^\infty(\mathbb{R}^2)$ and for every $k \in \mathbb{C}$ satisfying

$$\frac{\|(1 + |x|^2)^{\frac{1}{2}} q\|_{L^\infty(\mathbb{R}^2)}}{|k|} < \epsilon, \tag{19.78}$$

there exists a unique solution to

$$(\Delta - q)u = 0 \text{ in } \mathbb{R}^2 \tag{19.79}$$

such that

$$u = e^{izk}(1 + \psi(z, k)) \tag{19.80}$$

with $\psi \in L^p_\delta(\mathbb{R}^2)$ satisfying

$$\|\psi\|_{L^p_\delta(\mathbb{R}^2)} + \frac{1}{|k|}\|\nabla\psi\|_{L^p_\delta(\mathbb{R}^2)} \leq \frac{C}{|k|}\|q\|_{L^p_{\delta+1}(\mathbb{R}^2)}. \tag{19.81}$$

Furthermore,

$$\psi(x, k) = \frac{1}{4}\frac{(\overline{\partial}^{-1} q)}{ik} + b(x, k) \tag{19.82}$$

with

$$\|b\|_{L^p_\delta(\mathbb{R}^2)} + \frac{1}{|k|}\|\nabla b\|_{L^p_{\delta+1}(\mathbb{R}^2)} \leq \frac{C}{|k|^2}\|q\|_{L^p_{\delta+1}(\mathbb{R}^2)}. \tag{19.83}$$

Outline of Proof of Theorem 19.21. We note that in two dimensions for $z, k \in \mathbb{C} - \{0\}$,

$$e^{-izk} \Delta e^{izk} f = 4 \overline{\partial}(\partial + ik) f. \tag{19.84}$$

We first compute $(\overline{\partial}(\partial + ik))^{-1}$. We observe that

$$\partial(e^{i(kz + \overline{k}\overline{z})} f) = e^{i(kz + \overline{k}\overline{z})}(\partial + ik) f, \tag{19.85}$$

where $f \in C_0^\infty(\mathbb{R}^2)$.

We then define

$$(\partial + ik)^{-1} f = e^{-i(kz + \overline{k}\overline{z})} \partial^{-1}(e^{i(kz + \overline{k}\overline{z})} f), \tag{19.86}$$

where $f \in C_0^\infty(\mathbb{R}^2)$.

Because of Lemma 19.4 and (19.86) we have that $(\partial + ik)^{-1}$ extends as a bounded operator from $L_{\delta+1}^p(\mathbb{R}^2)$ and $L_\delta^p(\mathbb{R}^2)$ and

$$\|(\partial + ik)^{-1}\|_{L_{\delta+1}^p, L_\delta^p} \leq C_1 \tag{19.87}$$

with C_1 independent of k. Here $\| \cdot \|_{L_{\delta+1}^p, L_\delta^p}$ denotes the operator norm. Also using (19.86) we have that

$$\|D_z(\partial + ik)^{-1}\|_{L_{\delta+1}^p, L_{\delta+1}^p} \leq C_2 |k|, \tag{19.88}$$

where D_z denotes differentiation in any direction and C_2 is independent of k.

The following identities are easy to check.

Let $f \in L_{\delta+1}^p(\mathbb{R}^2)$

$$(\partial + ik)^{-1} f = \frac{1}{ik}(I - \partial(\partial + ik)^{-1}) f \tag{19.89}$$

and

$$(\partial + ik)^{-1} f = \frac{f}{ik} - \frac{\partial}{(ik)^2}(I - \partial(\partial + ik)^{-1})) f. \tag{19.90}$$

Therefore it follows from (19.89) that

$$\left(\overline{\partial}(\partial + ik)\right)^{-1} f = \frac{1}{ik}[\overline{\partial}^{-1} - \overline{\partial}^{-1}\partial(\partial + ik)^{-1}] f. \tag{19.91}$$

Using now (19.87) and (19.91) we get

$$\|(\overline{\partial}(\partial + ik))^{-1}\|_{L_{\delta+1}^p, L_\delta^p} \leq \frac{C_3}{|k|} \tag{19.92}$$

and

$$\|D_z(\overline{\partial}(\partial + ik))^{-1}\|_{L^p_{\delta+1}, L^p_{\delta+1}} \leq C_4 \tag{19.93}$$

with C_3, C_4 independent of k.

Now substituting (19.80) into (19.79) we get that ψ must satisfy

$$4\overline{\partial}(\partial + ik)\psi = q(\psi + 1) \tag{19.94}$$

or

$$\psi = \frac{1}{4}((\overline{\partial}(\partial + ik))^{-1})(q(1 + \psi)). \tag{19.95}$$

The existence of a unique ψ in $L^p_\delta(\mathbb{R}^2)$ follows easily from a contraction argument using (19.81) and (19.78). Also the estimate (19.81) follows from (19.92) and (19.93). To obtain (19.82) and (19.83) we note that (19.90) implies that

$$(\overline{\partial}(\partial + ik))^{-1}f = \frac{\overline{\partial}^{-1}f}{ik} - \frac{\overline{\partial}^{-1}}{(ik)^2}\partial(I - \partial(\partial + ik)^{-1}f). \tag{19.96}$$

According to (19.95),

$$\psi = \frac{1}{4}((\overline{\partial}(\partial + ik)^{-1}q + (\overline{\partial}(\partial + ik))^{-1}(q\psi)). \tag{19.97}$$

The first term in (19.97) is

$$\frac{1}{4}(\overline{\partial}(\partial + ik)^{-1})q = \frac{1}{4}\frac{\overline{\partial}^{-1}q}{ik} - \frac{\overline{\partial}^{-1}\partial(I - \partial(\partial + ik)^{-1})q}{4ik^2} \tag{19.98}$$

and the second term in (19.97) satisfies the estimate (19.83) because of (19.92), (19.93), (19.81), concluding the proof.

Outline of proof of Theorem 19.20. The proof follows using a "compactness" lemma for elements orthogonal to the product of solutions of the Schrödinger equation. Specifically

Lemma 19.22. *Let $\Omega \subseteq \mathbb{R}^2$ be a bounded domain with smooth boundary. Let $0 < s < 1$. Let q_i, $i = 1, 2$, satisfy*

$$\|q_i\|_{L^4(\Omega)} \leq M. \tag{19.99}$$

Then there exists a constant $C = C(\Omega, s, M)$ such that, if

$$\int_\Omega f u_1 u_2 = 0 \tag{19.100}$$

for all $u_i \in H^1(\Omega)$ solution of $\Delta u_i - q_i u_i = 0, i = 1, 2$, with $f \in L^2(\Omega)$, then $f \in H^s(\Omega)$ and

$$\|f\|_{H^s(\Omega)} \leq C\|f\|_{L^2(\Omega)}. \tag{19.101}$$

We now use the above lemma to conclude the proof of Theorem 19.20. Suppose that Theorem 19.20 is false. Then there is a sequence of pairs $\left\{ q_1^{(n)}, q_2^{(n)} \right\}$ with $q_1^{(n)} \neq q_2^{(n)}$ for all n approaching zero in $W^{1,\infty}(\Omega)$ and satisfying

$$\Lambda_{q_1^{(n)}} = \Lambda_{q_2^{(n)}}. \tag{19.102}$$

(The set of q's for which $\mathcal{C}_q = \Lambda_q$ is dense in $W^{1,\infty}(\Omega)$.) Now we have in this case

$$\int_\Omega (q_1^{(n)} - q_2^{(n)}) u_1^{(n)} u_2^{(n)} = 0 \tag{19.103}$$

for all $u_i^{(n)}$ solution of

$$(\Delta - q_i^{(n)}) u_i^{(n)} = 0 \text{ in } \Omega, \quad i = 1, 2. \tag{19.104}$$

Let

$$f_n = \frac{q_1^{(n)} - q_2^{(n)}}{\left\| q_1^{(n)} - q_2^{(n)} \right\|_{L^2(\Omega)}}. \tag{19.105}$$

It follows from Rellich's Lemma and (19.101) that the f_n have a convergent subsequence $f_{n(i)} \to f \in L^2(\Omega)$ with $\|f\|_{L^2(\Omega)} = 1$. Now, let u and v be arbitrary harmonic functions in Ω with boundary values of h and g. Let $u_1^{(n)}$ and $v_2^{(n)}$ denote the solutions to (19.104) with the same boundary values as u and v, respectively; then

$$\begin{aligned} \Delta \left(u - u_1^{(n)} \right) &= q_1^{(n)} \left(u - u_1^{(n)} \right) - q_1^{(n)} u, \\ \left(u - u_1^{(n)} \right) \Big|_{\partial\Omega} &= 0. \end{aligned}$$

Hence

$$\begin{aligned} \left\| u - u_1^{(n)} \right\|_{W^{2,2}(\Omega)} &\leq C_1 \left(\left\| q_1^{(n)} \left(u - u_1^{(n)} \right) \right\|_{L^2(\Omega)} + \left\| q_1^{(n)} \right\|_{L^\infty(\Omega)} \|u\|_{L^2(\Omega)} \right) \\ &\leq C_1 \left(\left\| q_1^{(n)} \right\|_{L^\infty(\Omega)} \left\| \left(u - u_1^{(n)} \right) \right\|_{L^2(\Omega)} + \left\| q_1^{(n)} \right\|_{L^\infty(\Omega)} \|u\|_{L^2(\Omega)} \right). \end{aligned} \tag{19.106}$$

For $\left\| q_1^{(n)} \right\|_{L^\infty(\Omega)}$ small enough

$$\left\| u - u_1^{(n)} \right\|_{W^{2,2}(\Omega)} \leq C_2 \left\| q_1^{(n)} \right\| \|u\|_{L^2(\Omega)}$$

so that

$$u_1^{(n)} \xrightarrow{W^{2,2}(\Omega)} u$$

and similarly

$$v_2^{(n)} \xrightarrow{W^{2,2}(\Omega)} v.$$

In particular, since convergence in $W^{2,2}(\Omega)$ implies convergence in $L^4(\Omega)$ (e.g., [21])

$$u_1^{(n)} v_2^{(n)} \xrightarrow{L^2(\Omega)} uv. \tag{19.107}$$

We know that

$$0 = \int_\Omega f_n u_1^{(n)} v_2^{(n)},$$

so it follows from (19.103) and (19.107) that

$$0 = \int_\Omega f uv, \tag{19.108}$$

for all harmonic u and v. By Calderón's argument we get $f = 0$, a contradiction.

Proof of Lemma 19.22. We begin by choosing u_1 and u_2 to be solutions to

$$\Delta u_i - q_i u_i = 0, \quad i = 1, 2,$$

in $\mathbb{R}^2$ of the form (19.80), with q_i extended to be zero outside Ω, and δ satisfying (19.77) with $p = 4$ of the form

$$\begin{aligned} u_1 &= e^{izk}(1 + \psi_1(x, k)), \\ u_2 &= e^{i\overline{z}\overline{k}}(1 + \psi_2(x, k)), \end{aligned}$$

and, in order to satisfy (19.78),

$$|k| \geq \frac{\|(1 + |x|^2)^{1/2} q\|_{L^\infty(\Omega)}}{\epsilon} =: R. \tag{19.109}$$

Hence, using the expansion (19.82) with $g_i = \overline{\partial}^{-1} q_i$, $w_i := \psi_{q_i}$ (19.100) becomes

$$\begin{aligned} -\int_{\mathbb{R}^2} e^{i(zk+\overline{z}\overline{k})} f &= \int_{\mathbb{R}^2} f w_1 w_2 e^{i(zk+\overline{z}\overline{k})} + \int_{\mathbb{R}^2} f \frac{(g_1+g_2)}{4ik} e^{i(zk+\overline{z}\overline{k})} \\ &+ \int_{\mathbb{R}^2} f (b_1 + b_2) e^{i(zk+\overline{z}\overline{k})}, \end{aligned} \tag{19.110}$$

where f is extended to be zero outside Ω. We denote by I_1, I_2, and I_3 the three terms on the right-hand side of (19.110). We have

$$
\begin{aligned}
|I_1| &\leq C_1 \|f\|_{L^2(\Omega)} \|\psi_1\psi_2\|_{L^2(\Omega)} \\
&\leq C_1 \|f\|_{L^2(\Omega)} \|\psi_1\|_{L^4(\Omega)} \|\psi_2\|_{L^4(\Omega)},
\end{aligned}
$$

which implies, in view of (19.81),

$$
|I_1| \leq \frac{C_2(\Omega)}{|k|^2} \|f\|_{L^2(\Omega)} \|q_1\|_{L^4_{\delta+1}((\mathbb{R}^2))} \|q_2\|_{L^4_{\delta+1}(\mathbb{R}^2)}
$$

so that, for $0 < s < 1$,

$$
\| \, |k|^s I_1 \|_{L^2(|k|\geq R)} \leq C_3(\Omega,s) \|q_1\|_{L^4_{\delta+1}(\mathbb{R}^2)} \|q_2\|_{L^4_{\delta+1}(\mathbb{R}^2)} \|f\|_{L^2(\Omega)}. \tag{19.111}
$$

In addition,

$$
I_2 = \frac{[f(g_1+g_2)]^\wedge(k)}{4ik} \quad \text{with } \widehat{f}(k) := \int_{\mathbb{R}^2} e^{i(zk+\bar{z}\bar{k})} f
$$

so that

$$
\begin{aligned}
\| \, |k|^s I_2 \|_{L^2(|k|\geq R)} &\leq \|[f(g_1+g_2)]^\wedge\|_{L^2(|k|\geq R)} \\
&\leq \|f(g_1+g_2)\|_{L^2(\mathbb{R}^2)} \\
&\leq \|f\|_{L^2(\Omega)} \|g_1+g_2\|_{L^\infty(\Omega)} \\
&\leq C_4 \|f\|_{L^2(\Omega)} \|q_1+q_2\|_{L^4_\delta(\Omega)}.
\end{aligned} \tag{19.112}
$$

In summary,

$$
\| \, |k|^s I_2 \|_{L^2(|k|\geq R)} \leq C_6 \|f\|_{L^2(\Omega)} (\|q_1\|_{L^4_{\delta+1}(\mathbb{R}^2)} + \|q_2\|_{L^4_{\delta+1}(\mathbb{R}^2)}). \tag{19.113}
$$

To estimate I_3, we use (19.83). We obtain

$$
\begin{aligned}
|k|^s |I_3| &\leq |k|^s \|f\|_{L^2(\Omega)} (\|b_1\|_{L^2_\delta(\mathbb{R}^2)} + \|b_2\|_{L^2_\delta(\mathbb{R}^2)}) \\
&\leq |k|^{s-2} \|f\|_{L^2(\Omega)} (\|q_1\|_{L^2_{\delta+1}(\mathbb{R}^2)} + \|q_2\|_{L^2_{\delta+1}(\mathbb{R}^2)}).
\end{aligned} \tag{19.114}
$$

Finally, we note that

$$
\|f\|_{H^s}^2 \leq 2(\|\widehat{f}\|_{L^2(|k|\leq R)}^2 (1+R^2)^s + \| \, |k|^s \widehat{f}\|_{L^2(|k|\geq R)}). \tag{19.115}
$$

Combining (19.115) with (19.110), (19.111), (19.113), and (19.114), where R is chosen in (19.109), gives (19.101).

19.6   First-Order Perturbations of the Laplacian

In this section we consider inverse boundary-value problems associated with first-order perturbations of the Laplacian. We consider two important cases arising in applications. We first consider the Schrödinger equation in the presence of a magnetic potential. The problem is to determine both the electric and magnetic potential of a medium by making measurements at the boundary of the medium. The second example involves an elliptic system. We consider an elastic body. The problem is to determine the elastic parameters of this body by making displacements and traction measurements at the boundary. Another important inverse boundary-value problem involving the system of Maxwell's equations is the determination of the electric permittivity, magnetic permeability, and electrical conductivity of a body by measuring the tangential component of the electric and magnetic field [51]. A global identifiability result and a reconstruction method can be found in [48]. This problem can also be reduced to construct complex geometrical optics solutions for first-order perturbations of the Laplacian [58]. Recently, in [49] it was shown that, by considering a larger system, one can reduce the problem of constructing the complex geometrical optics solutions for Maxwell's equations to a zeroth-order perturbation of the Laplacian. Then the solutions can be constructed as in §19.3.

19.6.1   The Schrödinger Equation with Magnetic Potential

Let Ω be a bounded domain in $\mathbb{R}^n$, $n \geq 3$, with smooth boundary. The Schrödinger equation in a magnetic field is given by

$$H_{\vec{A},q} = \sum_{j=1}^{n} \left(\frac{1}{i} \frac{\partial}{\partial x_j} + A_j(x) \right)^2 + q(x), \quad i = \sqrt{-1}, \qquad (19.116)$$

where $\vec{A} = (A_1, A_2, \ldots, A_n) \in C^1(\bar{\Omega})$ is the magnetic potential, and $q \in L^\infty(\Omega)$ is the electric potential. The magnetic field is the rotation of the magnetic potential, $\text{rot}(\vec{A})$.

We assume that $\vec{A}$ and q are real-valued function, and thus (19.116) is self-adjoint. We also assume that zero is not a Dirichlet eigenvalue of (19.116) on Ω (or as in §19.3, one can consider the Cauchy data associated to $H_{\vec{A},q}$), so that the boundary-value problem

$$\begin{cases} H_{\vec{A},q}\, u \;=\; 0 \quad \text{in } \Omega, \\ u|_{\partial\Omega} \;=\; f \;\in H^{\frac{1}{2}}(\partial\Omega), \end{cases} \qquad (19.117)$$

has a unique solution $u \in H^1(\Omega)$. The Dirichlet to Neumann map $\Lambda_{\vec{A},q}$, which maps $H^{\frac{1}{2}}(\partial\Omega)$ into $H^{-\frac{1}{2}}(\partial\Omega)$, is defined by

$$\Lambda_{\vec{A},q} : f \to \left.\frac{\partial u}{\partial \nu}\right|_{\partial\Omega} + i(\vec{A}\cdot\nu)f, \quad f \in H^{\frac{1}{2}}(\partial\Omega), \qquad (19.118)$$

where u is the unique solution to (19.117), and ν is the unit outer normal on $\partial\Omega$.

The inverse boundary-value problem for (19.116) is to recover information of $\vec{A}$ and q from knowledge of $\Lambda_{\vec{A},q}$.

It is easy to see in [54] that the the Dirichlet to Neumann map $\Lambda_{\vec{A},q}$ is invariant under a gauge transformation in the magnetic potential: $\vec{A} \to \vec{A} + \nabla g$, where $g \in C_\Omega^1$, where we denote

$$C_\Omega^s = \{f \in C^s(\mathbb{R}^n), \operatorname{supp} f \subset \Omega\}. \qquad (19.119)$$

In fact, if we consider u as in (19.117), then $v = e^{ig}u$ solves $H_{\vec{A}+\nabla g,q}\, v = 0$ in Ω and $\Lambda_{\vec{A},q} f = \Lambda_{\vec{A}+\nabla g,q} f$. Thus, $\Lambda_{\vec{A},q}$ carries information about the magnetic field instead of information about $\vec{A}$. The natural question is whether $\Lambda_{\vec{A},q}$ determines uniquely $\operatorname{rot}(\vec{A})$ and q. In [54], this question was answered affirmatively for $\vec{A}$ in the C_Ω^2 class and q in the $L^\infty(\Omega)$ class, under the assumption that $\operatorname{rot}(\vec{A})$ is small in the L^∞ topology.

The smallness assumption in Sun's result was used to construct complex geometrical optics solutions similar to (19.23) in this case. In §19.7 we will describe how Nakamura and Uhlmann [39] constructed these types of solutions without the smallness assumption on $\operatorname{rot}(\vec{A})$. This combined with the methods of [54] leads to the following result [42].

Theorem 19.23. *Let $\vec{A}_j \in C^\infty(\bar{\Omega})$, $q_j \in C^\infty(\bar{\Omega})$, $j = 1, 2$. Assume that zero is not a Dirichlet eigenvalue of $H_{\vec{A}_j,q_j}$, $j = 1, 2$. If*

$$\Lambda_{\vec{A}_1,q_1} = \Lambda_{\vec{A}_2,q_2},$$

then

$$\operatorname{rot}(\vec{A}_1) = \operatorname{rot}(\vec{A}_2) \text{ and } q_1 = q_2 \text{ in } \Omega.$$

If we assume $\vec{A}_j \in C_\Omega^\infty$, it was proved in [42] that Theorem 19.23 holds even for $q_j \in L^\infty(\Omega)$. We have

Theorem 19.24. *Let* $\vec{A}_j \in C_\Omega^\infty$, $q_j \in L^\infty(\Omega)$, $j = 1, 2$. *Assume that zero is not a Dirichlet eigenvalue of* $H_{\vec{A}_j, q_j}$, $j = 1, 2$. *If*

$$\Lambda_{\vec{A}_1, q_1} = \Lambda_{\vec{A}_2, q_2},$$

then

$$\mathrm{rot}(\vec{A}_1) = \mathrm{rot}(\vec{A}_2) \ and \ q_1 = q_2 \ in \ \Omega.$$

The inverse scattering problem at a fixed energy in this case has been considered in [19].

C. Tolmasky [66] reduced the regularity of $\vec{A}_j$ in Theorem 19.24 to just $\vec{A}_j \in C_\Omega^1$. He constructs the complex geometrical optics solutions under weaker regularity conditions. We shall outline his approach in the next section.

19.6.2 Inverse Boundary-Value Problems for Elastic Materials

We assume now that Ω is an elastic material, that is, if we deform Ω, it will try to come back to its original shape. Let $u(x)$ denote the displacement of the point x under the deformation. The undeformed domain is called the reference configuration space. The *linear strain tensor*,

$$\varepsilon_{ij} = \frac{1}{2}\left(\frac{\partial u_j}{\partial x_i} + \frac{\partial u_i}{\partial x_j}\right), \quad i, j = 1, \dots, n, \tag{19.120}$$

measures the rate of deformation with respect to the Euclidean metric for small deformations. Under the assumption of no-body forces acting on Ω, the equation of equilibrium in the reference configuration is given by the generalized Hooke's law (see [16] for an excellent treatment of elasticity theory),

$$L_C u = \mathrm{div}\ \sigma(u) = 0 \text{ in } \Omega, \tag{19.121}$$

where $\sigma(u)$ is a symmetric two-tensor called the *stress tensor*. The *elastic tensor* C is a fourth-order tensor which satisfies

$$\sigma_{ij}(u) = \sum_{k,l=1}^n C_{ijkl}(x)\varepsilon_{kl}(u), \quad i, j = 1, \dots, n. \tag{19.122}$$

We shall assume that the elastic tensor satisfies the hyperelasticity condition ([16], chap. 4)

$$C_{ijkl}(x) = C_{klij}(x) \quad \forall\ x \in \bar{\Omega}. \tag{19.123}$$

We also assume that C satisfies the strong convexity condition: there exists $\delta > 0$ such that

$$\sum_{i,j,k,l=1}^{n} C_{ijkl}(x)t_{ij}t_{kl} \geq \delta \sum_{i,j=1}^{n} t_{ij}^2, \quad x \in \bar{\Omega} \tag{19.124}$$

for any real-symmetric matrix $(t_{ij})_{1 \leq i,j \leq n}$. Condition (19.124) guarantees the unique solvability of the Dirichlet problem

$$\begin{cases} \operatorname{div} \sigma(u) = 0 & \text{in } \Omega, \\ u\big|_{\partial\Omega} = f. \end{cases} \tag{19.125}$$

The Dirichlet integral associated with (19.125) is given by

$$W_C(f) = \sum_{i,j,k,l=1}^{n} \int_\Omega C_{ijkl}(x) \frac{\partial u_k}{\partial x_l} \frac{\partial u_i}{\partial x_j} dx \tag{19.126}$$

with u the solution of (19.125). Physically, $W_C(f)$ measures the deformation energy produced by the displacement f at the boundary.

Applying the divergence theorem we have that

$$W_C(f) = \sum_{i=1}^{n} \int_\Omega (\Lambda_C(f))_i f_i \, dx, \tag{19.127}$$

where

$$(\Lambda_C(f))_i = \sum_{j,k,l=1}^{n} \nu^j C_{ijkl} \frac{\partial u_k}{\partial x_l}\Big|_{\partial\Omega}, \quad i = 1, \dots, n \tag{19.128}$$

with u the solution of (19.125) and ν the unit outer normal to $\partial\Omega$. In other words, Λ_C is the linear operator associated to the quadratic form W_C. The map

$$f \xrightarrow{\Lambda_C} \Lambda_C(f) \tag{19.129}$$

is the *Dirichlet* to *Neumann* map in this case. It sends the displacement at the boundary to the corresponding traction at the boundary. The inverse problem we consider in this subsection is whether we can determine C from Λ_C.

We shall assume that C is isotropic, i.e., satisfies

$$C_{ijkl}(x) = \lambda(x)\delta_{ij}\delta_{kl} + \mu(x)(\delta_{ik}\delta_{jl} + \delta_{il}\delta_{jk}), \tag{19.130}$$

where δ_{ik} denotes the Krönecker delta.

In this case the strain tensor takes the form

$$\sigma(u) = \lambda(x)(\text{trace } \varepsilon(u)) + 2\mu(x)\text{div } (u). \tag{19.131}$$

The strong-convexity condition (19.124) is equivalent to

$$n\lambda + 2\mu > 0, \quad \mu > 0 \text{ in } \bar{\Omega}. \tag{19.132}$$

The main known results for identifiability of C from Λ_C are

Theorem 19.25 ([39]). *Let $n \geq 3$, $C_j \in C^\infty(\bar{\Omega})$, be an isotropic elastic tensor, $j = 1, 2$. Assume*

$$\Lambda_{C_1} = \Lambda_{C_2}.$$

Then $C_1 = C_2$.

Theorem 19.26 ([41]). *Let $n = 2$ and C_j as in Theorem 19.9 with Lamé parameters λ_j, μ_j, $j = 1, 2$. There exists $\epsilon > 0$ such that if*

$$\|(\lambda_j, \mu_j) - (\lambda_0, \mu_0)\|_{W^{31,\infty}(\Omega)} < \epsilon, \quad j = 1, 2,$$

and $\Lambda_{C_1} = \Lambda_{C_2}$ then $(\lambda_1, \mu_1) = (\lambda_2, \mu_2)$. Here (λ_0, μ_0) denotes a constant and $\|u\|_{W^{31,\infty}(\Omega)} = \sup\limits_{\substack{x \in \Omega \\ |\alpha| \leq 31}} |\partial^\alpha u(x)|$.

The global uniqueness result Theorem 19.25 is the analog to Theorem 19.6 for the inverse conductivity problem. Theorem 19.26 is analogous to the local result, Theorem 19.20. We remark that there is no known global result in two dimensions similar to the one proven by Nachman [36] for the inverse conductivity problem. Theorem 19.25 has been extended in [38] to a class of nonlinear elastic materials, the so-called St. Venant-Kirchhoff's materials, using a linearization technique similar to Lemma 19.13.

We now indicate the main steps in the proofs of Theorems 19.25 and 19.26. The first observation is that under the hypothesis of Theorems 19.25 and 19.26 we can prove an identity involving $\lambda_1 - \lambda_2$, $\mu_1 - \mu_2$.

Lemma 19.27. *Let $u^{(i)}$, $i = 1, 2$ be solutions*

$$\text{div } \sigma(u^{(i)}) = 0 \text{ in } \Omega,$$

with $u^{(i)} \in H^1(\Omega)$, $i = 1, 2$. Assume $\Lambda_{C_1} = \Lambda_{C_2}$. Then

$$E\left(u^{(1)}, u^{(2)}\right) := \int_\Omega (\lambda_1 - \lambda_2)\text{div } u^{(1)} \cdot \overline{\text{div } u^{(2)}} dx \tag{19.133}$$

$$+ 2\int_\Omega (\mu_1 - \mu_2)\varepsilon(u^{(1)}) \cdot \overline{\varepsilon(u^{(2)})} dx = 0.$$

The proof of the Lemma follows readily by applying the divergence theorem and the boundary determination result of [40], which is the analog of Theorem 19.8. Namely, under the hypothesis of Theorems 19.23 and 19.24 we have that the Taylor series of λ_1, μ_1 and λ_2, μ_2 coincide. (It only needed $\partial^\alpha \mu, \partial^\alpha \lambda$, $|\alpha| \leq 1$.)

The problem is now to find "enough" solutions of $L_{C_i} u^{(i)} = 0$ in Ω to conclude that the Lamé parameters coincide in Ω.

It is quite difficult to construct solutions of the form (19.23) directly for the elasticity system. In the paper [39], a reduction to a system with principal part the biharmonic operator is made by multiplying L_C on the left by an explicit second-order system T_C to get

$$T_C L_C = \Delta^2 + M_1(x, D)\Delta + M_2(x, D) \tag{19.134}$$

with $M_i(x, D)$ an $n \times n$ system of order i, $i = 1, 2$. Then to construct solutions of $L_C u = 0$, it is enough to construct solutions of $Mu = (\Delta^2 + M_1(x, D)\Delta + M_2(x, D))u = 0$. By introducing a new dependent variable $v = \Delta u$ we want to find solutions of the $2n \times 2n$ system

$$\Delta \begin{pmatrix} u \\ v \end{pmatrix} + \begin{pmatrix} 0 & 0 \\ 0 & M_1 \end{pmatrix} \begin{pmatrix} u \\ v \end{pmatrix} + \begin{pmatrix} 0 & -I \\ 0 & M_2\Delta^{-1} \end{pmatrix} \begin{pmatrix} u \\ v \end{pmatrix} = \begin{pmatrix} 0 \\ 0 \end{pmatrix} \tag{19.135}$$

where Δ^{-1} denotes the inverse of the Laplacian. Notice that (19.135) is a first-order system perturbation of the Δ with the zeroth-order perturbation being a pseudodifferential operator.

Ikehata [21] reduced the elasticity system to an $(n+1) \times (n+1)$ differential system as follows.

Lemma 19.28 ([23]). Let $\begin{pmatrix} u \\ f \end{pmatrix}$ with $u = (u_1, \ldots, u_n)^t$ be a solution of the $(n+1) \times (n+1)$ system

$$\Delta I_{n+1} \begin{pmatrix} u \\ f \end{pmatrix} + V^1(x) \begin{pmatrix} \nabla f \\ \nabla \cdot u \end{pmatrix} + V^0(x) \begin{pmatrix} u \\ f \end{pmatrix} = \begin{pmatrix} 0 \\ 0 \end{pmatrix} \tag{19.136}$$

with

$$V^1(x) = \begin{pmatrix} 2\mu^{-\frac{1}{2}}(-\nabla^2 + I_n\Delta)\mu^{-1} & -\nabla \log \mu \\ 0 & \frac{\lambda+\mu}{\lambda+2\mu}\mu^{\frac{1}{2}} \end{pmatrix}, \tag{19.137}$$

$$V^0(x) = \begin{pmatrix} -\mu^{-\frac{1}{2}}(2\nabla^2 + I_n\Delta)\mu^{\frac{1}{2}} & 2\mu^{-\frac{5}{2}}(\nabla^2 - \Delta I_n)\mu\nabla\mu \\ -\frac{\lambda-\mu}{\lambda+2\mu}(\nabla\mu^{\frac{1}{2}})^t & -\mu\Delta\mu^{-1} \end{pmatrix}, \tag{19.138}$$

and I_k denotes the identity $k \times k$ matrix. Then

$$w = \mu^{-\frac{1}{2}}u + \mu^{-2}\nabla(\mu f) \tag{19.139}$$

satisfies

$$L_C w = 0. \tag{19.140}$$

Therefore we are reduced to finding "enough" solutions of the first-order system (19.136), which we rewrite as

$$(\Delta + P^{(1)}(x, D))v = 0 \tag{19.141}$$

with $P^{(1)}(x, D)$ a first-order $(n + 1) \times (n + 1)$ differential system.

We shall indicate in §19.7, how to construct these complex geometrical optics solutions.

19.7 Complex Geometrical Optics Solutions for First-order Perturbations of the Laplacian

In this section we outline the construction of complex geometrical optics solutions for first-order perturbations of the Laplacian. For simplicity we do this for scalar equations. A similar method applies to the first-order system (19.136) arising from the elasticity system [39] Let us consider an operator of the form

$$P(x, D) = \Delta I + P^{(1)}(x, D_x), \tag{19.142}$$

where $P^{(1)}(x, D_x)$ is a first-order scalar system with smooth coefficients in $\mathbb{R}^n$ and I denotes the identity matrix. Let $\rho \in \mathbb{C}^n$ with $\rho \cdot \rho = 0$, $\rho \neq 0$.

In this section, for $|\rho|$ sufficiently large we shall outline the method of [39] to construct solutions of

$$P(x, D)u = 0 \tag{19.143}$$

in compact sets of $\mathbb{R}^n$ of the form

$$u = e^{x \cdot \rho}v(x, \rho) \tag{19.144}$$

with a fairly precise control of the behavior of $v(x, \rho)$ as $|\rho| \to \infty$.

We will construct $v(x, \rho)$ as a solution of

$$P_\rho(x, D)v = 0, \tag{19.145}$$

where

$$P_\rho(x, D) = \Delta_\rho + P_\rho^{(1)}(x, D) \qquad (19.146)$$

and

$$\Delta_\rho u = e^{-x \cdot \rho} \Delta(e^{x \cdot \rho} u), \ P_\rho^{(1)}(x, D)u = e^{-x \cdot \rho} P^{(1)}(x, D)(e^{x \cdot \rho} u).$$

As shown in §19.3, we have precise estimates for Δ_ρ^{-1}. The problem is that derivatives of $\Delta_\rho^{-1} f$ do not decay for large ρ. Also $P_\rho^{(1)}$ involves terms growing in ρ. The goal is to get rid of the first-order terms in (19.146). Roughly speaking, we will construct invertible operators A_ρ, B_ρ, and an operator C_ρ of "lower order" so that

$$P_\rho A_\rho = B_\rho(\Delta_\rho + C_\rho). \qquad (19.147)$$

We will then construct solutions of $P_\rho v_\rho = 0$ of the form

$$v_\rho = A_\rho w_\rho$$

with w_ρ solution of $(\Delta_\rho + C_\rho)w_\rho = 0$.

We will accomplish (19.147) using the theory of pseudodifferential operators depending on the complex vector ρ. The main point is to regard the variables ξ and ρ in equal footing. We digress to discuss the main features of this theory. For more details, see for instance [50]. Let

$$Z = \{\rho \in \mathbb{C}^n; |\rho| \geq 1, \rho \cdot \rho = 0\}.$$

Definition 19.29. Let $l \in \mathbb{R}$, $0 \leq \delta < 1$, $U \subset \mathbb{R}^m$, U open. We say that $a_\rho \in S_\delta^l(U, Z) \Leftrightarrow \forall \rho \in Z$ fixed, $a_\rho(x, \xi) \in C^\infty(U \times \mathbb{R}^n)$; $\forall \alpha \in \mathbb{Z}_+^n$, $\forall \beta \in \mathbb{Z}_+^m$, $\forall K \subset U$, K compact $\exists C_{\alpha,\beta,K} > 0$ such that

$$\sup_{x \in K} |\partial_\xi^\alpha \partial_x^\beta a_\rho(x, \xi)| \leq C_{\alpha,\beta,k}(1 + |\xi| + |\rho|)^{l + \beta\delta - |\alpha|} \qquad \forall \rho \in Z, \xi \in \mathbb{R}^n.$$

Example 19.30. We have that

$$\widetilde{r}_\rho(x, \xi) = -|\xi|^2 + 2i\xi \cdot \rho \in S_0^2(\mathbb{R}^n \times Z)$$

since it is homogeneous of degree 2 in (ξ, ρ). Notice, however, that $\widetilde{r}_\rho$ is not elliptic.

In fact, if $n = 3$ and we take $\text{Re}\,\rho = s(1, 0, 0)$, $\text{Im}\,\rho = s(0, 1, 0)$, $s \in \mathbb{R}$, then the zeros of $\widetilde{r}_\rho(\xi)$ is a codimension two circle in the plane $\xi_1 = 0$ centered at the point $(0, -s, 0)$ of radius s.

Definition 19.31. Let $U \subseteq \mathbb{R}^n$, U open, $a_\rho \in S_\delta^l(U, Z)$. We define the operator $A_\rho \in L_\delta^l(U, Z)$ by

$$A_\rho f(x) = \frac{1}{(2\pi)^n} \int e^{ix\cdot\xi} a_\rho(x, \xi) \widehat{f}(\xi) d\xi, \quad f \in C_0^\infty(U). \tag{19.148}$$

The kernel of A_ρ is given by

$$k_{A_\rho}(x, y) = \frac{1}{(2\pi)^n} \int e^{i(x-y)\cdot\xi} a_\rho(x, \xi) d\xi, \tag{19.149}$$

where the integral in (19.149) is interpreted as an oscillatory integral. A_ρ extends continuously as a linear operator

$$A_\rho : \mathcal{E}'(\mathcal{U}) \to \mathcal{D}'(\mathcal{U}),$$

where $\mathcal{E}'(\mathcal{U})$ (resp. $\mathcal{D}'(\mathcal{U})$) denotes the space of compactly supported distributions (resp. distributions).

As usual, it is easy to check that if $a_\rho \in S_\delta^l(U, Z)$ for all l, then $A_\rho : \mathcal{E}'(\mathcal{U}) \to C^\infty(\mathcal{U})$, i.e., A_ρ is a smoothing operator.

Definition 19.32. We say that A_ρ is *uniformly properly supported* if supp k_{A_ρ} is contained in a fixed neighborhood V of the diagonal in $U \times U$ for all $\rho \in Z$, so that $\forall K \subset U$, K compact, V intersected with $\Pi^{-1}(K)$ is compact where Π denotes either one the projections of $U \times U$ onto U.

Proposition 19.33. *Let $A_\rho \in L_\delta^m(U, Z)$. Then we can write*

$$A_\rho = B_\rho + R_\rho$$

with $B_\rho \in L_\delta^m$ uniformly properly supported and R_ρ smoothing.

We shall assume from now on that all pseudodifferential operators are uniformly properly supported.

Definition 19.34. Let $A_\rho \in L_\delta^m(U, Z)$ as in (19.148).

(a) Then the full symbol of A_ρ is given by $\widetilde{\sigma}_m(A_\rho)(x, \xi) = a_\rho(x, \xi)$.

(b) The principal symbol of A_ρ is given by

$$\sigma_m(A_\rho)(x, \xi) = a_\rho(x, \xi) \bmod S_\delta^{m-1}(U, Z).$$

The functional calculus for pseudodifferential operators depending on a parameter is completely analogous to the standard calculus. Namely, we have

Theorem 19.35. *Let $A_\rho \in L_\delta^m(U, Z)$, $B_\rho \in L_\delta^{\widetilde{m}}(U, Z)$. Then*

(a) $A_\rho B_\rho \in L_\delta^{m+\tilde{m}}(U, Z)$

(b) $\tilde{\sigma}_{m+\tilde{m}}(A_\rho B_\rho) \sim \Sigma \frac{1}{\alpha!} D_\xi^\alpha \tilde{\sigma}_m(A_\rho) \partial_x^\alpha \tilde{\sigma}_{\tilde{m}}(B_\rho)$

(c) $\sigma_{m+\tilde{m}}(A_\rho B_\rho) = \sigma_m(A_\rho) \sigma_{\tilde{m}}(B_\rho)$

(d) $\sigma_{m+\tilde{m}}([A_\rho, B_\rho]) = H_{\sigma_m(A_\rho)} \sigma_{\tilde{m}}(B_\rho)$, where $[A_\rho, B_\rho]$ denotes the commutator and H_p denotes the Hamiltonian vector associated to p, i.e.,

$$H_p = \sum_{j=1}^n \left(\frac{\partial p}{\partial \xi_j} \frac{\partial}{\partial x_j} - \frac{\partial p}{\partial x_j} \frac{\partial}{\partial \xi_j} \right).$$

Finally we shall use the following continuity property of A_ρ's on Sobolev spaces (see [50]).

Theorem 19.36. *Suppose $l \leq 0$, K a compact subset of $\mathbb{R}^{2n}$ and $A_\rho \in L_\delta^l(\mathbb{R}^n, Z)$ with supp $K_{A_\rho} \subset K$, $\forall \rho \in Z \Rightarrow \forall k \in \mathbb{R}$, A_ρ is a bounded operator from $H^k(\mathbb{R}^n)$ to $H^k(\mathbb{R}^n)$ and $\exists C_{k,K} \geq 0$ such that*

$$\|A_\rho\|_{k,k} \leq C_{k,K} |\rho|^l \ \forall \ \rho \in Z,$$

where $\|A_\rho\|_{k,k}$ denotes the operator norm.

We define

$$\Lambda_\rho^s \in L_0^s(\mathbb{R}^n, Z) \tag{19.150}$$

by

$$\sigma(\Lambda_\rho^s) = (|\xi|^2 + |\rho|^2)^{\frac{s}{2}}. \tag{19.151}$$

We use this to get a first-order equation. Let $\tilde{P}_\rho = P_\rho \Lambda_\rho^{-1} = \tilde{\Delta}_\rho + \tilde{P}_\rho^{(1)}(x, D)$ with $\tilde{\Delta}_\rho = \Delta_\rho \Lambda_\rho^{-1}$, $\tilde{P}_\rho^{(1)} = P_\rho^{(1)} \Lambda_\rho^{-1}$. A key ingredient in the construction of complex geometrical optics solutions is the following result proven in [39].

Theorem 19.37 (Intertwining property). *For all positive integers $N \exists A_\rho, B_\rho \in L_\delta^0(\mathbb{R}^n, Z)$ invertible for $|\rho|$ large so that*

$$\forall \ \varphi_1 \in C_0^\infty(\mathbb{R}^n), \quad \exists \varphi_j \in C_0^\infty(\mathbb{R}^n), \quad j = 2, 3, 4$$

so that

$$\varphi_1 \tilde{P}_\rho A_\rho = (\varphi_1 B_\rho \varphi_2) \Lambda_\rho^{-1} (\Delta_\rho I + \varphi_3 R^{(-N)} \varphi_4) \tag{19.152}$$

with $R^{(-N)} \in L_\delta^{-N}(\mathbb{R}^n, Z)$, and $\forall s \in \mathbb{R}$

$$\|\varphi_3 R^{(-N)} \varphi_4\|_{s,s} \leq C_s |\rho|^{-N}.$$

Since the kernel of $\phi_3 R^{(-N)}\phi_4$ has compact support independent of ρ, the last estimate follows from Theorem 19.36. Then to find solutions of $P_\rho v_\rho = 0$ in compact sets, it is enough to find solutions of

$$(\Delta_\rho I + \varphi_3 R^{(-N)}\varphi_4)w_\rho = 0. \tag{19.153}$$

Then $\Lambda_\rho^{-1} A_\rho w_\rho$ solves $P_\rho v_\rho = 0$ on compact sets (take $\varphi_1 = 1$ on the compact set).

The proof of this result for the case $\delta = 0$ is given in [39]. We will write in this case $L_0^m(\mathbb{R}^n, Z) =: L^m(\mathbb{R}^n)$ and $S_0^m(U, Z) =: S^m(U, Z)$. We will just say a few words about the proof which is quite technical. The main problem is to construct A_ρ, B_ρ near the characteristic variety (i.e., the set of zeros of $\tilde{r}_\rho$ with $\tilde{r}_\rho$ as in Example 19.30). This is because away from the characteristic variety P_ρ and Δ_ρ are elliptic and therefore invertible modulo elements of $L^{-N}(\mathbb{R}^n, Z)$ for all $N \in \mathbb{N}$, and it is therefore easy to construct the intertwining operators A_ρ, B_ρ in that case. The characteristic variety is given by

$$r_\rho^{-1}(0) = \{(x, \xi) \in \mathbb{R}^{2n} : \operatorname{Re}\rho \cdot \xi = 0, |\xi + \operatorname{Im}\rho|^2 = |\operatorname{Im}\rho|^2\},$$

where

$$r_\rho(\xi) = (|\xi|^2 + |\rho|^2)^{-\frac{1}{2}}(-|\xi|^2 + 2i\rho \cdot \xi).$$

Near the characteristic variety we take $A_\rho = B_\rho$.

So we are looking for $A_\rho \in L^0(\mathbb{R}^n, Z)$ such that

$$\tilde{P}_\rho A_\rho = A_\rho \tilde{\Delta}_\rho \bmod L^{-N}(\mathbb{R}^n, Z),$$

i.e.,

$$(\tilde{\Delta}_\rho + \tilde{P}_\rho^{(1)}(x, D))A_\rho = A_\rho \tilde{\Delta}_\rho \bmod L^{-N}(\mathbb{R}^n, Z).$$

We proceed inductively. We choose

$$A_\rho = \sum_{j=0}^{M} A_\rho^{(j)}, A_\rho^{(j)} \in L^{-j}(\mathbb{R}^n, Z).$$

Let $\sigma_j(A_\rho^{(j)})$ be the principal symbol of $A_\rho^{(j)}$. Then by the calculus of pseudodifferential operators depending on a parameter we need to solve

$$\frac{1}{i}H_{r_\rho}\sigma_0(A_\rho^{(0)}) + \sigma_0(\tilde{P}_\rho^{(1)})\sigma_0(A_\rho^{(0)}) = 0 \tag{19.154}$$

and

$$\frac{1}{i}H_{r_\rho}(\sigma_j(A_\rho^{(j)}) + \sigma_0(\tilde{P}_0^{(1)})\sigma_j(A_\rho^{(M)}) = g_j(x, \xi)$$

with $g_j \in S^{-j}(\mathbb{R}^n, Z)$ so that

$$\widetilde{P}_\rho A_\rho = A_\rho \widetilde{\Delta}_\rho \bmod L^{-M}(\mathbb{R}^n, Z).$$

(Recall that this is all done near the characteristic variety.) We note that

$$H_{r_\rho} = L_{1,\rho} + i L_{2,\rho}$$

with $L_{1,\rho}, L_{2,\rho}$ real-valued vector fields $\mathbb{R}^{2n}$ so that $[L_{1,\rho}, L_{2,\rho}] = 0$. The vector fields $L_{1,\rho}, L_{2,\rho}$ are linearly independent near the characteristic variety. Therefore H_{r_ρ} can be reduced to a Cauchy-Riemann equation in two variables of the form $\frac{\partial}{\partial x_1} + i\frac{\partial}{\partial x_2}$. In fact, one can write down an explicit change of variables in (ξ, ρ) to accomplish this (see [39]). We also give conditions at ∞ on A_ρ to guarantee that it is invertible. Namely in the coordinates in which the vector field H_{r_ρ} is the Cauchy-Riemann equation we require, for $-1 < \alpha < 0$, that $\sigma(A_\rho) - I \in L^2_\alpha$, where I denotes the identity matrix.

To prove Theorem 19.24 with less regularity in the magnetic potentials Tolmasky [66] constructed complex geometrical optics solutions with the co-efficients of $P^{(1)}(x, D)$ in $C^{2/3+\epsilon}(\overline{\Omega})$ for any ϵ positive. Using techniques from the theory of pseudodifferential operators with nonsmooth symbols ([9], [17], [65]) one can decompose a nonsmooth symbol into a smooth symbol plus a less smooth symbol but of lower order. We describe below more precisely this result. First we introduce some notation. $C^s_*(\mathbb{R}^n)$ will denote the Zygmund class.

Definition 19.38. Let $\delta \in [0, 1]$:

(a) $p_\rho(x, \xi) \in C^s_* S^m_{1,\delta,\rho}(\mathbb{R}^n)$ if and only if

$$|D^\alpha_\xi p_\rho(x, \xi)| \le C_\alpha ((1 + |\xi|^2 + |\rho|^2)^{\frac{1}{2}})^{m-|\alpha|}$$

and

$$\|D^\alpha_\xi p_\rho(\cdot, \xi)\|_{C^s_*} \le C_\alpha ((1 + |\xi|^2 + |\rho|^2)^{\frac{1}{2}})^{m-|\alpha|+s\delta}$$

for any $\alpha \in \mathbb{Z}^n_+$.

(b) $p_\rho(x, \xi) \in C^s S^m_{1,\delta,\rho}(\mathbb{R}^n)$ if the conditions on (a) are satisfied and additionally:

$$\|D^\alpha_\xi p_\rho(\cdot, \xi)\|_{C^j} \le C_\alpha ((1 + |\xi|^2 + |\rho|^2)^{\frac{1}{2}})^{m-|\alpha|+j\delta}$$

for any $\alpha \in \mathbb{Z}^n_+$ and $\forall\, j \in \mathbb{N}$ such that $0 \le j \le s$.

Proposition 19.39. Let $p_\rho(x, \xi) \in C^s_* S^m_{1,0,\rho}$. Then we can write for any δ so that $0 \le \delta < 1$:

$$p_\rho(x, \xi) = p^\sharp_\rho(x, \xi) + p^\flat_\rho(x, \xi), \tag{19.155}$$

where

$$p_\rho^\sharp(x,\xi) \in S_{1,\delta,\rho}^m \tag{19.156}$$

and

$$p_\rho^\flat(x,\xi) \in C_*^{s-t} S_{1,0,\rho}^{m-t\delta}, \quad s, \, s-t > 0. \tag{19.157}$$

Here $S_{1,\rho,\delta}^m$ denotes the symbol class similar to Definition 19.29 with $|\alpha|$ replaced by $\rho|\alpha|$. Let $p_\rho(x,D)$, $p_\rho^\sharp(x,D)$, $p_\rho^\flat(x,D)$ denote the corresponding operators associated to $p_\rho(x,\xi)$, $p_\rho^\sharp(x,\xi)$, $p_\rho^\flat(x,\xi)$, respectively. Then we have the following estimates which are proved using a Littlewood-Paley decomposition of the phase space depending on the parameter ρ.

Theorem 19.40. *Let* $p_\rho(x,\xi) \in C_*^r S_{1,0,\rho}^m(\mathbb{R}^n)$. *Then*

$$p_\rho(x,D): \, H^{s+m,p}(\mathbb{R}^n) \longrightarrow H^{s,p}(\mathbb{R}^n)$$

with

$$\|p_\rho(x,D)\|_{s+m,s} \leq C((1+|\rho|^2)^{\frac{1}{2}})^{s+m}, \tag{19.158}$$

where $0 < s < r$, $p \in (1,\infty)$ *and* $\|\cdot\|_{s+m,s}$ *denotes the operator norm between Sobolev spaces.*

Now we describe how to construct complex geometrical optics solution of

$$P(x,D) = \Delta + P^{(1)}(x,D). \tag{19.159}$$

Using Theorem 19.37 (for the sake of exposition we will eliminate all the cutoff functions) we can find operators $A_\rho, B_\rho \in L_\delta^0(\mathbb{R}^n, Z)$ such that A_ρ, B_ρ are invertible for large ρ and

$$(\Delta_\rho + N_\rho^\sharp)A_\rho = B_\rho(\Delta_\rho + C_\rho) \tag{19.160}$$

with $C_\rho \in L_\delta^0(\mathbb{R}^n, Z)$ where $P_\rho^{(1)}(x,D) = N_\rho^\sharp(x,D) + N_\rho^\flat(x,D)$ using the decomposition (19.155). Then

$$\begin{aligned}
(\Delta_\rho + P_\rho(x,D))A_\rho &= B_\rho(\Delta_\rho + C_\rho) + N_\rho^\flat(x,D)A_\rho \tag{19.161} \\
&= B_\rho(\Delta_\rho + C_\rho + B_\rho^{-1} N_\rho^\flat(x,D)A_\rho).
\end{aligned}$$

Using the estimate (19.158) and the estimates for the operators A_ρ, B_ρ implied by Theorem 19.36 we conclude the following estimate under the regularity assumption that the coefficients of the first-order term are in $C^{2/3+\epsilon}$ with $\epsilon > 0$

$$\|(C_\rho + B_\rho^{-1} N_\rho^\flat(x,D)A_\rho)\Delta_\rho^{-1}\|_{L^2(\Omega), L^2(\Omega)} \leq C|\rho|^{-\beta} \tag{19.162}$$

for some $\beta = \beta(\epsilon) > 0$.

19.8   Anisotropic Conductors

In §§19.1–19.5 we considered isotropic conductivities, i.e., the electrical properties of Ω do not depend of direction. Examples of anisotropic media are
muscle tissue. In this section we consider the inverse conductivity problem
for anisotropic medium. The problem is well understood in two dimensions.
Using isothermal coordinates [2] one can in fact reduce, by a change of variables, the anisotropic conductivity equation to an isotropic one and therefore
one can apply the two-dimensional results of §19.5. Of course, this is not
available in dimension $n > 2$. In fact, in this case the problem is equivalent
to the problem of determining a Riemannian metric from the Dirichlet to
Neumann map associated to the Laplace-Beltrami operator [34]. Only the
real-analytic case is understood at present [34].

   In this section we will consider the case of a quasilinear anisotropic conductivity. We outline recent results [57] proving identical results to the linear
case. One needs to go further than the linearization procedure of Lemma
19.13 for isotropic non-linear conductivities. In fact, we show that one can
reduce the problem question about the density of product of solutions for the
linear anisotropic conductivities by using a second linearization.

   We assume that $\gamma(x, t) \in C^{1,\alpha}(\overline{\Omega} \times \mathbb{R})$ be a symmetric, positive definite
matrix function satisfying

$$\gamma(x,t) \geq \epsilon_T I, \quad (x,t) \in \overline{\Omega} \times [-T, T], \quad T > 0, \tag{19.163}$$

where $\epsilon_T > 0$ and I denotes the identity matrix.

   It is well known (see, e.g., [21]) that, given $f \in C^{2,\alpha}(\overline{\Omega})$, there exists a
unique solution of the boundary-value problem

$$\begin{cases} \nabla \cdot (\gamma(x,u)\nabla u) = 0 & \text{in } \Omega, \\ u\big|_{\partial\Omega} = f. \end{cases} \tag{19.164}$$

We define the Dirichlet to Neumann map $\Lambda_\gamma : C^{2,\alpha}(\partial\Omega) \to C^{1,\alpha}(\partial\Omega)$ as the
map given by

$$\Lambda_\gamma : f \to \nu \cdot \gamma(x, f)\nabla u\big|_{\partial\Omega}, \tag{19.165}$$

where u is the solution of (19.164) and ν denotes the unit outer normal of
$\partial\Omega$.

   Physically, $\gamma(x, u)$ represents the (anisotropic, quasilinear) conductivity
of Ω and $\Lambda_\gamma(f)$ the current flux at the boundary induced by the voltage f.

   We study the inverse boundary-value problem associated to (19.164): how
much information about the coefficient matrix γ can be obtained from knowledge of the Dirichlet to Neumann map Λ_γ?

The uniqueness, however, is false in the case where γ is a general matrix function as it is also in the linear case [33]: if $\Phi : \overline{\Omega} \to \overline{\Omega}$ is a smooth diffeomorphism which is the identity map on $\partial\Omega$, and if we define

$$(\Phi_*\gamma)(x,t) = \frac{(D\Phi)^T\gamma(\cdot,t)(D\Phi)}{|D\Phi|} \circ \Phi^{-1}(x), \qquad (19.166)$$

then it follows that

$$\Lambda_{\Phi_*\gamma} = \Lambda_\gamma, \qquad (19.167)$$

where $D\Phi$ denotes the Jacobian matrix of Φ and $|D\Phi| = \det(D\Phi)$.

The main results of [57] concern with the converse statement. We have

Theorem 19.41. *Let $\Omega \subset \mathbb{R}^2$ be a bounded domain with $C^{3,\alpha}$ boundary, $0 < \alpha < 1$. Let γ_1 and γ_2 be quasilinear coefficient matrices in $C^{2,\alpha}(\overline{\Omega} \times \mathbb{R})$ such that $\Lambda_{\gamma_1} = \Lambda_{\gamma_2}$. Then there exists a $C^{3,\alpha}$ diffeomorphism $\Phi : \overline{\Omega} \to \overline{\Omega}$ with $\Phi\big|_{\partial\Omega} = $ identity such that $\gamma_2 = \Phi_*\gamma_1$.*

Theorem 19.42. *Let $\Omega \subset \mathbb{R}^n$, $n \geq 3$, be a bounded simply connected domain with real-analytic boundary. Let γ_1 and γ_2 be real-analytic quasilinear coefficient matrices such that $\Lambda_{\gamma_1} = \Lambda_{\gamma_2}$. Assume that either γ_1 or γ_2 extends to a real-analytic quasilinear coefficient matrix on $\mathbb{R}^n$. Then there exists a real-analytic diffeomorphism $\Phi : \overline{\Omega} \to \overline{\Omega}$ with $\Phi\big|_{\partial\Omega} = $ identity, such that $\gamma_2 = \Phi_*\gamma_1$.*

Theorems 19.41 and 19.42 generalize all known results for the linear case [64]. In this case and when $n = 2$, with a slightly different regularity assumption, Theorem 19.41 follows using a reduction theorem of Sylvester [60], using isothermal coordinates, and Theorem 19.14 for the isotropic case.

In the linear case and when $n \geq 3$, Theorem 19.42 is a consequence of the work of Lee and Uhlmann [34], in which they discussed the same problem on real-analytic Riemannian manifolds. The assumption that one of the coefficient matrices can be extended analytically to $\mathbb{R}^n$ can be replaced by a convexity assumption on the Riemannian metrics associated to the coefficient matrices. Thus Theorem 19.42 can also be stated under this assumption, which we omit here. We mention that, in the linear case, complex geometrical optics solutions have not been constructed for the Laplace-Beltrami operator in dimensions $n \geq 3$. The proof of Theorem 19.1 in the linear case follows a different approach.

A. Linearization. The proof of linearization Lemma 19.13 is also valid in the anisotropic case. We shall use γ^t to denote the function of x obtained by freezing t in $\gamma(x,t)$.

Under the assumptions of Theorem 19.6, using Lemma 19.13 we have that

$$\Lambda_{\gamma_1^t} = \Lambda_{\gamma_2^t}, \quad \forall\, t \in \mathbb{R}. \tag{19.168}$$

Since Theorems 19.41 and 19.42 hold in the linear case, it follows that there exists a diffeomorphism Φ^t, which is in C_α^3 when $n = 2$ and is real-analytic when $n \geq 3$, and the identity at the boundary such that

$$\gamma_2^t = \Phi_*^t \gamma_1^t. \tag{19.169}$$

It is proven in [57] that Φ^t is uniquely determined by γ_l^t, and thus by γ_l, $l = 1, 2$. We then obtain a function

$$\Phi(x, t) = \Phi^t(x) : \overline{\Omega} \times \mathbb{R} \to \overline{\Omega} \times \mathbb{R}, \tag{19.170}$$

which is in $C^{3,\alpha}(\overline{\Omega})$ for each fixed t in dimension two and real analytic in dimension $n \geq 3$. It is also shown in [57] that Φ is also smooth in t. More precisely we have, in every dimension $n \geq 2$, that $\frac{\partial \Phi}{\partial t} \in C^{2,\alpha}(\overline{\Omega})$.

In order to prove Theorems 19.41 and 19.42, we must then show that Φ^t is independent of t. Without loss of generality, we shall only prove

$$\left. \frac{\partial \Phi}{\partial t} \right|_{t=0} = 0 \quad \text{in } \overline{\Omega}. \tag{19.171}$$

It is easy to show, using the invariance (19.167) that we may assume that

$$\Phi(x, 0) \equiv x, \text{ that is, } \Phi^0 = \text{ identity.} \tag{19.172}$$

Let us fix a solution $u \in C^{3,\alpha}(\overline{\Omega})$ of

$$\nabla \cdot A \nabla u = 0, \quad u|_{\partial \Omega} = f, \tag{19.173}$$

where we denote $A = \gamma_1^0 = \gamma_2^0$.

For every $t \in \mathbb{R}$ and $l = 1, 2$, we solve the boundary-value problem (19.173) with γ^t replaced by γ_l^t. We obtain a solution $u_{(l)}^t$:

$$\begin{cases} \nabla \cdot \gamma_l^t \nabla u_{(l)}^t = 0 \quad \text{in } \Omega \\ u_{(l)}^t \big|_{\partial \Omega} = f, \qquad l = 1, 2. \end{cases} \tag{19.174}$$

It is easy to check that

$$u_{(1)}^t(x) = u_{(2)}^t(\Phi^t(x)), \qquad x \in \overline{\Omega}.$$

Differentiating this last formula in t and evaluating at $t = 0$ we obtain

$$\left(\frac{\partial u_{(1)}^t}{\partial t} - \frac{\partial u_{(2)}^t}{\partial t}\right)\Bigg|_{t=0} - X \cdot \nabla u = 0, \qquad x \in \overline{\Omega}, \tag{19.175}$$

where

$$X = \frac{\partial \Phi^t}{\partial t}\Bigg|_{t=0}. \tag{19.176}$$

It is easy to show that $X \cdot \nabla u = 0$ for every solution of (19.173) implies $X = 0$. So we are reduced to prove

$$\left(\frac{\partial u_{(1)}^t}{\partial t} - \frac{\partial u_{(2)}^t}{\partial t}\right)\Bigg|_{t=0} = 0. \tag{19.177}$$

Using (19.174) we get

$$\nabla \cdot (\gamma_1(x,t)\nabla u_{(1)}^t) - \nabla \cdot (\gamma_2(x,t)\nabla u_{(2)}^t) = 0. \tag{19.178}$$

Differentiating (19.178) in t at $t = 0$ we conclude

$$\nabla \cdot \left[\left(\frac{\partial \gamma_1}{\partial t} - \frac{\partial \gamma_2}{\partial t}\right)\Bigg|_{t=0} \nabla u\right] + \nabla \cdot \left[A\nabla \left(\frac{\partial u_{(1)}^t}{\partial t} - \frac{\partial u_{(2)}^t}{\partial t}\right)\Bigg|_{t=0}\right] = 0. \tag{19.179}$$

We claim that in order to prove (19.177) it is enough to show that

$$\nabla \cdot \left[\left(\frac{\partial \gamma_1}{\partial t} - \frac{\partial \gamma_2}{\partial t}\right)\Bigg|_{t=0} \nabla u\right] = 0. \tag{19.180}$$

This is the case since we get from (19.179) and (19.180)

$$\nabla \cdot \left[A\nabla \left(\frac{\partial u_{(1)}^t}{\partial t} - \frac{\partial u_{(2)}^t}{\partial t}\right)\Bigg|_{t=0}\right] = 0.$$

The claim now follows since the operator $\nabla \cdot A\nabla : H_0^2(\Omega) \cap H^1(\Omega) \to L^2(\Omega)$ is an isomorphism and therefore

$$\left(\frac{\partial u_{(1)}^t}{\partial t} - \frac{\partial u_{(2)}^t}{\partial t}\right)\Bigg|_{t=0}\Bigg|_{\partial\Omega} = 0.$$

B. Second linearization and products of solutions. In order to show (19.180) we now study the second linearization. We introduce, for every

$t \in \mathbb{R}$, the map $K_{\gamma,t} : C^{2,\alpha}(\partial\Omega) \to H^{\frac{1}{2}}(\partial\Omega)$ which is defined implicitly as follows (see [53]): for every pair $(f_1, f_2) \in C^{2,\alpha}(\partial\Omega) \times C^{2,\alpha}(\partial\Omega)$,

$$\int_{\partial\Omega} f_1 K_{A,t}(f_2) dS = \int_{\Omega} \nabla u_1 \frac{\partial A}{\partial t} \nabla u_2^2 dx \qquad (19.181)$$

with u_l, $l = 1, 2$, as in (19.174) with f replaced by f_l, $l = 1, 2$. We have

Proposition 19.43 ([53]). *Let $\gamma(x, t)$ be a positive definite symmetric matrix in $C^2(\overline{\Omega} \times \mathbb{R})$, satisfying (19.163). Then for every $f \in C^{2,\alpha}(\partial\Omega)$ and $t \in \mathbb{R}$,*

$$\lim_{s \to 0} \left\| \frac{1}{s} \left[\frac{1}{s} \Lambda_A(t + sf) - \Lambda_{A^t}(f) \right] - K_{A,t}(f) \right\|_{H^{\frac{1}{2}}(\partial\Omega)} = 0.$$

Under the assumptions of Theorems 19.41 and 19.42, using Proposition 19.43 with $t = 0$, we obtain

$$K_{\gamma_1,0}(f) = K_{\gamma_2,0}(f), \qquad \forall f \in C^{3,\alpha}(\partial\Omega).$$

Thus, by (19.181) we have

$$\int_{\Omega} \nabla u_1^T \frac{\partial \gamma_1}{\partial t} \bigg|_{t=0} \nabla u_2^2 \, dx = \int_{\Omega} \nabla u_1^T \frac{\partial \gamma_2}{\partial t} \bigg|_{t=0} \nabla u_2^2 \, dx, \qquad (19.182)$$

with u_1, u_2 solutions of (19.174). By writing

$$B = \left(\frac{\partial \gamma_1}{\partial t} - \frac{\partial \gamma_2}{\partial t} \right) \bigg|_{t=0} \qquad (19.183)$$

and replacing in (19.182) u_1 by u, and u_2^2 by $(u_1 + u_2)^2 - u_1^2 - u_2^2$, we obtain

$$\int_{\Omega} \nabla u \cdot B(x) \nabla(u_1 u_2) dx = 0 \qquad (19.184)$$

with u, u_1 and u_2 solutions of (19.174).

To continue from (19.184), we need the following two lemmas.

Lemma 19.44. *Let $h(x) \in C^1(\overline{\Omega})$ be a vector-valued function. If*

$$\int_{\Omega} h(x) \nabla(u_1 u_2) dx = 0$$

for arbitrary solutions u_1 and u_2 of (19.174), then $h(x)$ lies in the tangent space $T_x(\partial\Omega)$ for all $x \in \partial\Omega$.

Lemma 19.45. *Let $A(x)$ be a positive definite, symmetric matrix in $C^{2,\alpha}(\overline{\Omega})$. Define*

$$D_A = \text{Span}_{L^2(\Omega)} \{ uv : u, v \in C^{3,\alpha}(\overline{\Omega}), \nabla \cdot A \nabla u = \nabla \cdot A \nabla v = 0 \}.$$

Then the following are valid:

(a) If $l \in C^\omega(\overline{\Omega})$ and $l \perp D_A$, then $l = 0$ in $\overline{\Omega}$,

(b) If $n = 2$, then $D_A = L^2(\Omega)$.

Now we finish the proof of (19.180) concluding the proofs of Theorems 19.41 and 19.42.

By Lemma 19.44 we have that $\nu \cdot B(x)\nabla u \equiv 0$ in $\partial\Omega$. Integrating by parts in (19.184), we obtain

$$\int_\Omega [\nabla \cdot B(x)\nabla u] u_1 u_2 dx = 0. \qquad (19.185)$$

We now apply Lemma 19.44 to (19.185). If $n \geq 3$, we have that γ_1 and γ_2 are real-analytic on $\overline{\Omega} \times \mathbb{R}$. Thus $B \in C^\omega(\overline{\Omega})$. Since the solutions u solve an elliptic equation with a real-analytic coefficient matrix, we have that u is analytic in Ω. If u is analytic on $\overline{\Omega}$, we can conclude from Lemma 19.45 that

$$\nabla \cdot (B(x)\nabla u) = 0, \qquad x \in \overline{\Omega}. \qquad (19.186)$$

We shall prove that (19.186) holds independent of whether u is analytic up to $\partial\Omega$ or not. This is due to the Runge approximation property of the equation. Using the assumptions of Theorem 19.42, we extend A analytically to a slightly larger domain $\widetilde{\Omega} \supset \overline{\Omega}$. For any solution $u \in C^{3,\alpha}(\overline{\Omega})$ and an open subset $\mathcal{O}$ with $\overline{\mathcal{O}} \subset \Omega$, we can find a sequence of solutions $\{u_m\} \subset C^\omega(\widetilde{\Omega})$, which solves (19.184) on $\widetilde{\Omega}$, and $u_m\big|_{\mathcal{O}_1} \xrightarrow[m\to\infty]{} u\big|_{\mathcal{O}_1}$ in the L^2 sense, where $\overline{\mathcal{O}}_1 \subset \Omega$, $\overline{\mathcal{O}} \subset \mathcal{O}_1$. By the local regularity theorem of elliptic equations this convergence is valid in $H^2(\mathcal{O})$. Since (19.186) holds with $u = u_m$, letting $m \to \infty$ yields the desired result for u on $\mathcal{O}$. Thus (19.184) holds. If $n = 2$, Lemma 19.44(b) implies that $\nabla \cdot (B(x)\nabla u) = 0$ for any solution $u \in C^{3,\alpha}(\overline{\Omega})$.

The proof of Lemma 19.44 follows an argument of Alessandrini [4], which relies on the use of solutions with isolated singularities. It turns out that in our case, only solutions with Green's function type singularities are sufficient in the case $n \geq 3$, while in the case $n = 2$, solutions with singularities of higher order must be used. There are additional difficulties since we are dealing with a vector function h. We refer the readers to [57] for details.

The proof of part (a) of Lemma 19.45 follows the proof of Theorem 1.3 in [3] (which also follows the arguments of [31]). Namely, one constructs solutions u of (19.173) in a neighborhood of Ω with an isolated singularity of arbitrary given order at a point outside of Ω. We then plug this solution into the identity

$$\int_\Omega lu^2 dx = 0.$$

By letting the singularity of u approach to a point x in $\partial\Omega$, one can show that any derivative of l must vanish on x and thus by the analyticity of l, $l \equiv 0$ in $\overline{\Omega}$. For more details, see [57].

To prove the part (b) of Lemma 19.45, we first reduce the problem to the Schrödinger equation.

Using isothermal coordinates (see [2]), there is a conformal diffeomorphism $F : (\overline{\Omega}, g) \to (\overline{\Omega}', e)$, where g is the Riemannian metric determined by the linear coefficient matrix A with $g_{ij} = A_{ij}^{-1}$. One checks that F transforms the operator $\nabla \cdot A\nabla$ (on Ω) to an operator $\nabla \cdot A'\nabla$ (on Ω') with A' a scalar matrix function $\beta(x)I$. Therefore the proof of the part (b) is reduced to the case where $A = \beta I$, with $\beta(x) \in C^{2,\alpha}(\overline{\Omega})$. By approximating by smooth solutions, we see that the $C^{3,\alpha}$ smoothness can be replaced by H^2 smoothness. Thus we have reduced the problem to showing that

$$D_\beta = \text{Span}_{L^2}\{uv : u,v \in H^2(\Omega); \nabla \cdot \beta\nabla u = \nabla \cdot \beta\nabla v = 0\} = L^2(\Omega).$$

We make one more reduction by transforming, as in §19.3, the equation $\nabla \cdot \beta\nabla u = 0$ to the Schrödinger equation

$$\Delta v - qv = 0$$

with

$$u = \beta^{-\frac{1}{2}}v, \quad q = \frac{\Delta\sqrt{\beta}}{\sqrt{\beta}} \in C^\alpha(\overline{\Omega}). \tag{19.187}$$

This allows us to reduce the proof to showing that

$$D_q = \text{Span}_{L^2}\{v_1 v_2 : v_i \in H^2(\Omega), \Delta v_i - qv_i = 0, i = 1,2\} = L^2(\Omega) \tag{19.188}$$

for potentials q of the form (19.187).

Statement (19.188) was proven by Novikov [46]. In [57] it was shown that it is enough to use the Proposition below which is valid for any potential $q \in L^\infty(\Omega)$. This result uses some of the techniques of [61,62] similar to the proof of Theorem 19.20.

Proposition 19.46. *Let $q \in L^\infty(\Omega)$, $n = 2$. Then D_q has a finite codimension in $L^2(\Omega)$.*

It is an interesting open question whether $D_q = L^2(\Omega)$ in the two-dimensional case.

References

[1] M. Ablowitz, D. Bar Yaacov, and A. Fokas, *On the inverse scattering transform for the Kadomtsev-Petviashvili equation*, Studies Appl. Math. 69 (1983), 135–143.

[2] L. Ahlfors, *Quasiconformal mappings*, Van Nostrand, 1966.

[3] G. Alessandrini, *Stable determination of conductivity by boundary measurements*, App. Anal. 27 (1988), 153–172.

[4] ———, *Singular solutions of elliptic equations and the determination of conductivity by boundary measurements*, J. Diff. Equations 84 (1990), 252-272.

[5] D. Barber and B. Brown, *Applied potential tomography*, J. Phys. E 17 (1984), 723–733.

[6] R. Beals and R. R. Coifman, *Transformation Spectrales et équation d'évolution non linéares*, Seminaire Goulaouic-Meyer-Schwarz, exp. 21, 1981-1982.

[7] ———, *Multidimensional inverse scattering and nonlinear PDE*, Proc. Symp. Pure Math. 43, American Math. Soc., Providence (1985), 45–70.

[8] ———, *The spectral problem for the Davey-Stewarson and Ishimori hierarchies*, in *Nonlinear evolution equations: Integrability and spectral methods*, pp. 15-23 (1988), Manchester University Press.

[9] G. Bourdaud, *L^p estimates for certain non-regular pseudodifferential operators*, Comm. PDE 7 (1982), 1023–1033.

[10] R. Brown, *Global uniqueness in the impedance imaging problem for less regular conductivities*, SIAM J. Math. Anal. 27 (1996), 1049–1056.

[11] ———, *Recovering the conductivity at the boundary from the Dirichlet to Neumann map: A pointwise result*, preprint.

[12] R. Brown and G. Uhlmann, *Uniqueness in the inverse conductivity problem with less regular conductivities in two dimensions*, Comm. PDE 22 (1997), 1009-1027.

[13] A. P. Calderón, *On an inverse boundary-value problem*, Seminar on Numerical Analysis and its Applications to Continuum Physics, Soc. Brasileira de Matemática, Rio de Janeiro (1980), 65–73.

[14] ———, *Boundary value problems for elliptic equations*, Outlines of the joint Soviet-American symposium on partial differential equations, 303-304, Novisibirsk (1963).

[15] S. Chanillo, *A problem in electrical prospection and an n-dimensional Borg-Levinson theorem*, Proc. AMS 108 (1990), 761–767.

[16] P. Ciarlet, *Mathematical elasticity, Vol. I*, Elsevier Science (1988).

[17] R. Coifman and Y. Meyer, *Au delà des opérateurs pseudo-différentiels*, Astérisque 57 (1978).

[18] V. L. Druskin, *The unique solution of the inverse problem in electrical surveying and electrical well logging for piecewise-constant conductivity*, Physics of the Solid Earth 18 (1982), no. 1, 51–53.

[19] G. Eskin and J. Ralston, *Inverse scattering problem for the Schrödinger equation with magnetic potential at a fixed energy*, Comm. Math. Phys. 173 (1995), 199-224.

[20] L. Faddeev, *Growing solutions of the Schrödinger equation*, Dokl. Akad. Nauk SSSR 165 (1965), 514–517 (translation in Sov. Phys. Dokl. 10, 1033).

[21] D. Gilbarg and N. Trudinger, *Elliptic Partial Differential Equations*, Interscience Publishers (1964).

[22] P. Hähner, *A periodic Faddeev-type solution operator*, J. Diff. Equations 128 (1996), 300–308.

[23] M. Ikehata, *A remark on an inverse boundary-value problem arising in elasticity*, preprint.

[24] D. Isaacson and E. Isaacson, *Comment on Calderón's paper: "On an inverse boundary-value problem,"* Math. Comput. 52, 553–559.

[25] D. Isaacson, J. C. Newell, J. C. Goble, and M. Cheney, *Thoracic impedance images during ventilation*, Annual Conference of the IEEE Engineering in Medicine and Biology Society, vol. 12 (1990), 106–107.

[26] V. Isakov, *On uniqueness of recovery of discontinuity conductivity coefficient*, Comm. Pure Appl. Math. 41 (1988), 865–877.

[27] ———, *On uniqueness in inverse problems for semilinear parabolic equations*, Arch. Rat. Mech. Anal. 124 (1993), 1–12.

[28] ———, *Completeness of products of solutions and some inverse problems for PDE*, J. Diff. Equations 92 (1991), 305–317.

[29] V. Isakov and A. Nachman, *Global uniqueness for a two-dimensional semilinear elliptic inverse problem*, Trans. of AMS 347 (1995), 3375–3390.

[30] V. Isakov and J. Sylvester, *Global uniqueness for a semilinear elliptic inverse problem*, Comm. Pure Appl. Math. 47 (1994), 1403–1410.

[31] R. Kohn and M. Vogelius, *Determining conductivity by boundary measurements*, Comm. Pure App. Math. 38 (1985), 643–667.

[32] ———, *Determining conductivity by boundary measurements II. Interior results*, Comm. Pure Appl. Math. 38 (1985), 644–667.

[33] ———, *Identification of an unknown conductivity by means of measurements at the boundary*, in Inverse Problems, edited by D. McLaughlin, SIAM-AMS Proc. no. 14, Amer. Math. Soc., Providence (1984), 113–123.

[34] J. Lee and G. Uhlmann, *Determining anisotropic real-analytic conductivities by boundary measurements*, Comm. Pure Appl. Math 42 (1989), 1097–1112.

[35] A. Nachman, *Reconstructions from boundary measurements*, Annals of Math. 128 (1988), 531–587.

[36] ———, *Global uniqueness for a two-dimensional inverse boundary value problem*, Annals of Math (1996), 71–96.

[37] A. Nachman and M. Ablowitz, *A multidimensional inverse scattering method*, Studies in App. Math. 71 (1984), 243–250.

[38] G. Nakamura and Z. Sun, *An inverse boundary-value problem for St. Venant-Kirchhoff materials*, Inverse Problems 10 (1994), 1159–1163.

[39] G. Nakamura and G. Uhlmann, *Global uniqueness for an inverse boundary value problem arising in elasticity*, Invent. Math. 118 (1994), 457–474.

[40] ———, *Inverse problems at the boundary for an elastic medium*, SIAM J. Math. Anal. 26 (1995), 263–279.

[41] ———, *Identification of Lamé parameters by boundary measurements*, American Journal of Math. 115 (1993), 1161–1187.

[42] G. Nakamura, G. Uhlmann, and Z. Sun, *Global identifiability for an inverse problem for the Schrödinger equation in a magnetic field*, Math. Ann. 303 (1995), 377–388.

[43] J. C. Newell, D. Isaacson, M. Cheney, G. J. Saulnier, and D. G. Gisser, *Acute pulmonary edema assessed by electrical impedance imaging*, Proceedings IEEE-EMBS Conf. 14 (1993), 92-93.

[44] L. Nirenberg and H. Walker, *Null spaces of elliptic partial differential equations in* $\mathbb{R}^n$, J. Math. Anal. Appl. 42 (1973), 271–301.

[45] R. Novikov, *Multidimensional inverse spectral problems for the equation* $-\Delta\psi + (v(x) - Eu(x))\psi = 0$, Funktsionalny Analizi Ego Prilozheniya, vol. 22, no. 4 (1988), 11-12, Translation in Functional Analysis and Its Applications, vol. 22, no 4 (1988), 263–272.

[46] ———, $\bar{\partial}$-*method with nonzero background potential. Application to inverse scattering for the two-dimensional acoustic equation*, Comm. PDE 21 (1996), 597–618.

[47] R. Novikov and G. Henkin, $\bar{\partial}$-*equation in the multidimensional inverse scattering problem*, Uspekhi Mat. Nauk, vol. 42 (1987), 93–1526 translation in Russian Mathematical Surveys, vol. 42, no. 4 (1987), 109–180.

[48] P. Ola, L. Päivärinta, and E. Somersalo, *An inverse boundary-value problem arising in electrodynamic*, Duke Math. J. 70 (1993), 617–653.

[49] P. Ola and E. Somersalo, *Electromagnetic inverse problems and generalized Sommerfeld potentials*, SIAM J. Appl. Math. 56 (1996), 1129–1145.

[50] M. Shubin, *Pseudodifferential Operators and Spectral Theory*, Springer Series in Soviet mathematics, Springer-Verlag (1987).

[51] E. Somersalo, D. Isaacson, and M. Cheney, *A linearized inverse boundary-value problem for Maxwell's equations*, Journal of Comp. and Appl. Math. 42 (1992), 123-136.

[52] P. Stefanov, *Stability of the inverse scattering in potential scattering at a fixed energy*, Ann. Inst. Fourier, Grenoble 40 (1990), 867–884.

[53] Z. Sun, *On a quasilinear boundary-value problem*, Math. Z. 221 (1996), 293–305.

[54] ———, *An inverse boundary-value problem for Schrödinger operators with vector potentials*, Trans. of AMS 338 (1993), 953–969.

[55] Z. Sun and G. Uhlmann, *Recovery of singularities for formally determined inverse problems*, Comm. Math. Phys. 153 (1993), 431–445.

[56] ———, *Generic uniqueness for an inverse boundary value problem*, Duke Math. Journal 62 (1991), 131–155.

[57] ———, *Inverse problems in quasilinear anisotropic media*, Amer. J. Math. 119 (1997), 771–797.

[58] ———, *An inverse boundary-value problem for Maxwell's equations,* Archive Rat. Mech. Anal. 119 (1992), 71–93.

[59] L. Sung, *An inverse scattering transform for the Davey-Stewartson II equations, I,II,III,* J. Math. Anal. Appl. 183 (1994), 121–154, 289–325, 477–494.

[60] J. Sylvester, *An anisotropic inverse boundary-value problem,* Comm. Pure Appl. Math. (1990), 201–232.

[61] J. Sylvester and G. Uhlmann, *A global uniqueness theorem for an inverse boundary-value problem,* Ann. of Math. 125 (1987), 153–169.

[62] ———, *A uniqueness theorem for an inverse boundary-value problem in electrical prospection,* Comm. Pure. App. Math. 39 (1986), 91–112.

[63] ———, *Inverse boundary-value problems at the boundary-continuous dependence,* Comm. Pure Appl. Math. 41 (1988), 197–221.

[64] ———, *Inverse problems in anisotropic media,* Contemp. Math. 122 (1991), 105–117.

[65] M. Taylor, *Pseudodifferential Operators and Nonlinear PDE,* Progress in Mathematics 100 (1991), Birkhäuser.

[66] C. Tolmasky, *Global uniqueness for an inverse boundary-value problem with non smooth coefficients,* Ph.D. thesis, University of Washington (1996).

[67] T. Y. Tsai, *The Schrödinger equation in the plane,* Inverse Problems 9 (1993), 763–787.

[68] G. Uhlmann, *Inverse boundary-value problems and applications,* Astérisque 207 (1992), 153–211.

[69] A. Wexler, B. Fry, and M. Neumann, *Impedance-computed tomography algorithm and system,* Applied Optics 24 (1985), 3985–3992.

[70] M. S. Zhdanov and G. V. Keller *The geoelectrical methods in geophysical exploration,* Methods in Geochemistry and Geophysics, 31, Elsevier (1994).

Remarks Addressed to the Conference

Alberto P. Calderón

Dear Friends:

Thank you for attending this conference. I have turned seventy-five and I am starting to realize that the cardinality of this number is larger than I felt it was.

Turning now to the purpose of this conference, I believe that a few mathematical comments of mine before the end of the conference would not be totally out of place. Well, here they are. They concern the relationship between singular integral operators and pseudodifferential operators. The latter are a very refined mathematical tool whose full power is yet to be exploited. Singular integral whose kernels are infinitely differentiable off the diagonal are a very special case of pseudodifferential operators. However, singular integral operators with kernels which are not infinitely differentiable as above are not pseudodifferential. Similarly, linear partial differential operators are pseudodifferential operators only if the coefficients are infinitely differentiable. Such differential operators with nonsmooth (i.e., noninfinitely differentiable) coefficients have important applications. They appear in problems in physics and engineering, and they can be studied by using singular integral operators with nonsmooth kernels. It is not difficult to construct an algebra of singular integral operators K in such a way that differential operators of a given order m with coefficients with specified regularity be contained in the module, over this algebra, of operators of the form $K\Lambda^m$, where K is an operator in the algebra and Λ is the operator defined by $(\Lambda f)^\frown(x) = \hat{f}(x)\varphi(x)$, where $\frown$ denotes Fourier transform and $\varphi(x)$ is a positive infinitely differentiable function such that $\varphi(x) = |x|$ for $|x| \geq 1$. The study of these algebras is not yet complete. There are still open questions about best possible assumptions.

Thank you.

Contributors

Alexandra Bellow
Department of Mathematics
Northwestern University
Evanston, IL 60208

Earl Berkson
Department of Mathematics
University of Illinois, Urbana
Urbana, IL 61801

Jean Bourgain
School of Mathematics
Institute for Advanced Study
Princeton, NJ 08540

Donald L. Burkholder
Department of Mathematics
University of Illinois, Urbana
Urbana, IL 61801

Luis A. Caffarelli
Department of Mathematics
University of Texas, Austin
Austin, TX 78712

Alberto P. Calderón (1920–1998)
Department of Mathematics
University of Chicago
Chicago, IL 60637

Sun-Yung A. Chang
Department of Mathematics
Princeton University
Princeton, NJ 08544

Michael Christ
Department of Mathematics
University of California, Berkeley
Berkeley, CA 94720

A. B. Cruzeiro
Grupo de Fisica–Matematica
Universidade de Lisboa
1649-003 Lisboa
Portugal

Ingrid Daubechies
Program in Applied and Computational Mathematics
and Department of Mathematics
Princeton University
Princeton, NJ 08544

Guy David
Departement de Mathématiques, Bâtiment 425
Université de Paris-Sud
91405 Orsay Cedex
France

Charles Fefferman
Department of Mathematics
Princeton University
Princeton, NJ 08544

Robert A. Fefferman
Department of Mathematics
University of Chicago
Chicago, IL 60637

David Jerison
Department of Mathematics
Massachussetts Institute of Technology
Cambridge, MA 02139

Carlos E. Kenig
Department of Mathematics
University of Chicago
Chicago, IL 60637

Gilles Lebeau
  Centre de Mathématiques
  Ecole Polytechnique
  F-91128 Plaiseau Cedex
  France

Stéphane Maes
  IBM T. J. Watson Research Center
  Human Language Technologies
  Acoustic Processing Department
  Yorktown Heights, NY 10598

P. Malliavin
  10 rue Saint Louis en l'Isle
  75004 Paris
  France

Yves Meyer
  CMLA-CNRS URA 1611
  ENS de Cachan
  61 Avenue du President Wilson
  94235 Cachan Cedex
  France

Louis Nirenberg
  Courant Institute
  New York University
  New York, NY 10012

Maciej Paluszyński
  Institute of Mathematics
  University of Wroclaw
  Poland

Cora Sadosky
  Department of Mathematics
  Howard University
  Washington, DC 20059

Robert Seeley
  Department of Mathematics
  University of Massachussetts, Boston
  Boston, MA 02125

Stephen Semmes
    Department of Mathematics
    Rice University
    Houston, TX 77005

Elias M. Stein
    Department of Mathematics
    Princeton University
    Princeton, NJ 08544

Gunther Uhlmann
    Department of Mathematics
    University of Washington
    Seattle, WA 98195

Guido Weiss
    Department of Mathematics
    Washington University
    St. Louis, MO 63130

Index

353